W9-AMT-620

ADVANCES IN CHEMICAL PHYSICS

VOLUME 116

Advances in
CHEMICAL PHYSICS

Edited by

I. PRIGOGINE

Center for Studies in Statistical Mechanics and Complex Systems
The University of Texas
Austin, Texas
and
International Solvay Institutes
Université Libre de Bruxelles
Brussels, Belgium

and

STUART A. RICE

Department of Chemistry
and
The James Franck Institute
The University of Chicago
Chicago, Illinois

VOLUME 116

AN INTERSCIENCE® PUBLICATION
JOHN WILEY & SONS, INC.
NEW YORK • CHICHESTER • WEINHEIM • BRISBANE • SINGAPORE • TORONTO

For ordering and customer service, call 1-800-CALL-WILEY

Library of Congress Catalog Number: 58-9935

ISBN 0-471-40541-8

Printed in the United States of America.

10 9 8 7 6 5 4 3 2 1

CONTRIBUTORS TO VOLUME 116

BIMAN BAGCHI, Solid State and Structural Chemistry Unit, Indian Institute of Science, Bangalore, India

SARIKA BHATTACHARYYA, Solid State and Structural Chemistry Unit, Indian Institute of Science, Bangalore, India

MICHAEL E. CATES, Department of Physics and Astronomy, University of Edinburgh, Edinburgh, United Kingdom

J. W. HALLEY, School of Physics and Astronomy, University of Minnesota, Minneapolis, MN

JOSEPH KLAFTER, School of Chemistry, Tel Aviv University, Tel Aviv, Israel

RALF METZLER, School of Chemistry, Tel Aviv University, Tel Aviv, Israel and Department of Physics, Massachusetts Institute of Technology, Cambridge, MA

D. L. PRICE, Department of Physics, University of Memphis, Memphis, TN

WOLFFRAM SCHRÖER, Institut für Anorganische und Physikalische Chemie, Universität Bremen, Bremen, Germany

PETER SOLLICH, Department of Mathematics, King's College, University of London, London, United Kingdom

S. WALBRAN, Forschungszentrum Jülich GmbH, Institut fuer Werkstoffe und Verfahren der Energietechnik (IVW-3), Jülich, Germany

PATRICK B. WARREN, Unilever Research Port Sunlight, Bebington, Wirral, United Kingdom

HERMANN WEINGÄRTNER, Physikalische Chemie II, Ruhr-Universität Bochum, Bochum, Germany

INTRODUCTION

Few of us can any longer keep up with the flood of scientific literature, even in specialized subfields. Any attempt to do more and be broadly educated with respect to a large domain of science has the appearance of tilting at windmills. Yet the synthesis of ideas drawn from different subjects into new, powerful, general concepts is as valuable as ever, and the desire to remain educated persists in all scientists. This series, *Advances in Chemical Physics*, is devoted to helping the reader obtain general information about a wide variety of topics in chemical physics, a field that we interpret very broadly. Our intent is to have experts present comprehensive analyses of subjects of interest and to encourage the expression of individual points of view. We hope that this approach to the presentation of an overview of a subject will both stimulate new research and serve as a personalized learning text for beginners in a field.

I. PRIGOGINE
STUART A. RICE

CONTENTS

ADVANCES IN CHEMICAL PHYSICS

VOLUME 116

CRITICALITY OF IONIC FLUIDS

HERMANN WEINGÄRTNER

Physikalische Chemie II, Ruhr-Universität Bochum, Bochum, Germany

WOLFFRAM SCHRÖER

Institut für Anorganische und Physikalische Chemie, Universität Bremen, Bremen, Germany

CONTENTS

Advances in Chemical Physics, Volume 116, edited by I. Prigogine and Stuart A. Rice.
ISBN 0-471-40541-8 © 2001 John Wiley & Sons, Inc.

I. INTRODUCTION

This chapter deals with critical phenomena in simple ionic fluids. Prototypical ionic fluids, in the sense considered here, are molten salts and electrolyte solutions. Ionic states occur, however, in many other systems as well; we quote, for example, metallic fluids or solutions of complex particles such as charged macromolecules, colloids, or micelles. Although for simple atomic and molecular fluids thermodynamic anomalies near critical points have been extensively studied for a century now [1], for a long time the work on ionic fluids remained scarce [2, 3]. Reviewing the rudimentary information available in 1990, Pitzer [4] noted fundamental differences in critical behavior between ionic and nonionic fluids.

In preparing the present account, we have been impressed by how much the field has changed since Pitzer's review and a similar review published by us in 1995 [5]. Therefore, a comprehensive account of the present status of the field seems timely. Thus, we presume that the reader is aware of the fundamentals of critical phenomena, as described in many reviews [6–8] and monographs [9–12]. Earlier reviews of one or another aspect of ionic criticality by Pitzer [4,13], Levelt Sengers and Given [14], Fisher [15,16], Stell [17,18], and ourselves [5] are notable.

The organization of this review is as follows: In Section II we describe the theoretical and experimental background of the field. Section III reviews experimental work on the criticality of ionic fluids. Section IV presents the basic theoretical methods for describing ionic phase transitions at the mean-field level. Results obtained by these techniques are reviewed in Section V. Section VI reviews the theoretical work concerned with the nature of the critical point. The review closes in Section VII with a brief summary and outlook.

II. BACKGROUND

Liquid–vapor and liquid–liquid coexistence curves of nonionic fluids have an approximately cubic top in the temperature–density plane [6, 8, 9]. Any analytical equation of state such as the van der Waals equation predicts a parabolic top, which simply follows from the analytical expansion of the free energy in powers of temperature and density. This anomaly is but one example for the universal nonanalytic behavior of fluid properties near critical points. In the 1960s and 1970s the theoretical concepts for describing this universal behavior have been developed to a mature state [10], culminating in the renormalization group (RG) analysis of critical phenomena [11, 19, 20].

Nonanalytical divergences at the critical point result from fluctuations of the order parameter M, which is a measure of the dissimilarity of the coexisting phases. In pure fluids one identifies M with the density difference of the coexisting phases, for mixtures M is related to some concentration variable. The magnitude and spatial extend of these fluctuations is described by their correlation length ξ, which upon approach to the critical point on an isochoric path diverges as

$$\xi = \xi_0 \tilde{T}^{-\nu} + \cdots \qquad \text{as } \tilde{T} \to 0, \qquad (1)$$

where $\tilde{T} = |(T - T_c)/T_c|$ describes the temperature distance from the critical temperature T_c, ν is a universal exponent, and ξ_0 a nonuniversal critical amplitude. The diverging fluctuations result in divergences of other thermophysical properties which along specified paths are described by asymptotic power laws with universal exponents μ and nonuniversal amplitudes X_0:

$$X = X_0 \tilde{T}^{\mu} + \cdots \qquad \text{as } \tilde{T} \to 0. \qquad (2)$$

Table I compiles the scaling laws of interest here. Only two exponents are independent, and the others are related by the exponent equalities given in Table I.

TABLE I
Asymptotic Scaling Laws and Critical Exponents [a,b]

Property	Application	Scaling Law	Path[c]	Exponent[d] Mean-Field	Ising
Correlation length	Either	$\xi = \xi_0 \tilde{T}^{-\nu}$	$d = d_c, X = X_c$	1/2	0.63
Isothermal compressibility	One-component	$\tilde{K}_T = \Gamma_0 \tilde{T}^{-\gamma}$	$d = d_c$	1	1.24
Osmotic susceptibility	Two-component	$\chi = \chi_0 \tilde{T}^{-\gamma}$	$X = X_c$		
Heat capacity	One-component	$C_V = A_0 \tilde{T}^{-\alpha}$	$d = d_c$	0	0.11
	Two-component	$C_{p,x} = A_0 \tilde{T}^{-\alpha}$	$X = X_c$		
Pressure	One-component	$\Delta P = D_0 (\Delta d)^\delta$	$T = T_c$	3	4.8
Chemical potential	Two-component	$\Delta\mu = D_0'(\Delta d)^\delta$	$T = T_c$		
Order parameter	Either	$M = B_0 \tilde{T}^{-\beta}$	coexistence	1/2	0.326
Viscosity	Either	$\eta/\eta_{bg} = H\tilde{T}^{-y}$	$d = d_c, X = X_c$	0(?)	0.04
Correlation function	Either	$g(r) \propto r^{-(D+2-\eta)}$	$d = d_c, X = X_c$	0	0.033
Scaling corrections Δ	Either				0.54

[a] The isothermal compressibility is made dimensionless by defining $\tilde{K}_T = K_T/P_c$, where P_c is the critical pressure. The osmotic susceptibility is defined as $(\partial X_i/\partial\mu_i)_{p,T}$, where X_i is the mole fraction of component i and μ_i its chemical potential.
[b] The following exponent equalities are valid: $\gamma = \beta(\delta - 1); 2 - \alpha = \beta(\delta + 1); \gamma = \nu(2 - \eta);$ $D\nu = (2 - \alpha)$. The last equation is only valid for Ising systems.
[c] X stands for a suitably defined composition variable. For details see text.
[d] For values of the exponents given with more decimal places see Ref. 23.

According to RG theory [11, 19, 20], universality rests on the spatial dimensionality D of the systems, the dimensionality n of the order parameter (here $n = 1$), and the short-range nature of the interaction potential $\phi(r)$. In $D = 3$, short-range means that $\phi(r)$ decays as r^{-p} with $p \geq D + 2 - \eta \cong 4.97$ [21], where $\eta \cong 0.033$ is the exponent of the correlation function $g(r)$ of the critical fluctuations [22] (cf. Table I). Then, the critical exponents map onto those of the Ising spin-1/2 model, which are known from RG calculations [23], series expansions [11, 12, 24] and simulations [25, 26]. For insulating fluids with a leading term of $\phi(r) \propto r^{-6}$ [6–8] and for liquid metals [27–29] the experimental verification of Ising-like criticality is unquestionable.

The *bare* Coulombic interaction ($p = 1$) and interactions of charges with rotating dipoles ($p = 4$) do not fall into this class, and it has been argued for a long time [30] that in this case one expects analytical ("classical") behavior. This implies that the system can be described by a mean-field Hamiltonian, in which the interaction of a particle is ascribed to the mean field of all other particles, thus ignoring local fluctuations [10]. In real ionic fluids the screening of the effective electrostatic interactions to shorter range by counterions—that is, the so-called Debye shielding [31, 32]—may, however, restore Ising behavior. Thus, one has to discriminate between mean-field behavior and $(D=3, n=1)$ Ising criticality. The critical exponents are compared in Table I.

We note that even short-range interactions may, however, allow a mean-field scenario, if the system has a tricritical point, where three phases are in equilibrium. A well-known example is the ^{3}He–^{4}He system, where a line of critical points of the fluid–superfluid transition meets the coexistence curve of the ^{3}He–^{4}He liquid–liquid transition at its critical point [33]. In $D=3$, tricriticality implies that mean-field theory is exact [11], independently from the range of interactions. Such a mechanism is quite natural in ternary systems. For one or two components it would require a further line of hidden phase transitions that meets the coexistence curve at or near its critical point.

Starting with a study on the liquid–vapor coexistence of ammonium chloride (NH$_4$Cl) [34], there have been repeated reports on classical ionic criticality [4], but none of these studies allows unambiguous conclusions [14]. In 1990 more decisive results were reported by Singh and Pitzer [35], who observed a parabolic liquid–liquid coexistence curve of an electrolyte solution. This apparent classical behavior was the stimulus for most theoretical and experimental work reported here.

A prerequisite for understanding these phenomena is a proper description of the molecular forces driving criticality. This puts the problem at the very heart of traditional electrolyte theory. The "restrictive primitive model" (RPM) of equisized, charged hard spheres in a dielectric continuum forms the simplest model to deal with these issues [36, 37]. The existence of a critical point of the RPM was for a long time taken for granted [38, 39], and it was proved in the 1970s by Monte Carlo (MC) simulations [40, 41] and statistical–mechanical theories [42]. By a comparative analysis of the theory of ionic and neutral fluids, Hafskjold and Stell [36] asserted in 1982 that the RPM shows Ising behavior. There are, however, problems in this regard [15], and a decisive RG analysis is still lacking. Some time ago, Fisher [15] and Stell [17] discussed the status of the theory, but did not agree on firm conclusions.

There are other scenarios for an apparent mean-field criticality [15, 17]. The most likely one is crossover from asymptotic Ising behavior to mean-field behavior far from the critical point, where the critical fluctuations must vanish. For the vicinity of the critical point, Wegner [43] worked out an expansion for nonasymptotic corrections to scaling of the general form

$$X = X_0 \tilde{T}^\mu \left(1 + X_1 \tilde{T}^\Delta + X_2 \tilde{T}^{2\Delta} + \cdots\right), \tag{3}$$

where Δ is the universal exponent for corrections to scaling (cf. Table I), and the X_i are nonuniversal correction amplitudes. In nonionic fluids, the lowest-order corrections of this expansion prove to be sufficient for describing the near-critical behavior [6–8], because the crossover region is wide and mean-field behavior is never attained [44, 45]. An apparent mean-field behavior of ionic fluids may mean that the asymptotic range is more narrow, as was indeed observed in early work on the liquid–liquid phase transition in solutions of sodium in ammonia [46]. About 40 years ago, Ginzburg [47] derived a criterion for the temperature distance $\tilde{T}_{\times}$ from the critical point up to which mean-field theory remains self-consistent, thus allowing us to estimate the location of the crossover regime. Contemporary crossover theories incorporate Ginzburg's ideas into an RG treatment, as discussed in Section VI.E.

III. SURVEY OF EXPERIMENTAL RESULTS

A. Liquid–Vapor Transitions in One-Component Ionic Fluids

The critical points of alkali halides such as NaCl are located at temperatures above 3000 K [48–50], while experimental data do not extend beyond 2000 K [48]. Critical point estimates are, however, often needed for comparative purposes. Matching results of molecular dynamics (MD) simulations to the available experimental data, Guissani and Guillot [51] developed an equation of state (EOS) for NaCl which predicts $T_c = 3300$ K and a critical mass density of $d_c = 0.18\,\mathrm{g\,cm^{-3}}$. Pitzer [13] recommended a lower critical density, but, as discussed later, some MC data used in his assessment are questionable.

Comparison with nonionic fluids is possible through the corresponding-states principle [37]. If, as usual, the reduced temperature T^* is defined as the ratio of the thermal energy $k_B T$ to the depth of the potential, one finds for a symmetrical Coulomb system with charges $q = z_+ e = |z_-|e$

$$T^* = \varepsilon_0 \sigma / \beta q^2, \tag{4}$$

where $\beta = 1/k_B T$. ε_0 is the dielectric constant of the background. $\sigma = (\sigma_+ + \sigma_-)/2$ is the mean diameter of the ions. The reduced density is defined via the excluded volume as

$$\rho^* = \rho \sigma^3, \tag{5}$$

where $\rho = \rho_+ + \rho_- = (N_+ + N_-)/V$ is the total number density of the particles. With $\varepsilon_0 = 1$ and $\sigma \cong 0.276\,\mathrm{nm}$ for NaCl, one finds from Guissani and

Guillot's EOS $T_c^* \cong 0.058$ and $\rho_c^* \cong 0.08$. As discussed later in Section V.A, the best available MC results for the charge- and size-symmetrical hard-sphere ionic fluid yield [52, 53]

$$T_c^* = 0.048 - 0.05, \qquad \rho_c^* = 0.07 - 0.08, \qquad (6)$$

in surprisingly good agreement with these values. Both T_c^* and ρ_c^* are much lower than observed for simple nonionic fluids; for example $T_c^* = 1.31$ and $\rho_c^* \cong 0.32$ for the Lennard-Jones fluid [54]. The low values are signatures for phase transitions driven by Coulombic interactions [37].

A direct observation of critical points of molten salts is possible, if specific interactions reduce T_c. One such system is NH_4Cl, where decomposition in the gaseous phase into HCl and NH_3 lowers T_c to about 1150 K [34]. Liquid NH_4Cl is a highly conducting ionic melt even near criticality [55]. Guillot and Guissani [56, 57] performed an extensive theoretical analysis of the effect of this decomposition on the EOS. Buback and Franck [34] determined in 1972 the liquid–vapor coexistence curve of NH_4Cl.

Guided by comparison with the lattice gas model, the order parameter of one-component systems is associated with the mass density d; that is, $M \equiv \tilde{d} = (d - d_c)/d_c$ (in real fluids a better choice would be a linear combination of density and entropy, which accounts for the liquid–vapor asymmetry [58]). Then, the scaling law in Table I predicts a straight line in a double-logarithmic plot of the density difference between the coexisting liquid and gaseous phases, $d_L - d_G$, vs. $T - T_c$ with a slope given by the exponent β. Figure 1 shows that for NH_4Cl such a representation yields $\beta \cong 0.5$, distinctly different from $\beta \cong 0.33$, generally found for nonionic fluids, as exemplified in Fig. 1 by data for xenon and ammonia. Because, at the extreme conditions the critical point of NH_4Cl could be reached only to within 10 K, it is easy to object against the observed mean-field behavior [59]. Nevertheless, Buback and Franck's results indicate that ionic fluids may differ appreciably in their effective critical behavior from nonionic fluids.

On the other hand, Ising behavior is observed for $BiCl_3$ ($T_c \cong 1178$ K) [60]. $BiCl_3$ is, however, little conducting in the liquid phase, and it resembles nonionic rather than ionic fluids. This may warn that specific interactions may destroy the peculiar effects of ionic criticality. Likewise, water ($T_c = 647$ K) shows Ising behavior [61]. At the critical point, autoprotolysis of water is enhanced by three orders of magnitude relative to ambient conditions [62].

For comparison, we briefly consider results for metallic fluids. Early experiments for sodium ($T_c \cong 2573$ K) [63] indicate a parabolic coexistence curve, but the more accurate results for rubidium ($T_c \cong 2017$ K), cesium ($T_c \cong 1924$ K) [27, 28] and mercury ($T_c \cong 1751$ K) [29] ensure an Ising-like

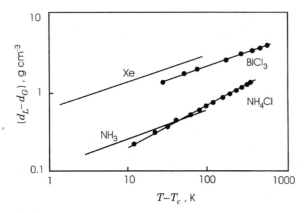

Figure 1. Log–log plot of the density difference $d_L - d_G$ between the liquid and vapor phase of ammonium chloride [34] and bismuth chloride [60] versus the temperature distance $T - T_c$ from the critical temperature T_c. For comparison, data are also shown for xenon and ammonia. The slope of the straight line for NH_4Cl is $\beta = 0.5$. The slopes of the other lines are $\beta = 0.326$. Redrawn with permission from M. Buback, Thesis, Karlsruhe 1969.

criticality. A rationale is that the electrons screen the Coulombic interactions between the cores to short range. Calculations of Maggs and Ashcroft [64] and Goldstein et al. [65] show the screened interactions to decay as r^{-6}, just as in nonionic fluids.

Another interesting aspect of results for metallic fluids concerns the liquid–vapor asymmetry. The "law of the rectilinear diameter" implies that the diameter $(d_L + d_G)/2$ of the coexistence curve is a linear function of $\tilde{T}$. In contrast to nonionic fluids, where deviations from this law are detectable only in experiments of high accuracy [6–8], large departures occur in metallic fluids [27–29]. This indicates substantial differences in the effective interaction potentials for the vapor and liquid phases [66, 67], obviously related to a transition from nonconducting to conducting states. We will come back to this issue, when considering results for electrolyte solutions, where large differences are found in the conductances of the coexisting phases [68] as well.

B. Liquid–Liquid Demixings in Binary Electrolyte Solutions

1. Forces Driving the Phase Separation

In discussing the results for NH_4Cl of Buback and Franck, Friedman [69] suggested in 1972 that some electrolyte solutions with liquid–liquid demixing near room temperature, as discovered by him earlier [3], may be more suitable for studies of ionic criticality than molten salts. Although already observed in 1903 for $KI + SO_2$ [2], for a long time the examples for liquid–liquid immiscibilities in electrolyte solutions remained rare [3, 70]. One difficulty in finding such systems arises from the interference of crystallization, driven by high melting points of salts. Low-melting salts with large organic cations and anions enable the systematic design of more suitable systems [71, 72]. The use of some apparently "exotic" systems in studies of ionic criticality is dictated by this need for low-melting salts.

The presence of the solvent may introduce a wide spectrum of short-range ion–solvent interactions. An intriguing hypothesis was that mean-field-like criticality is restricted to systems, where Coulombic interactions prevail [5, 35, 72]. The critical parameters (6) may serve as a criterion for the dominance of Coulombic interactions, because the corresponding states principle can be extended to solutions, if ε_0 in Eq. (4) is interpreted as the dielectric constant ε_s of the solvent [37]. One should, however, appreciate that, owing to uncertainties in this approximation, and also in estimating the diameters of large, internally flexible ions, a mapping of experimental critical parameters onto the reduced variables defined by Eqs. (4) and (5) can only be done in an approximate way. A general criterion evolving from such a reasoning is that "Coulombic" immiscibilities occur at low ion densities in solvents of low dielectric constant.

Such immiscibilities were, for example, observed for aqueous solutions comprising salts with multivalent ions such as $BaCl_2$ [73] or UO_2SO_4 [74] at high temperatures, where ε_s is low. Because these systems were never applied in studies of ionic criticality, we restrict the discussion to salts with univalent ions.

For salts with univalent ions, Eq. (4) predicts critical points near room temperature for systems with $\varepsilon_s \cong 5$ [72]. Liquid–liquid immiscibilities in several electrolyte solutions are known to satisfy this criterion [5, 71, 72]. Note that these gaps do not necessarily possess an upper critical solution temperature (UCST). Theory can rationalize a lower critical solution temperature (LCST) as well, if the product $\varepsilon_s T$ decreases with increasing temperature.

The critical parameters (6) exclude immiscibilities in aqueous solutions near room temperature, where $T^* > 0.5$. Nevertheless, liquid–liquid coexistence curves were found at such conditions for some tetraalkylammonium salts with large cations and large anions [75–77]. In early debates of ionic criticality, this observation led to some confusion. Originally studied near the UCST [77], such gaps later proved to be closed loops with the LCST suppressed by crystallization [72, 78]. In one case, the LCST could be reached [79, 80].

Due to the obvious non-Coulombic nature, the mechanism of these transitions was investigated in great detail [72, 76, 81], also with regard to an unusual pressure dependence of the miscibility gaps [78, 80, 82]. All evidence suggests that the hydrophobic nature of the cation in an aqueous environment drives these immiscibilities, just as in many aqueous solutions of hydrophobic nonelectrolytes. This interpretation is confirmed by an analysis of the solution thermodynamics in terms of a semiempirical EOS [76] and more refined statistical mechanical calculations [81]. Some well-studied aqueous solutions of weak organic electrolytes such as isobutyric acid + water [6, 8] seem to represent such hydrophobic demixings as well.

In passing, we note that such hydrophobic interactions enforce cation–anion and cation–cation pairing [72, 76, 81]. In contrast to expectations from electrostatic arguments, cation–anion pairing increases with increasing ion size— for example, in the anion series $F^- < Cl^- < Br^- < NO_3^- < I^- < ClO_4^-$. In biophysical chemistry, this anion series is known as "lyotropic series" or "Hofmeister series" [83], which governs salting-out effects and many other properties of biomolecules in aqueous solutions [84]. The tendency for hydrophobic phase separation increases along this anion series as well [72].

The dichotomy of Coulombic versus hydrophobic interactions in driving criticality was probably first clearly stated in Ref. 76. Becoming aware that mechanisms of the hydrophobic type also occur in some nonaqueous solvents of high cohesive energy density e.g., in ethylene glycol, formamide, etc., Weingärtner et al. [72] created the term *solvophobic demixing*, which is now widely in use for discriminating these phase equilibria from Coulombic transitions [5, 15, 17].

Another subtle case, where specific interactions may obscure the effects of Coulombic criticality, is ethylammonium nitrate $(EtNH_3NO_3)$ +1-octanol $(T_c \cong 315 \, K)$ [85]. In contrast to all other known examples, the critical point is located in the salt-rich regime at a critical mole fraction of $X_c \cong 0.77$. Electrical conductance data indicate strong ion pairing, presumably caused by a hydrogen bond between the cation and anion which stabilizes the pairs in excess to what is expected from the Coulombic interactions [85]. This warns that, beyond the Coulombic/solvophobic dichotomy widely discussed in the literature, additional mechanisms may affect the phase separation [5].

2. Coexistence Curves

Expansion of the asymptotic power law for the order parameter in terms of a Wegner series (3) leads to [43]

$$M = B_0 \tilde{T}^\beta \left(1 + B_1 \tilde{T}^\Delta + B_2 \tilde{T}^{2\Delta} + \cdots\right). \qquad (7)$$

Note that the correction amplitudes B_i are not independent, but related to B_1 through $B_2 \propto B_1^2$, $B_3 \propto B_1^3$, and so on [43]. Moreover, their signs are well-defined: A converging series requires $B_1 > 0$ followed by alternating signs for the higher terms [86]. One should therefore be careful when trying to assess the physical significance of the sign and magnitude of correction amplitudes obtained in fits with freely adjustable amplitudes.

It is usually stated that there is some ambiguity in the order parameter in Eq. (7), because, on thermodynamic or experimental grounds, many choices exist, and there is *a priori* no reason for selecting a preferred variable [8]. Experience for nonionic fluids shows that all reasonable choices (e.g., the mole fraction or volume fraction) yield the same value of β [6, 8]. In contrast, the size of the asymptotic range, the numerical values of the Wegner corrections, and the behavior of the diameter of the coexistence curve depend on the choice of M. If one accepts, however, the principle of isomorphism between one- and two-component systems, one can argue that, among all variables of practical relevance, the volume fraction ϕ_i of component i is the best choice [58]. In fact, in terms of ϕ_i, a symmetrization of coexistence curves is often achieved [6, 8]. According to these arguments, the mole fraction X_i is only suitable in data analysis, when the molar volumes of the components are of similar size. The refractive index n, which often forms the primary experimental quantity, or better the Lorenz–Lorentz function $(n^2 - 1)/(n^2 + 2)$, may also form reasonable choices, because both are almost linearly related to ϕ_i. However, other views can be found [8].

Figure 2 shows an example of the order parameter dependence of the coexistence curve data for tetra-*n*-butylammonium picrate (Bu$_4$NPic) + 1-dodecanol [87] plotted in terms of mole and volume fractions, respectively. If represented in the mole fraction scale, the phase diagram is highly skewed and located in the solvent-rich regime. As emphasized by Fisher [15], such highly skewed phase diagrams resemble those of polymer solutions in poor solvents [88], suggesting that we look for theoretical analogies. The volume fraction leads to a more symmetric phase diagram, and the asymptotic range becomes larger. No variable was found that generates a fully symmetric coexistence curve [87].

Table II summarizes the existing studies of ionic criticality and lists the critical parameters. In the following, we will focus on results for immiscibilities which seem to be primarily driven by Coulombic interactions, as exemplified by Pitzer's system *n*-hexyl-triethylammonium *n*-hexyl-triethyl-borate (HexEt$_3$N$^+$HexEt$_3$B$^-$) + diphenylether [35], solutions of Bu$_4$NPic in alcohols [87], and solutions of Na in NH$_3$ [46].

We note that some coexistence curves of ionic systems with pronounced non-Coulombic interactions were also investigated with great care. These include aqueous solutions of tetraalkylammonium salts [77, 79] and solutions

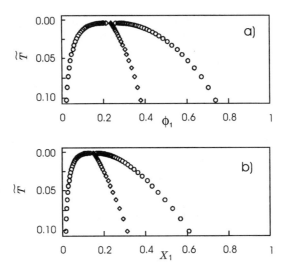

Figure 2. Liquid–liquid coexistence curve of the system tetra-n-butylammonium picrate + 1-dodecanol [87] plotted (a) in terms of the volume fraction ϕ_1 and (b) in terms of the mole fraction x_1 of the salt.

of $EtNH_3NO_3$ in 1-octanol [89]. In these cases, plain Ising behavior was found. The same is true for many phase-separating aqueous solutions of organic compounds with OH, amine, and COOH groups that are slightly conducting, as exemplified by isobutyric acid + water. These are, however, usually considered as nonionic fluids; for reviews and references see Refs. 6, 8, and 90. Plain Ising behavior was also found in other cases: ethyl-trimethyl-ammonium bromide ($EtMe_3NBr$) + $CHCl_3$ doped with ethanol [91] to suppress crystallization; tetra-n-butylammonium naphtylsulfonate (Bu_4NNS) + toluene [92]; n-heptyl-tri-n-butylammonium dodecylsulfate ($HepBu_3$-NDS) + cyclohexane [92, 93]. The observation of Ising behavior in the latter systems is interesting, because the solvents are fairly inert, but in total, these systems are little characterized, and the role of specific interactions remains unclear. When specific interactions drive the phase separation, results may bear little relevance for the problem of ionic criticality.

Turning now to phase transitions primarily driven by Coulombic forces, Pitzer's group [94] reported in 1985 experiments for Bu_4NPic + 1-chloroheptane with an UCST at $T_c \cong 414$ K. The experiments indicate a parabolic coexistence curve, but their limited precision gave rise to doubts [14, 77]. Singh and Pitzer were therefore quick to search for a more suited Coulombic system [71], and in 1990 they reported a system comprising the low-melting salt $HexEt_3NHexEt_3B$ dissolved in diphenylether with $T_c \cong 317$ K [35].

TABLE II
Survey of Work on Liquid–Liquid Phase Separations in Binary Electrolyte Solutions

System [a]	T_c, K [b]	Composition	Method [c]
HexEt$_3$NHexEt$_3$B + diphenylether	317.69	$X_c = 0.053$	C [35,71]
	311.18	$X_c = 0.049$	C [96]
	294.82	$X_c = 0.049$	T [95]
	308.69	$X_c = 0.049$	T,S [95], D [107]
	312.49	$X_c = 0.049$	V [125]
	309.50	$X_c = 0.049$	I [134]
Bu$_4$NPic + 1-chloroheptane	414.4	$X_c = 0.085$	C [94]
Bu$_4$NPic + 1-tetradecanol	351.09	$w_c = 0.272$	C [87], T,S [93], D [111]
Bu$_4$NPic + 1-tridecanol	341.73	$w_c = 0.298$	S,D [97]
	342.35	$w_c = 0.284$	C [87], T,S [93], D [111]
	341.79	$w_c = 0.298$	T [108-110]
	341.85	$w_c = 0.298$	V [122]
Bu$_4$NPic + 1-dodecanol	335.91	$w_c = 0.295$	C [87], T,S [93], D [111]
	334.8	$w_c = 0.298$	C [89]
	331.99	$w_c = 0.304$	T [108-110]
	334.38	$w_c = 0.304$	E [133]
Bu$_4$NPic + 1-undecanol	326.53	$w_c = 0.309$	T [108–110]
Bu$_4$NPic + 1-decanol	318.29	$w_c = 0.315$	C [87], T,S [93], D [111]
Bu$_4$NPic + 2-propanol	318.29	$w_c = 0.350$	C [87], T,S [93], D [111]
Bu$_4$NPic + 1,2-propanediol	322.19	$w_c = 0.410$	T [108–110]
Bu$_4$NPic + 1,4-butanediol	332.43	$w_c = 0.469$	T [108–110]
	331.90	$w_c = 0.469$	E [133]
EtNH$_3$NO$_3$ + 1-octanol	315.19	$X_c = 0.773$	T,S,D [116]
	314.94	$X_c = 0.766$	C [89]
	313	$X_c = 0.760$	S [118]
	302.31 [d]	$X_c = 0.760$	N [117]
	313.56	$X_c = 0.766$	V [126]
EtMe$_3$NBr + chloroform [e]	298.84	$X_c = 0.053$	C,S [91]
Bu$_4$NNS + toluene	319	$X_c = 0.106$	C,S [92]
HepBu$_3$NDS + cyclohexane	356	$X_c = 0.140$	C,S [92,111]
Pe$_4$NBr + H$_2$O	404.90	$X_c = 0.0305$	C [77]
Bu3PrNI + H$_2$O (UCST)	345.72	$w_c = 0.4377$	C [79], C,S,D,T [5]
Bu3PrNI + H$_2$O (LCST)	331.72	$w_c = 0.4457$	C [79], C,S,D,T [5]
Na + NH$_3$	231.55	$X_c = 0.0415$	C [46]
	231.55	$X_c = 0.0415$	N [112,113]

[a] Abbreviations: Me, methyl; Et, ethyl; Pr, n-propyl; Bu, n-butyl; Pe, n-pentyl; He, n-hexyl; Hep, n-heptyl; Pic, picrate; NS, naphtylsulfonate; DS, dodecyl-sulfate.

[b] Some samples show a temperature drift with time, so that T_c is only approximately given, or there are different samples with different T_c values within a series of measurements. For details we refer to the original papers.

[c] Abbreviations: C, coexistence curve; S, static light scattering; D, dynamic light scattering; T, turbidity; N, coherent neutron scattering; V, shear viscosity; E, electrical conductance; I, interfacial phenomena (ellipsometry).

[d] Deuterated 1-octanol.

[e] Sample contained 1 wt% ethanol.

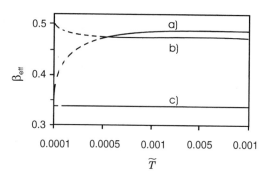

Figure 3. Effective exponent β_{eff} of Pitzer's system n-hexyl-triethylammonium n-hexyl-triethylborate + diphenylether. Curves a and b are derived from Singh and Pitzer's data presuming asymptotic mean-field behavior and asymptotic Ising behavior [35], respectively. Curve c is derived from the data of Wiegand et al. [96].

Refractive indices of the coexisting phases were measured at $\tilde{T} \geq 2 \times 10^{-5}$. Figure 3 shows the effective exponent $\beta_{\text{eff}} = d \ln M/d \ln \tilde{T}$, defined by the local slope in the double-logarithmic plot of M versus $\tilde{T}$. At $\tilde{T} > 10^{-4}$ the behavior is mean-field-like. At lower $\tilde{T}$, the data are consistent with mean-field behavior; but at the cost of large amplitudes of the Wegner corrections, they can also be forced to reflect asymptotic Ising behavior. We anticipate here that Sing and Pitzer's results were immediately confirmed by the turbidity experiments of Zhang et al. [95].

Due to the extraordinary implications of these observations in these two studies, Pitzer's system was reexamined with regard to the coexistence curve and to other critical properties. Wiegand et al. [96] remeasured the refractive indices of the coexisting phases in the range $10^{-4} < \tilde{T} < 0.04$. The results disagree with Singh and Pitzer's data, yielding $\beta = (0.34 \pm 0.01)$ close to the Ising value. Crossover is not detectable. Figure 3 compares the effective exponents resulting from these data with Pitzer's original results. This brings, of course, the status of Singh and Pitzer's data into question.

We note that Pitzer's value of $T_c \cong 317\,\text{K}$ could not be reproduced in later work which, depending on the sample, yielded values between 288 and 309 K [95, 96]. These differences seem to indicate a chemical instability of the salt, so that decomposition products displace T_c. Weingärtner et al. [72, 97] suggested that solutions comprising Bu_4NPic as a low-melting salt may behave more reproducibly, because the salt is chemically more stable.

Extending earlier work [72], Schröer, Weingärtner and coworkers [87] were able to observe the critical points of Bu_4NPic in a homologous series of alcohols with dielectric constants between 16.8 (2-propanol) and 3.6

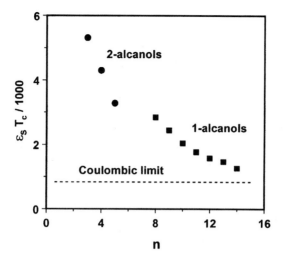

Figure 4. Dielectric constant–temperature product $\varepsilon_s T_c$ at the critical temperature for tetra-n-butylammonium picrate + alcohols as a function of the chain length n of the alcohols [87]. The dashed line reflects the RPM prediction with $\sigma = 0.6\,\text{nm}$.

(1-tetradecanol). Figure 4 shows the critical temperatures as a function of the chain length n of the alkyl residues. For a given salt, Eq. (4) predicts the product $T_c \cdot \varepsilon_s$ to be constant; one finds $T_c \cdot \varepsilon_s \cong 800\text{–}1000$. Figure 4 shows that, when increasing the apolar part of the molecule, one clearly moves from a distinctly non-Coulombic to an essentially Coulombic mechanism for phase separation.

Some of these systems were subsequently employed in precise studies of the coexistence curves [87]. Figure 5 shows the effective exponents β_{eff} based on volume fractions derived from these data. When reducing the solvophobic character of the systems by increasing the chain length n of the alcohols, deviations of β_{eff} from the Ising value become visible which show a trend toward the mean-field value.

The results substantiate earlier observations for the liquid–liquid phase transition of Na + NH$_3$. This system shows a transition to metallic states in concentrated solutions; but in dilute solutions and near criticality, ionic states prevail [98], and the gross phase behavior seems to be in accordance with a Coulombic transition [37]. Crossover was found at $\tilde{T} = 10^{-2}$ [46], and it seems to be much more abrupt than in the picrate systems. However, much depends on the subtle details of the data evaluation. Das and Greer [99] could smoothly represent the data by a Wegner series.

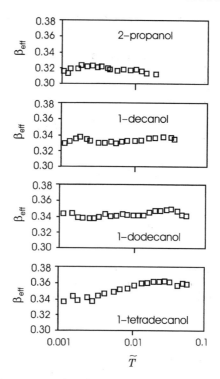

Figure 5. Effective exponents β_{eff} of tetra-n-butylammonium picrate dissolved in several alcohols [87] as a function of the reduced temperature.

The diameter $(M_1 + M_2)/2$ of the coexistence curve is represented by the series [100, 101]

$$(M_1 + M_2)/2 = M_c + D_0\tilde{T}^{1-\alpha}(1 + D_1\tilde{T}^{\Delta} + D_2\tilde{T}^{2\Delta} + \cdots) + A\tilde{T} + C\tilde{T}^{2\beta} \tag{8}$$

involving the exponent α of the heat capacity. The linear term contains both a critical contribution and a regular contribution. Moreover, there is a spurious 2β contribution [8, 102], when an improper order parameter is chosen in data evaluation. In a mean-field system with $\alpha = 0$ and $\beta = 1/2$ the diameter is "rectilinear." For a long time, the weak diameter anomaly in nonionic systems was a matter of controversy, because the deviations from rectilinear behavior are small [6, 8], and the 2β contribution obscures the results.

A careful investigation of the picrate systems yields a substantial diameter anomaly [87] observed with all reasonable choices of the order parameter (see, however, somewhat contradicting results in Ref. 89). The data are consistent with the predicted $(1 - \alpha)$ anomaly. Large diameter anomalies are expected, when the intermolecular interactions depend on the density, as expected in these cases: The dilute phase is essentially nonconducting and mainly composed of neutral ion pairs, while the concentrated phase is a highly conducting ionic melt [68]. However, any general conclusion is weakened by the fact that with Pitzer's system no such anomaly was observed [96].

3. Scattering and Turbidity

Measurements of static light or neutron scattering and of the turbidity of liquid mixtures provide information on the osmotic compressibility χ and the correlation length ξ of the critical fluctuations and, thus, on the exponents γ and v. Owing to the exponent equality $\gamma = v(2 - \eta) \cong 2v$, data about γ and v are essentially equivalent. In the classical case, $\gamma = 2v$ holds exactly. Dynamic light scattering yields the time correlation function of the concentration fluctuations which decays as $\exp(-Dk^2t)$, where k is the wave vector and D is the diffusion coefficient. Kawasaki's theory [103] then allows us to extract the correlation length, and hence the exponent v.

Following Singh and Pitzer's report on the parabolic coexistence curve [35], there was immediate interest in the exponent γ, because Khodolenko and Beyerlein [104, 105] claimed that the RPM is described the spherical model. The spherical model predicts indeed a parabolic coexistence curve with $\beta = 1/2$ [106], but the other exponents differ largely from the classical values (e.g., $\gamma = 2$ instead of 1). By light scattering measurements for $Bu_4NPic + 1$-tridecanol, Weingärtner et al. [97] immediately ruled out the spherical model. Turbidity measurements in Pitzer's system by Zhang et al. [95] yielded the same conclusion.

Later, both Pitzer's system and several picrate systems were reexamined. Let us first consider results for Pitzer's system. In the original turbidity study, Zhang et al. [95] performed experiments at $10^{-4} < \tilde{T} < 10^{-1}$ and found $\gamma = 1.01 \pm 0.01$. Crossover did not appear in this temperature range. Of course, at the cost of sufficiently large amplitudes of the Wegner coefficients, the data can be forced to asymptotic Ising-like criticality. However, turbidity measurements performed later in the same group with a new sample differed considerably [96], and they were indicative for Ising-like behavior. With the new sample, an Ising-like criticality was also obtained by dynamic scattering experiments [107]. Thus, there is the same situation as encountered in coexistence curve experiments: There are several sets of highly accurate data which, however, contradict fundamentally. The differences in critical temperatures obtained with the different samples point toward impurity effects that apparently change the critical behavior. Details are, however, poorly understood.

Again, more consistent results were obtained by different groups with the picrate systems. Following the early light scattering study of Weingärtner et al. [97] on Bu_4NPic dissolved in 1-tridecanol, Narayanan and Pitzer [108–110] performed turbidity experiments for this system and also for solutions of Bu_4NPic in the homologous solvents 1-undecanol and 1-dodecanol. The data show quite sharp crossovers from mean-field criticality away from T_c to asymptotic Ising criticality that occurs almost within one decade of $\tilde{T}$. This crossover is closest to T_c for the highest homologue, where Coulombic interactions are expected to be strongest.

Phase separations could also be observed for Bu_4NPic in alkanediols and glycerol, but the latter solvents possess a high cohesive energy density, and the phase transition seems to be of solvophobic nature [72]. In fact, Narayanan and Pitzer [108–110] showed that addition of 1,4-butanediol to 1-dodecanol shifts the crossover region away from T_c. For pure 1,4-butanediol, plain Ising behavior is observed.

Very recently, Kleemeier repeated the turbidity measurements for some systems [93], and he supplemented the experiments by static [93] and dynamic [111] light scattering experiments. His results confirm some essential features of Pitzer's data—for example, that crossover becomes more relevant at higher chain length of the alcohol. However, according to Kleemeier's data, crossover seems to be more smooth than observed by Narayanan and Pitzer, as also suggested by the coexistence curve data discussed in the preceding section. Because any firm conclusions concerning the sharpness of crossover depend on subtle details of the data evaluation, a final answer is not yet available.

For $Na + NH_3$, light scattering is unsuitable, owing to strong light absorption over the entire spectral range, but neutron scattering is the method of choice. In such experiments [112, 113], sharp crossover was noted to occur for both the osmotic susceptibility and correlation length, in a range consistent with that of the coexistence curve. In contrast, neutron scattering results for $K + KBr$ ($T_c \cong 1000\,K$) [113, 114] and similar systems [115] imply Ising-like behavior. It remains to note that for the other cases mentioned in Table 2, the Ising-like character of the phase transition deduced from the coexistence curves has been reconfirmed by light scattering measurements [5, 92, 116–118].

In concluding this section, we draw attention to the amplitudes ξ_0 derived from the scattering experiments. As shown later, ξ_0 enters into theoretical expressions for the crossover temperature. Large ξ_0 favor a small Ising regime. In simple nonelectrolyte mixtures, ξ_0 is of the order of the molecular

CRITICALITY OF IONIC FLUIDS

diameter, 0.1–0.4 nm say. In some ionic fluids, considerably larger amplitudes were observed. While it would now appear that such large ξ_0 values are
a signature of ionic systems, the detailed results for the amplitudes are
obviously not related to the mean-field versus Ising problem in a transparent
way.

4. Viscosity

Perhaps a more decisive discrimination between Ising and mean-field behavior could be provided by the investigation of "weak" anomalies [6] as predicted for the specific heat. Such weak anomalies are absent in the mean-field
case (cf. Table I). Except for the diameter anomalies already mentioned, no
thermodynamic investigations of weak anomalies were reported so far. However, dynamical properties such as the shear viscosity and electrical conductance may show weak anomalies as well.

The viscosity of Ising-like systems is known to exhibit a weak divergence
of the form [119]

$$\eta/\eta_{bg} = (Q_0\xi)^z = (Q_0\xi_0)^z \tilde{T}^{-y} \qquad \text{as } \tilde{T} \to 0, \qquad (9)$$

where η_{bg} is the background viscosity, Q_0 is a system-dependent wave vector,
$(Q_0\xi_0)^z$ is a system-dependent critical amplitude, y is the viscosity exponent,
and $z = v \cdot y$. For an Ising system, mode coupling theory and dynamic RG
theory both yield $z = 0.065$, that is $y = 0.04$ [120], in agreement with the
best experimental data [121]. For mean-field systems there seems to be no
decisive answer, but probably mean-field behavior excludes an anomalous
viscosity enhancement [122, 123].

Figure 6 shows data for the critical contribution to the viscosity of the
system Bu$_4$NPic + 1-tridecanol [122], obtained after an appropriate
treatment of the background term. The data ensure scaling behavior with
an Ising exponent. Further away from T_c, the anomaly was found to vanish, in
accordance with crossover theory [124]. With ξ_0 taken from light scattering
data, the wave vector Q_0 was found to fall in the broad range of values
reported for nonionic fluids. Wiegand et al. [125] and Oleinikova and Bonetti
[126] observed similar anomalies for Pitzer's system and for EtNH$_3$NO$_3$ +
1-octanol, respectively. These experiments clearly prove the Ising-like
character of the critical points.

5. Electrical Conductance

Theory suggests that the electrical conductance Λ exhibits an anomalous contribution $\Lambda \propto \tilde{T}^{\Theta}$. With regard to the critical exponent Θ, one may think of
several scenarios. Scaling behavior with $\Theta = 1 - \alpha$ is expected for short-
range fluctuations [127] and also for a proton-hopping mechanism [128]. A

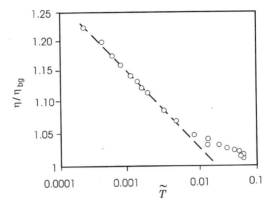

Figure 6. Anomalous contribution η/η_{bg} to the viscosity of the system tetra-*n*-butylammonium picrate + 1-tridecanol [122]. The dashed line represents the theoretical limiting slope for the Ising model.

percolation theory yields $\Theta = 2\beta$ [129, 130]. Yet another scenario is scattering of ions by concentration fluctuations with $\Theta = \nu$ [131]. All these theories place the critical exponent between $\nu = 0.63$ and $1 - \alpha = 0.89$. In view of weak effects obscured by large analytical background contributions, discrimination among these scenarios is extremely difficult.

Until recently, experiments were limited to aqueous solutions of nonelectrolytes which are weakly ionized such as isobutyric acid or phenol, or which were doped with ions to achieve conductance. Andersen and Greer [132] critically assessed many earlier data and concluded that a $(1 - \alpha)$ singularity is most probable.

Oleinikova and Bonetti [133] extended this type of measurements to Bu_4NPic dissolved in 1-dodecanol, in 1,4-butanediol, and in mixtures of these solvents. From all systems employed in conductance measurements, Bu_4NPic + 1-dodecanol may come closest to a Coulombic fluid. The anomaly becomes notable near $\tilde{T} = 10^{-2}$. The anomalous contribution never exceeded a few percent of the background contribution, which makes any decisive determination of the critical exponent extremely difficult. If correction exponents were included, the fits gave the best results for a $(1 - \alpha)$ anomaly. No essential differences between results for 1-dodecanol and 1,4-butanediole systems seem to be present.

It is interesting to compare conductance behavior with that of the shear viscosity, because conventional hydrodynamic conductance theories relate Λ to the frictional resistance of the surrounding medium. At first glance, one would expect from the Stokes–Einstein equation a critical anomaly of the conductance which is closely related to that of the shear viscosity. This is, however, neither predicted theoretically nor found experimentally.

6. Interfacial Properties

Caylor et al. [134] performed ellipsometric measurements at the liquid–liquid interface of Pitzer's system. The experiments yielded a decrease in ellipticity, when approaching the critical point. This is in clear contrast to a $(\beta - v)$ divergence predicted theoretically [135] and observed experimentally [136] for nonionic Ising mixtures. The results point toward an anisotropic interface caused by an orientation of ion pairs perpendicular to the interface. A rough model for such an interface captured some of the observed features.

7. The Ion Distribution Near Criticality

In concluding this section on liquid–liquid phase transitions, we briefly consider the available experimental information on the ion distribution near criticality. In the absence of scattering experiments, most experimental data come from electrical conductance data [72, 137, 138]. Moreover, there are some data on the concentration dependence of the dielectric constant in low-ε solutions [139] and near criticality [138].

We focus on the conductance data. In systems of the type considered here, the conductance is primarily determined by the degree of ion pair association α. However, at higher ion densities, substantial mobility effects come into play. In the absence of sufficiently accurate conductance theories for the region of interest, a reliable measure for estimating α is the conductance–viscosity product $\Lambda\eta$ which is often denoted as the "Walden product."

Figure 7 shows isotherms for the Walden product at $T \cong T_c$ for Bu$_4$NPic + 1-tridecanol [72] and Bu$_4$NPic + 1-chloroheptane [137] as a function of the

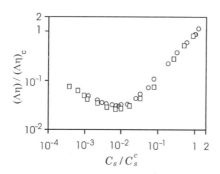

Figure 7. Conductance–viscosity products $\Lambda\eta$ of tetra-n-butylammonium picrate dissolved in 1-tridecanol [72] (○) and 1-chloroheptane (□) [137] as a function of the salt concentration along near-critical isotherms normalized to the values at the critical points.

molar concentration C_S of the salt. The data are normalized to the values at the critical points [138]. Both sets of data fall onto a common curve, which clearly indicates corresponding states behavior of the degree of dissociation. Moreover, a major feature is a conductance minimum at a concentration about two orders of magnitude below the critical concentration. This minimum is generally found in solvents of low dielectric constant, at $T^* < 0.12$ say [37, 68]. Along the low-concentration branch, neutral ion pairs dominate, but begin to break up, when infinite dilution is approached, as predicted by the mass action law for ion pair formation [140]. Thus, the dilute phase remains conducting as $\rho^* \to 0$.

The high-concentration branch, which encompasses the critical point, is characterized by an increase of Λ with increasing salt concentration. Below T_c, coexistence occurs between a weakly conducting vapor-like phase and a highly conducting liquid-like phase. The density dependence of Λ found in conductance experiments for NH_4Cl [55] resembles the increasing branch in Figure 7. The conductance increase indicates the reappearance of charged species. In electrolyte solution theory, the origin of this redissociation has been of long-standing and controversial debate. Heuristically, one may argue that such a redissociation must occur, because eventually the solution has to rearrange to obtain the fully dissociated structural pattern of the molten salt. Apart from the coexistence curve itself, the conductance isotherms and the location of the conductance minimum form major targets for testing theories of the species population near criticality [68, 137, 141].

C. Liquid–Vapor Transitions in Aqueous Electrolyte Solutions

Apart from liquid–liquid transitions, liquid–vapor transitions in aqueous electrolyte solutions have played a crucial role in debates on ionic criticality [142–144]. The liquid–vapor transition is usually associated with a mechanical instability with diverging density fluctuations, while liquid–liquid transitions are associated with a material instability with diverging concentration fluctuations. This requires, however, that both regimes are well-separated. Their interference can lead to complex phase behavior with continuous transitions from liquid–liquid demixing to liquid–gas condensation [9, 145, 146]. It is then not trivial to define the order parameter [147–149].

To single out the peculiarities in the phase behavior of ionic fluids, it is convenient to consider first the behavior of nonionic (e.g., van der Waals-like) mixtures. We note, however, that the subsequent considerations ignore liquid–solid phase equilibria, which in real electrolyte solutions can lead to far more complex topologies of the phase diagrams than discussed here [150].

In the standard classification of Scott and Konynenburg [145,146] the simplest case is type I behavior, where liquid–liquid demixing is absent.

Then, the vapor pressure curves of the two components are terminated by critical points which, in turn are connected by a continuous liquid–gas (L–G) critical line. For nonionic fluids, type I is only found for components with critical temperatures that are very close to one another—for example, neighboring compounds in homologous series.

For larger differences between the critical temperatures, one expects liquid–liquid miscibility gaps that are well-separated from the liquid–gas transition (type II). When further increasing the dissimilarity, these liquid–liquid immiscibilities are displaced to higher temperatures, and eventually the corresponding liquid–liquid–gas (L_1–L_2–G) three-phase line interferes with the L–G critical line (type III, IV, and V). Then, the L–G critical line starting from the critical point of the more volatile component (here water) is broken at a so-called upper critical end point (UCEP), where it meets the L_1–L_2–g three-phase line of the liquid–liquid equilibrium.

For electrolyte solutions such as NaCl + water the critical temperatures of the pure components differ by about a factor of five. From the perspective of nonelectrolyte thermodynamics, the absence of a liquid–liquid immiscibility then comes as a great surprise. It is a major challenge for theory to explain why this salt, as well as similar salts such as KCl or $CaCl_2$, seems to show a continuous critical line. Perhaps there is a slight indication for a transition toward an interrupted critical curve in Marshall's study [151] of the critical line of NaCl + H_2O. Marshall observed a dip in the $T_c(X_S)$ curve some K away from the critical point of pure water, which at first glance seems obscure. It was suggested [152] that the vicinity to an upper critical end point leaves its mark by this dip.

We note that for aqueous solutions of some salts with ions of higher valence such as $BaCl_2$ [73] or UO_2SO_4 [74] such a liquid–liquid immiscibility is indeed observed. Moreover, for some other salts such as $MgSO_4$ a metastable liquid–liquid immiscibility at elevated temperatures [76] seems to be suppressed by a retrograde salt solubility that rapidly decreases as the temperature is increased.

The continuous critical line for systems such as NaCl + H_2O offers a temperature window for studying the behavior of electrolyte solutions near their liquid–vapor transition. Pitzer [4,13,142,144] compiled much evidence that the nonclassical fluctuations in pure water are apparently suppressed when adding electrolytes. Thus, from the application's point of view, a classical EOS may be quite useful. The pressing question is to what degree these observations withstand more quantitative analysis.

A key role in this debate was played by experiments by Bischoff and Rosenbauer [153], who reported accurate data on isothermal vapor–liquid coexistence curves as a function of pressure near the critical line of NaCl + H_2O. Far from the critical point of pure water, one expects the compositions of the coexisting phases to be given by

$$X_L - X_G = |P - P_c|^\beta, \qquad T = \text{const.} \qquad (10a)$$

Bischoff and Rosenbauer found $\beta = 1/2$. Closer to the critical point of pure water they found however $\beta > 1/2$.

In comments, Harvey and Levelt Sengers [154] and Pitzer and Tanger [155] drew attention to a subtle ambiguity in this analysis: At high dilution of the salt the system changes its character from a two-component to a one-component system and the mole fraction becomes an increasingly unsuitable order parameter. This behavior is well-investigated in work on supercritical nonelectrolyte mixtures, for which we refer to a pertinent review [156]. It can be shown that in the limit of vanishing salt concentration X_S, Eq. (10a) goes over into

$$X_L - X_G = |P - P_c|^{1+\beta}, \qquad T = \text{const.,} \quad X_S \to 0; \qquad (10b)$$

that is, the exponent changes from β to $1 + \beta$. Pitzer and Tanger [155] showed that this transition affects the entire region covered by Bischoff and Rosenbauer's data. Distinguishing between $1 + \beta = 1.32$ and $1 + \beta = 1.5$ is an appreciable task. The problem might be circumvented by using the density rather than the composition as an order parameter. Data for coexisting densities (in place of mole fractions) as a function of pressure are scarcely available.

Another claim for an apparent mean-field behavior of ionic fluids came from measurements of heat capacities. The weak Ising-like divergences of the heat capacities C_V of the pure solvent and $C_{P,x}$ of mixtures should vanish in the mean-field case (cf. Table I). The divergence of C_V is firmly established for pure water. Accurate experiments for aqueous solutions of NaCl [157] and NaOH [158] near the L–G critical line show that the heat capacity loses its weak divergence as small amounts of electrolytes are added.

Anisimov et al. [157] attributed this behavior to crossover with a rapidly shrinking asymptotic regime. There is, however, another possible mechanism [158] based on Fisher renormalization. Fisher renormalization [159] is a subtle renormalization effect along a path toward the critical point which, due to the presence of a hidden variable, differs from the path along which the exponent α is originally defined. This renormalization occurs in liquid mixtures on a path that is asymptotically parallel to a critical line.

Clearly, the simultaneous presence of crossover from Ising to mean-field criticality, a transition from two-component behavior of solutions to one-component behavior, and the possible presence of Fisher renormalization renders any analysis difficult.

D. Liquid–Liquid Demixings in Multicomponent Systems

There is a large body of experimental work on ternary systems of the type salt + water + organic cosolvent. In many cases the binary water + organic solvent subsystems show "reentrant phase transitions," which means that there is more than one critical point. Well-known examples are closed miscibility loops that possess both a LCST and a UCST. Addition of salts may lead to an expansion or shrinking of these loops, or may even generate a loop in a completely miscible binary mixture. By judicious choice of the salt concentration, one can then achieve very special critical states, where two or even more critical points coincide [90, 160, 161]. This leads to very peculiar critical behavior—for example, a doubling of the critical exponent γ. We shall not discuss these aspects here in detail, but refer to a comprehensive review of reentrant phase transitions [90]. We note, however, that for reentrant phase transitions one has to redefine the reduced temperature $\tilde{T}$, because near a given critical point the system's behavior is also affected by the existence of the second critical point. An improper treatment of these issues will obscure results on criticality.

In the literature there have been repeated reports on an apparent mean-field-like critical behavior of such ternary systems. To our knowledge, this has first been noted by Bulavin and Oleinikova in work performed in the former Soviet Union [162], which only more recently became accessible to a greater community [163]. The authors measured and analyzed refractive index data along a near-critical isotherm of the system 3-methylpyridine (3-MP) + water + NaCl. The shape of the refractive index isotherm is determined by the exponent δ. Bulavin and Oleinikova found the mean-field value $\delta = 3$ (cf. Table I). Viscosity data for the same system indicate an Ising-like exponent, but a shrinking of the asymptotic range by added NaCl [164].

The observation of crossover has later been substantiated by several other studies. In particular, Jacob et al. [165] performed light scattering measurements on the system 3-MP + water + NaBr. The data indicate comparatively sharp crossover in the range $10^{-4} < \tilde{T} < 10^{-2}$, which becomes more pronounced when increasing the salt concentration. It is intriguing to characterize this crossover by a suitably defined crossover temperature $\tilde{T}_{\times}$, defined here by the point of inflection in the $\tilde{T}$-dependence of the effective exponent γ_{eff}. Figure 8 shows $\tilde{T}_{\times}$ as a function of the amount of added NaBr. Eventually, plain mean-field behavior is obtained in a solution containing about 16.8 mass% NaBr.

As viewed from the Coulombic versus solvophobic dichotomy, the mean-field-like behavior is quite unexpected. Thermodynamic properties of the salt-free systems clearly point toward a hydrophobic mechanism for phase

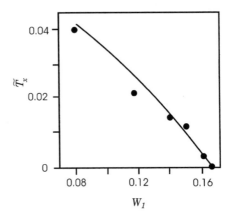

Figure 8. Crossover temperature $\tilde{T}_x$ in the ternary system 3-methylyridine + water + sodium bromide as a function of the salt concentration given in weight-percent. For details see text. Redrawn from Ref. 165 with permission.

separation. One would then expect the salt just to enhance this mechanism, which implies that the critical behavior remains Ising-like. Clearly, this gives rise to the question of whether the source for crossover is the same as in the binary salt-solvent systems. As discussed later in Section VI.E, tricriticality indeed forms a more probable scenario for explaining these observations.

E. Summary

While the early work on molten NH_4Cl gave only some qualitative hints that the effective critical behavior of ionic fluids may be different from that of nonionic fluids, the possibility of apparent mean-field behavior has been substantiated in precise studies of two- and multicomponent ionic fluids. Crossover to mean-field criticality far away from T_c seems now well-established for several systems. Examples are liquid–liquid demixings in binary systems such as Bu_4NPic + alcohols and Na + NH_3, liquid–liquid demixings in ternary systems of the type salt + water + organic solvent, and liquid-vapor transitions in aqueous solutions of NaCl. On the other hand, Pitzer's conjecture that the asymptotic behavior itself might be mean-field-like has not been confirmed.

In binary systems there is now some understanding of the role of Coulombic and solvophobic forces in driving the phase separation. The effect of this interplay is now well-established for solutions of Bu_4NPic in alcohols. When the chain length of the alcohol is increased, so that ε_S is lowered and the ionic forces become stronger, the crossover range is displaced toward T_c. An interesting speculation is that in the limit of infinite chain length the systems approach mean-field behavior. This would imply that in other cases specific interactions such as solvophobic effects destroy the mean-field criticality. From this perspective it is not finally decided what behavior could be expected for a generic Coulombic system such as the RPM.

Finally, a recurring result indicates that crossover is completed within a quite small temperature range, and in some cases is even nonmonotonous. Actually, accounting for the uncertainties necessarily involved in the primary experimental data and in data evaluation, the sharpness of crossover may be debated in some cases. However, at least at the qualitative level, sharper crossover than observed with nonionic fluids seems to be established. Such a sharp crossover has severe consequences for theoretical interpretations.

IV. THEORETICAL METHODS AT THE MEAN-FIELD LEVEL

A. Models for Ionic Fluids

Theories of ionic fluids usually start with simple Hamiltonians, in which only the essential features of real fluids are retained. Work on uncharged fluids (e.g., through the lattice-gas version of the Ising model) show that discrete-state lattice models have distinct theoretical advantages for treating criticality [166] (cf. Section VI.B). In trying to understand real ionic fluids, simple fluid models such as continuum models seem, however, to be more appropriate [15]. Thus, most theories have relied on "primitive models," which consider ions in a dielectric continuum, interacting by Coulombic forces. Assuming pairwise additivity of these forces, the ion–ion potential is given by

$$\varphi_{\alpha\beta}(r) = \begin{cases} \infty, & r \le \sigma_{\alpha\beta} \\ q_\alpha q_\beta / \varepsilon_0 r, & r > \sigma_{\alpha\beta} \end{cases} \tag{11}$$

with $\sigma_{\alpha\beta} = (\sigma_\alpha + \sigma_\beta)/2$. The model is often applied in its *restricted* version (RPM) in which all ions have the same diameter σ and absolute charge q. In fact, this is just the model behind many electrolyte theories, including Debye–Hückel theory [37]. Primitive models satisfy the corresponding states principle [37] which justifies the use of reduced variables defined by Eqs. (4) and (5).

For molten salts one sets $\varepsilon_0 = 1$. For electrolyte solutions $\varphi_{\alpha\beta}$ is a solvent-averaged potential [37]. Then, in real fluids, ε_0 in Eq. (11) depends on the ion density [167]. Usually, one sets $\varepsilon_0 = \varepsilon_S$, where ε_S is the dielectric constant of the solvent. A further assumption inherent in all primitive models is $\varepsilon_{in} = \varepsilon_S$, where ε_{in} is the dielectric constant *inside* the ionic spheres. This deficit can be compensated by a "cavity" term that, for electrolyte solutions with $\varepsilon_S > \varepsilon_{in}$, is repulsive. At zero ion density this cavity term decays as r^{-4} [17, 168]. At finite ion densities it is, however, shielded to shorter range by Debye screening [169–171].

More refined continuum models—for example, the well-known Fumi–Tosi potential with a soft core and a term for attractive van der Waals interactions [172]—have received little attention in phase equilibrium calculations [51]. Refined potentials are, however, vital when specific ion–ion or ion–solvent interactions in electrolyte solutions affect the phase stability. One can retain the continuum picture in these cases by using modified solvent-averaged potentials—for example, the so-called Friedman–Gurney potentials [81, 168, 173]. Specific interactions are then represented by additional terms in $\varphi_{\alpha\beta}(r)$ that modify the ion distribution in the desired way. Finally, there are models that account for the discrete molecular nature of the solvent—for example, by modeling the solvent as dipolar hard spheres [174, 175].

B.   Monte Carlo Simulations

Once the potential is chosen, Monte Carlo (MC) and molecular dynamics (MD) simulations form important tools for calculating phase equilibria [176]. With one notable exception [51], only MC techniques were employed. Methodological developments in MC techniques were addressed in a previous volume of this series [177], so that we summarize here only some aspects important for the treatment of ionic fluids.

In isochoric–isothermal MC simulations the phase transition lines can be determined from the free energy as a function of the state variables, using the criteria for thermodynamic equilibrium, mechanical stability, and critical points [178]. The Gibbs ensemble technique [54] now allows direct simulations of phase equilibria by running two simulations in physically detached but thermodynamically connected boxes that are representative of the coexisting phases. Particle transfer and volume exchanges between the boxes lead to an establishment of phase equilibrium. Originally, Panagiotopoulos [179] and Caillol [180, 181] shifted ions with a correction to maintain charge neutrality. Because the vapor contains almost exclusively ion pairs, this leads to severe problems in the convergence of the simulations. It seems more economical to transfer ion pairs [52, 53]. Moreover, the efficiency of grand canonical simulations can be improved [52] by employing a histogram reweighting technique [182]. Because at a given state, fluctuations contain information on neighboring states, histograms of fluctuating observables allow us, after appropriate reweighting, to extract thermodynamic properties of neighboring states.

One major problem in determining phase transition lines is associated with the use of finite systems, so that near criticality the correlation length of the fluctuations begins to exceed the size of the simulation box. Finite-size scaling theory [183, 184] provides expressions for the rounding of the critical point at finite size of the system, thus remedying these defects of the simulations. Special versions are now available for off-lattice fluids that lack symmetry between the coexisting phases [185].

C. Analytical Theories of the Restricted Primitive Model

1. General Issues

From the many tools provided by statistical mechanics for determining the EOS [36, 173, 186–188] we consider first integral equation theories for the pair correlation function $g_{\alpha\beta}(\mathbf{r}_\alpha, \mathbf{r}_\beta)$ of spherical ions which relates the density of ion β at location $\mathbf{r}_\beta$ to that of α at $\mathbf{r}_\alpha$. In most theories $g_{\alpha\beta}(\mathbf{r}_\alpha, \mathbf{r}_\beta)$ enters in the form of the total correlation function $h_{\alpha\beta}(\mathbf{r}_\alpha, \mathbf{r}_\beta) = g_{\alpha\beta}(\mathbf{r}_\alpha, \mathbf{r}_\beta) - 1$. The Ornstein–Zernike (OZ) equation splits up $h_{\alpha\beta}(\mathbf{r}_\alpha, \mathbf{r}_\beta)$ into the direct correlation function $c_{\alpha\beta}(\mathbf{r}_\alpha, \mathbf{r}_\beta)$ for pair interactions plus an indirect term that reflects these interactions mediated by all other particles γ:

$$h_{\alpha\beta}(\mathbf{r}_\alpha, \mathbf{r}_\beta) = c_{\alpha\beta}(\mathbf{r}_\alpha, \mathbf{r}_\beta) + \sum_\gamma \int c_{\alpha\beta}(\mathbf{r}_\alpha, \mathbf{r}_\gamma) \rho_\gamma(\mathbf{r}_\gamma) h_{\gamma\beta}(\mathbf{r}_\gamma, \mathbf{r}_\beta) d\mathbf{r}_\gamma. \quad (12a)$$

Approximations for $c_{\alpha\beta}(\mathbf{r}_\alpha, \mathbf{r}_\beta)$ are then used as a "closure" to solve the OZ equation for $h_{\alpha\beta}(\mathbf{r}_\alpha, \mathbf{r}_\beta)$.

In this section we consider isotropic, homogeneous fluids. Then $g_{\alpha\beta}(\mathbf{r}_\alpha, \mathbf{r}_\beta)$ depends only on the radial distance:

$$h_{\alpha\beta}(\mathbf{r}_{\alpha\beta}) = c_{\alpha\beta}(\mathbf{r}_{\alpha\beta}) + \sum_\gamma^{\rho_\gamma} \int c_{\alpha\gamma}(\mathbf{r}_{\alpha\gamma}) h_{\beta\gamma}(\mathbf{r}_{\beta\gamma}) d\mathbf{r}_\gamma \quad (12b)$$

with $\mathbf{r}_{\alpha\beta} = |\mathbf{r}_\beta - \mathbf{r}_\alpha| \equiv r$.

From the various possible closures, the mean spherical approximation (MSA) [189] has found particularly wide attention in phase equilibrium calculations of ionic fluids. The Percus–Yevick (PY) closure is unsatisfactory for long-range potentials [173, 187, 190]. The hypernetted chain approximation (HNC), widely used in electrolyte thermodynamics [168, 173], leads to an increasing instability of the numerical algorithm as the phase boundary is approached [191]. There seems to be no decisive relation between the location of this numerical instability and phase transition lines [192–194]. Attempts were made to extrapolate phase transition lines from results far away, where the HNC is soluble [81, 194].

Once $g_{\alpha\beta}(r)$ is known, the EOS can be extracted from equations for the internal energy, pressure, or compressibility, respectively. The approximate nature of $h_{\alpha\beta}(r)$ leads, however, to severe thermodynamic inconsistencies of the various routes for calculating the EOS. In the MSA, only the energy route is physically meaningful. The compressibility EOS yields only properties of the underlying hard-sphere fluid, and the pressure route yields unreasonable results as well [189]. Høye, Lebowitz, and Stell [195] have, however, constructed a thermodynamically consistent version termed "Generalized MSA" (GMSA) in which an *ad hoc* term is added to $c_{\alpha\beta}(r)$ to gain consistency.

Analytical solutions for the RPM are conveniently given in terms of the excess part Φ^{ex} of the reduced free energy density $\Phi = A\sigma^3/k_B T V = \Phi^{id} + \Phi^{ex}$, where A is the Helmholtz free energy and Φ^{id} the ideal gas contribution. For the MSA one finds for the ion–ion contribution [189]

$$\Phi^{II} = \left[2 + 6x + 3x^2 - 2(1 + 2x)^{3/2}\right]/12\pi\sigma^3. \tag{13}$$

Thus, Φ^{II} is only a function of the inverse Debye screening length Γ_D, defined by

$$\Gamma_D^2 = 4\pi\beta q^2\rho/\varepsilon_0, \tag{14}$$

which enters into Eq. (13) in form of the dimensionless "reduced Debye length"

$$x = \Gamma_D\sigma = (4\pi\rho^*/T^*)^{1/2}. \tag{15}$$

Alternatively, there has been a revival of Debye–Hückel (DH) theory [196–199] which provides an expression for the free energy of the RPM based on macroscopic electrostatics. Ions j are assumed to be distributed around a central ion i according to the Boltzmann factor $\exp(-\beta q_j\phi_j(r))$, where $\phi_j(r)$ is the mean local electrostatic potential at ion j. By linearization of the resulting Poisson–Boltzmann (PB) equation, one finds the Coulombic interaction to be screened by the well-known DH screening factor $\exp(-\Gamma_D r)$. The ion–ion contribution to the excess free energy then reads

$$\Phi^{II} = \left[\ln(1 + x)) - x + x^2/2\right]/4\pi\sigma^3, \tag{16}$$

which differs from the MSA result (13) in order x^2. Note that in Eq. (16) the excluded volume of the ions is only reflected by the electrostatic boundary condition. Even without an excluded volume term, both Eqs. (13) [189] and (16) [200] predict a two-phase regime with an upper critical point. This feature is lost if these expressions are reduced to the well-known DH limiting law at low ρ^*, $\Phi^{ex} = \Gamma_D^3/12\pi\beta$, which is common to both theories. If the charges are, however, switched off, the free energy reduces to that of an ideal gas. To remedy this defect, one has to add a hard-sphere term— for example, based on the Carnahan–Starling EOS [201] for hard spheres.

As a major deficit, in both DH and MSA theory the Mayer functions $f_{\alpha\beta} = \exp\{-\beta\varphi_{\alpha\beta}(r)\} - 1$ are linearized in β. This approximation becomes unreasonable at low T^* and near criticality. Pairing theories discussed in the next section try to remedy this deficit. Attempts were also made to solve the PB equation numerically without recourse to linearization [202–204]. Such PB theories were also applied in phase equilibrium calculations [204–206].

2. Pairing Theories

The proper treatment of ionic fluids at low T^* by appropriate pairing theories is a long-standing concern in standard ionic solution theory which, in the light of theories for ionic criticality, has received considerable new impetus. Pairing theories combine statistical–mechanical theory with a chemical model of ion pair association. The statistical–mechanical treatment is restricted to terms of the Mayer f-functions which are linear in β, while the higher terms are taken care by the mass action law

$$\frac{2(1-\alpha)}{\alpha^2}\frac{1}{\rho} = \frac{\rho_p}{\rho_+\rho_-} = K(T)\frac{\gamma_+\gamma_-}{\gamma_p},\tag{17}$$

where $K(T)$ is the pair association constant, α is the degree of dissociation, the ρ_i are the number densities of the free cations, anions, and the pairs (p), and the γ_i are the activity coefficients.

There is some arbitrariness in the definition of the ion pair, and hence the association constant. Often a structural definition of the ion pairs is preferred—for example, by adopting a cutoff distance such as $r_C = 2\sigma$ [141, 207] or similar choices [208, 209]. In contrast, Bjerrum (Bj) theory [140] uses an energetic criterion by defining ions as being associated, when their interaction energy is twice the thermal energy k_BT. Bjerrum theory yields

$$K_{Bj} = \int_\sigma^{b/2} \exp\left(\frac{b}{r}\right)4\pi r^2 dr = \int_\sigma^{\sigma/2T^*} \exp\left(\frac{\sigma}{rT^*}\right)4\pi r^2 dr,\tag{18}$$

where $b = \beta q^2/\varepsilon_0 = \sigma/T^*$ is the famous Bjerrum length. Bjerrum suggested $r_C = b/2$ for the upper cutoff, which corresponds to the minimum of the integrand.

A somewhat more subtle ion pair definition was introduced by Ebeling (Eb) [210–212]. Ebeling's definition of the association constant ensures consistency up to the level of the second ionic virial coefficient between the chemical picture of associating ions and the physical picture resulting from the direct evaluation of the Mayer functions. Ebeling's expression has the same low-temperature expansion as Eq. (18), and near criticality their predictions practically agree [199, 213]. At high T^* both Bjerrum's and Ebeling's definitions possess deficits [213]. Note also a slightly different definition of $K(T)$ by Zhou et al. [214, 215]. The Fuoss association theory [216], quite popular in electrolyte solution work, and applied to describe the ionic phase transition by McGahay and Tomozawa [200], is little adequate for RPM-like systems [199, 217].

Given the expression for $K(T)$, one can construct an EOS by modeling the excess free energy density by $\Phi^{ex} = \Phi^{HS} + \Phi^{II} + \Phi^{ID} + \Phi^{DI} + \Phi^{DD} + \Phi_p^{int}$, where Φ^{ex} is summed over contributions from hard-sphere (HS), ion-ion (II), ion–dipole (ID), dipole–ion (DI), and dipole–dipole interactions (DD), respectively. Φ^{ex} also contains the contribution due to the internal partition function of the ion pair, $\Phi_p^{int} = -\rho_p^* \ln K(T)$. Pairing theories differ in the terms retained in the expression for Φ^{ex}.

In the simplest case, only the interionic term Φ^{II} (with or without a hard core term) is retained, so that $\Phi^{ex} = \Phi^{HS} + \Phi^{II} + \Phi_p^{int}$. This implies that ion pairs form thermodynamically ideal species with $\gamma_p = 1$, as widely assumed in solution thermodynamics. In DHBj (or DHEb) theory [198, 199], Φ^{II} is then given by Eq. (16) with x determined by the density of the *free* ions. At the same level of approximation, an MSA-based pairing theory with $\gamma_p = 1$ was developed earlier by Ebeling and Grigo [210–212] (MSAEb) and was reconsidered with more appropriate hard core terms by Guillot and Guissani [56] and Yeh et al. [218]. Similar in spirit is the so-called "pairing MSA" of Zhou et al. [215] in its version PMSA1, in which some approximations are involved to gain a simple theory.

Because ion pairs possess high dipole moments [138, 139], DI interactions between ion pairs and free ions and DD interactions between pairs are by no means negligible. For a long time, these interactions were suspected to contribute to Φ^{ex}, favoring a redissociation of ion pairs [219, 220] at high ion densities. Fisher and Levin [198, 199] were the first to include DI interactions into DH theory which provides an expression for Φ^{DI} at the same level of approximation as used for Φ^{II}. This internal consistency is not reached when DI interactions are built into Ebeling's MSA [218] or into Stell's PMSA (version PMSA2 and PMSA3) [215].

Guillot and Guissani [56] and Weiss and Schröer [221] went one step further. Guillot and Guissani considered the effect of unscreened DD interactions on the FL theory. They also performed an approximate treatment of unscreened DD interactions in the framework of the MSA. In contrast, Weiss and Schröer theory (WS) considers ionic screening of the DD interactions by the remainder of the fluid, the change of the dielectric permittivity caused by the ion pair concentration, and the related effect on the Eb (Bj) association constant.

MSA theory including screened DD interactions is presently not available. There is, however, an analytical solution for the MSA of a hard-ion–hard-dipole system [174, 175] which involves DI and DD interactions in a natural way. If the dipole is taken as an ion pair controlled by the mass action law, the theory could be adopted for the RPM. However, the expected complexity makes such a theory unattractive [56].

One characteristic feature of theories that incorporate DD interactions is a density-dependent dielectric constant $\varepsilon(\rho^*) \geq \varepsilon_S$, induced by a varying concentration of dipolar ion pairs. Friedman [3] suggested such a variation of the dielectric constant to be responsible for phase separation. As a by-product, WS theory provides a generalization of the well-known Onsager expression for the dielectric permittivity of dipolar fluids to a system comprising free ions [221].

D. Lattice Theories

The lattice analogue of the RPM is the Coulomb gas with cations and anions on lattice sites. Then, one can exploit the isomorphism with the sine–Gordon field theory [222, 223], which enables the long-range Coulombic interaction to be exactly represented in terms of a nearest-neighbor Hamiltonian via the so-called sine–Gordon transformation. However, this mapping is only exact for a Coulomb gas of *point charges* in the grand canonical ensemble. The hard-core interactions have to be introduced *post facto* by imposing some cutoff on the momentum space integrals at high wave numbers. To what extent the treatment of the hard core affects the results is still unclear. Field theories yield the fugacity $z_\pm = \exp(\beta\mu_\pm)$, where $\mu_\pm$ is the chemical potential. To compare *e.g.* with MC data the results have to be transformed from the $z_\pm - T^*$ to the $\rho^* - T^*$ plane which may give rise to problems.

Attempts to exploit this isomorphism for the study of phase transitions go back to Saito [224]. Later, several authors have developed a systematic field theory [225–228] for treating ionic phase transitions. Once a field-theoretical approach is available, one can also explore the lattice analogues of DH theory and its extensions, because the lattice analogue of DH theory should result from a truncated expansion in sine–Gordon theory [229]. Netz and Orland [227] have discussed in great detail the status of DH theory with regard to field theory. Their work also includes a treatment of asymmetry effects when dealing with highly asymmetrical electrolytes in the UPM.

E. Beyond the Primitive Models

Few attempts have been made to develop theories that explicitly account for the molecular nature of the solvent. In the simplest case the system is treated as a mixture of charged hard spheres and *neutral* hard spheres. In MSA-based theories such an extension is straightforward, because the hard-sphere part is decoupled from the electrostatic part. Both the solution for the pure MSA [230–232] and for the PMSA [149] of such charged-hard-sphere–neutral-hard-sphere mixtures are available.

With regard to real electrolytes, mixtures of charged hard spheres with *dipolar* hard spheres may be more appropriate. Again, the MSA provides an established formalism for treating such a system. The MSA has been solved analytically for mixtures of charged and dipolar hard spheres of equal [174, 175] and of different size [233, 234]. "Analytical" means here that the system of integral equations is transformed to a system of nonlinear equations, which makes applications in phase equilibrium calculations fairly complex [235].

Solvent-induced effects on phase equilibria have also been described by models based on solvent-averaged Friedman–Gurney potentials using the HNC approximation [81]. The difficulty in extracting phase transition lines from HNC calculations has been noted earlier, but only the HNC seems to be flexible enough to account for specific interactions (e.g., present in solvophobic mechanisms).

F. Mean-Field Theories of Inhomogeneous Fluids and Fluctuations

One of the hallmarks of criticality is the divergence of fluctuations. In ionic fluids, density and charge fluctuations are of relevance. Density–density correlations in the RPM are reflected by the sum combination [187]

$$h_{\rho\rho}(r) = \{h_{++}(r) + h_{+-}(r)\}, \qquad (19)$$

where symmetry dictates $h_{++}(r) = h_{--}(r)$ and $h_{+-}(r) = h_{-+}(r)$. Charge–charge correlations are reflected by the difference combination

$$h_{zz}(r) = \{h_{++}(r) - h_{+-}(r)\}. \qquad (20)$$

Due to the reduced symmetry, more complex definitions are necessary for the UPM [17].

The spatial range of the fluctuations is described by their correlation length. Because in one-component fluids such as the RPM the total ion density is an appropriate order parameter, the correlation length for the density fluctuations, ξ, is at the same time the correlation length of the order parameter fluctuations introduced in Eq. (1). The correlation length for the charge fluctuations will be denoted by ξ_z. Thus, there is a need for theories that allow us to estimate the correlation lengths ξ and ξ_z. This is particularly important when estimating the range of validity of mean-field theories by the

Ginzburg criterion. Moreover, theories for the correlation length are basic ingredients for developing theories of inhomogeneous fluids, as needed in the treatment of interfacial phenomena.

One possibility for deriving an expression for ξ is to analyze the divergence of the pair correlation function $g(r)$. More traditionally, however, one uses the so-called Landau–Ginzburg theory [236]. As is well known, Landau–Ginzburg theory describes the free energy density of nonuniform systems, $\Phi = \Phi_{hom} + \Phi_{inhom}$, where the homogenous term represents analytical contributions that are not affected by the fluctuations. The inhomogeneous term reflects local densities and their gradients. This part of the local free energy density is then expanded in a power series of the reduced temperature and reduced density plus the density gradient. By symmetry, only the terms of even power in the density and in the gradient term $\nabla \tilde{\rho}$ survive. In theories, where the expansion yields uneven powers, an appropriate tansformation allows us to remove these terms [45]. Thus, the essential terms in the Landau–Ginzburg expansion are of the form

$$\Phi_{inhom}\big(\tilde{T}, \tilde{\rho}(\mathbf{r})\big) = \tfrac{1}{2}a_0\tilde{T}[\tilde{\rho}(\mathbf{r})]^2 + \tfrac{1}{4!}u_0[\tilde{\rho}(\mathbf{r})]^4 + \tfrac{1}{2}c_0[\nabla\tilde{\rho}(\mathbf{r})]^2 + \cdots \quad (21)$$

Equation (21) is usually denoted as Landau–Ginzburg expansion, but in calculations of interfacial phenomena the square gradient approach goes back to van der Waals [237].

It remains to work out expressions for the system-dependent coefficients a_0, u_0, and c_0 in terms of the molecular parameters of the RPM or other model potentials. a_0 and u_0 are determined by derivatives of the free energy of the homogeneous system. Moreover, theory shows that the correlation length of the fluctuations is given by

$$\xi = (c_0/a_0)^{1/2}\,\tilde{T}^{-1/2}, \quad (22)$$

so that $\xi_0 = (c_0/a_0)^{1/2}$. The exponent ν in Eq. (1) is given by the classical value $\nu = 1/2$.

A natural way is to calculate the square gradient term from appropriate functional derivatives of the free energy, as done in the generalized DH theory (GDH) of Lee and Fisher [238]. GDH theory provides an extension of conventional DH theory by accounting for the spatial variation of the total ion density. Standard DH theory assumes only a constant ion density.

An alternative, already implicit in van der Waals' theory of the surface tension [237], exploits an interconnection between the coefficient c_0 and the pair correlation function $g(\mathbf{r}_\alpha, \mathbf{r}_\beta)$ of the inhomogeneous system. As the local density distribution determining $g(\mathbf{r}_\alpha, \mathbf{r}_\beta)$ is unknown, one has to resort to a so-called local density approximation. To do so, Weiss and Schröer [239–242] considered several simple approximations for RPM-based theories. One can also attempt to relate c_0 to the direct correlation function $c(\mathbf{r}_\alpha, \mathbf{r}_\beta)$ associated with $g(\mathbf{r}_\alpha, \mathbf{r}_\beta)$. The latter function may be sufficiently short-ranged to ignore the local density variation in an approximate theory and to use instead $c(r)$ of the homogenous system. Groh et al. [243] constructed a modified direct correlation function for the MSA. Lee and Fisher [238], searching for consistency to other approaches, also used an approximation for $c(r)$, and they implemented the DH result for $h(r)$ into the HNC formalism.

Once the correlation length is known, one is able to estimate the range of validity of mean-field theory, as first demonstrated by Ginzburg [47]. By considering the magnitude of the fluctuations, Ginzburg derived a criterion for the temperature distance from the critical point up to which mean-field theory remains self-consistent. Ginzburg theory predicts that classical theory is only valid if

$$\tilde{T} \gg N_{\text{Gi}} = \frac{u_0^2 v_0^2}{64\pi^2 a_0^4 \xi_0^6}, \qquad (23)$$

where v_0 is a molecular volume, and ξ_0 is the amplitude of the order parameter fluctuations already defined in Eq. (22). N_{Gi} is called the Ginzburg number. There is quite a variety of other approaches [45, 86, 244, 245] to derive this criterion; these essentially result in different numerical factors entering into Eq. (23). We recall that ξ_0, and hence N_{Gi}, can also be derived from a direct analysis of the pair correlation function.

G. Summary

From the theoretical perspective, the need to assess the nature of the Coulombic phase transition has led to many activities. Thus, most theories have relied on the RPM as a generic model for the ionic phase transition. From the various theoretical tools for deriving the EOS, only MSA- and DH-based approaches have found wide application. Applications of the HNC, which is a standard theory in general electrolyte thermodynamics, have remained scarce because of numerical problems when approaching phase transitions. However, pure DH and MSA theory are linear theories that fail at low T^*. It is known for a long time that, at least in parts, this failure can be remedied by accounting for ion pair formation. More recently, it has become clear that at near- and subcritical temperatures, free-ion–ion-pair and ion-pair–ion-pair interactions play a crucial role. Just in this regard, DH theory seems to provide a particularly flexible and transparent scheme for such theoretical extensions.

V. RESULTS FROM MEAN-FIELD THEORIES

A. The Restricted Primitive Model

1. Critical Point and Coexistence Curve

Table III compiles MC results obtained over the years for the critical temperature and critical density of the RPM. Table III includes also results from the cluster calculations of Pitzer and Schreiber [141]. In a critical assessment of earlier work [40, 141, 179–181, 246], Fisher deduced in 1994 that $T_c^* = 0.052–0.056$ and $\rho_c^* = 0.023–0.035$ represent the best values [15]. Since then, however, the situation has substantially changed. Caillol et al. [53, 247] performed simulations of ions on the surface of a four-dimensional hypersphere and applied finite-size corrections. Valleau [248] used his thermodynamic-scaling MC for systems with varying particle numbers to extract the infinite-size critical parameters. Orkoulas and Panagiotopoulos [52] performed grand canonical simulations in conjunction with a histogram technique. All studies indicate an insufficient treatment of finite-size effects in earlier work. While their results do not agree perfectly, they are sufficiently close to estimate $T_c^* = 0.048–0.05$ and $\rho_c^* = 0.07–0.08$, as already quoted in Eq. (6). Critical points of some real Coulombic systems match quite well to these figures [5]. The coexistence curve derived by Orkoulas and Panagiotopoulos [52] is displayed in Fig. 9.

The critical pressure $P^* = P\sigma^4 \varepsilon_0 / q^2$ is more difficult to evaluate. In the earlier literature there is a large spread of values [17]. The recent MC simulations of Orkoulas and Panagiotopoulos [52] yield $P_c^* \cong 8 \times 10^{-5}$ near the lower limit of earlier estimates, along with a critical compressibility factor of $Z_c = P_c / (\rho_c T_c) \cong 0.024$ which is one order of magnitude lower than observed for nonionic fluids (e.g., $Z_c = 3/8 = 0.375$ for the van der Waals fluid).

TABLE III
MC Results for the Critical Parameters of the Restricted Primitive Model

	T_c^*	ρ_c^*
Vorontsov-Veliaminov et al., 1970 [40]	0.094	0.3–0.4
Pitzer and Schreiber, 1987[a] [141]	0.068	$\cong 0.05$
Valleau, 1991 [246]	0.07	0.07
Panagiotopoulos, 1992 [179]	0.056	0.04
Caillol, 1994 [180, 181]	0.057	0.04
Valleau, 1996 [248]	0.049	0.08
Caillol et al., 1996, 1997 [53, 247]	0.0488	0.08
Orkoulas and Panagiotopoulos, 1999 [52]	0.049	0.07

[a] Cluster calculations.

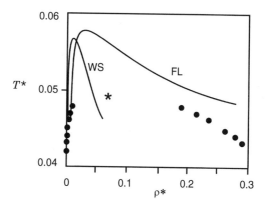

Figure 9. MC coexistence curves for the RPM reported by Orkoulas and Panagiotopoulos [52] in comparison with coexistence curves from FL and WS theory.

It should be emphasized that the comparatively large change obtained in more recent work is mainly caused by the application of finite-size scaling. Under these circumstances, one certainly needs to reconsider how far the results of analytical theories, which are basically mean-field theories, should be compared with data that encompass long-range fluctuations. For the van der Waals fluid the mean-field and Ising critical temperatures differ markedly [249]. In fact, an overestimate of T_c is expected for theories that neglect nonclassical critical fluctuations. Because of the asymmetry of the coexistence curve this overestimate may be correlated with a substantial underestimate of the critical density.

The high population of ion pairs near criticality motivated Shelley and Patey [250] to compare the RPM coexistence curve with that of a dipolar fluid. It is now known that a critical point does not develop in a system of dipolar hard spheres [251]. However, ion pairs resemble dumbbell molecules comprising two hard spheres at contact with opposite charges at their centers. Shelley and Patey found that the coexistence curves of these "charged dumbbells" are indeed very similar in shape and location to the RPM coexistence curve, but very different from the coexistence curve of "dipolar dumbbells" with a point dipole at the tangency of the hard-sphere contact.

Table IV summarizes the critical parameters obtained from RPM-based analytical theories. We have referred all values for DH pairing theories to expressions including a Carnahan (CS) hard-sphere term [68]. To obtain a well-balanced view, it is certainly not sufficient to refer to critical point predictions only. Nevertheless, some conclusions from such an analysis are pertinent:

TABLE IV
Critical Parameters of the RPM Derived from Analytical Theories

Theory	T_c^*	ρ_c^*
DH [200]	0.625	0.00497
DHEb (DHBj)[a] [68][b]	0.0625	0.04517
DHEb + DI (FL)[a] [68][b]	0.0574	0.02778
DHBj + DI + DD (WS) [68][b]	0.0562	0.00863
MSA [42]	0.079	0.014
MSAEb[c] [212]	0.0837	0.018
MSAEb[d] [218]	0.0789	0.026
MSAEb+DI [218]	0.0716	0.027
PMSA (PMSA1) [215]	0.0748	0.025
PMSA+DI (PMSA2) [215]	0.0733	0.0229
PMSA+DI (PMSA3) [215]	0.0744	0.0245

[a] Results from Bjerrum and Ebeling theory do not differ near criticality.
[b] Values with Carnahan–Starling hard-core term; for Fisher's original treatment see Ref. 198.
[c] Ebeling's original version.
[d] Ebeling theory with an improved hard-core term.

First, pure DH theory yields a reasonable prediction of T_c^*, while pure MSA theory and all MSA-based pairing theories tend to predict values that are high by about 50%. The origin of this deficit remains unclear. Both pure DH theory and MSA theory predict far too low critical densities. The HNC prediction fails to reproduce both the critical temperature and the critical density.

Second, predictions of ρ_c^* are substantially improved when account is made for ion pairs. The increase of the critical density is easily understood: A certain free-ion density is needed for driving criticality. If pairs are formed, this free-ion density can only be achieved at a higher overall ion density. Nevertheless, all theories yield too low values if assessed by the more recent MC data. As mentioned, one reason for low critical densities may result from comparison with MC data that encompass long-range fluctuations. It will, however, be shown in the subsequent section that all available analytical theories seem to overestimate the degree of dissociation. Such an overestimate almost invariably leads to an underestimate of the critical density.

Third, in DH-type theories a reasonable shape of the coexistence curves is only obtained when dipole–ion interactions are included. FL theory that includes a DI term yields probably the best representation of the MC coexistence curve near criticality, although the critical density is still low. However, addition of the missing DD term by WS theory lowers the critical density, opposite to the need dictated by the MC results. This may caution that the good performance of FL theory is to some extent fortuitous. In contrast, the critical parameters are not much changed when extending the MSA pairing theories to include DI interactions (MSAEb+DI, PMSA2, and PMSA3). Reasons for this insensitivity remain unclear.

Fourth, the significance of a proper hard-core term near the critical density remains subject to debate. Near the critical density, the MSA theory is more sensitive to the choice of the hard-core term than is the DH theory [198, 199]. At high ion densities the need for an appropriate choice of the hard-core term is unquestionable—for example, to prevent the coexistence curve from reaching states beyond the close packing of the b.c.c. solid phase of the RPM ($\rho^*_{bcc} \cong 1.3$ [252]). Figure 9 compares the coexistence curves of the FL and WS models with the simulation results of Orkoulas and Panagiotopoulos [52].

While there are conflicting views about the performances and benefits of all these versions, there are some criteria that may serve for a critical assessment. These include (a) Onsager's well-known lower bound for the mean electrostatic energy per ion [253] in its reformulation by Totsuji [254] and (b) Gillan's upper bound for the free energy [255]. Moreover, the condition for thermal stability requires the configurational isochoric heat capacity to be positive.

Zuckerman, Fisher, and Lee [213] reexamined these criteria in a comprehensive critique of RPM-based DH and MSA theories. Violations were found with almost all versions of DH- and MSA-based theories. There is, however, no evidence for the often-quoted bad performance of DH as compared to MSA theory. At high T^*, thermal stability is not satisfied by the existing pairing theories [213], which reflects a general deficit of such theories.

2. The Ion Distribution of the RPM Near Criticality

The nature of the low-conducting states at low ion densities has been treated both by cluster theories [141, 207, 256, 257] and MC simulations [208, 258, 259]. There is no doubt that in this regime neutral (1,1) pairs prevail, and there is some evidence for neutral (2, 2) clusters (a cluster s, t comprises s cations and t anions). Near the critical density, higher clusters come into play, and eventually the cluster representation becomes inappropriate [141]. Simulations indicate the importance of intercluster interactions that are unsatisfactorily described [208] by MSA-based estimates [141].

Weingärtner et al. [68] have calculated the degree of dissociation α for the DHBj (DHEb), FL, and WS models over a wide range of conditions from $T^* = 0.04$ up to $T^* = 0.15$. Figure 10 compares the calculated density dependence of α for the DHBj, FL and WS theories along the critical isotherms of the DHEb, FL and WS theories. Detailed analysis [68] shows that DI and DD interactions are an essential ingredient for rationalizing the degree of dissociation and, hence, the conductance behavior near the two-

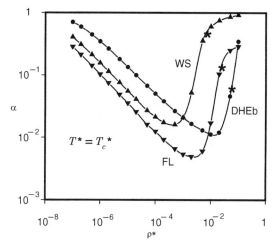

Figure 10. Degree of dissociation α of the RPM as a function of the total ion density along the critical isotherms of Debye–Hückel–Ebeling (or Bjerrum) theory (DHEb), Fisher–Levin theory (FL), and Weiss–Schröer theory (WS) [138]. The asterisks show the critical points of the three models.

phase regime. DHBj theory fails at $T^* < 0.08$, which is distinctly above T_c^* [68].

At a quantitative level, near criticality the FL theory overestimates dissociation largely, and WS theory deviates even more. The same is true for all versions of the PMSA. In WS theory the high ionicity is a consequence of the increase of the dielectric constant induced by dipolar pairs. The direct DD contribution of the free energy favors pair formation [221]. One can expect that an account for neutral (2,2) quadruples, as predicted by the MC studies, will improve the performance of DH-based theories, because the coupled mass action equilibria reduce dissociation. Moreover, quadrupolar ionic clusters yield no direct contribution to the dielectric constant, so that the increase of ε and the diminution of the association constant becomes less pronounced than estimated from the WS approach. Such an effect is suggested from dielectric constant data for electrolyte solutions at low T^* [138, 139], but these arguments may be subject to debate [215]. We note that according to all evidence from theory and MC simulations, charged triple ions [260], often assumed to explain conductance minima, do not seem to play a major role in the ion distribution.

Finally, we note that results for the RPM have been repeatedly discussed with regard to an insulator–conductor transition in the homogeneous regime that could interfere with the coexistence curve. Figure 10 shows that the degree of dissociation reflects a minimum, as also observed for the conductance in Fig. 7. The existence of a minimum of α implies that in the limit

$\rho^* \to 0$, one restores a conducting rather than insulating state, as dictated by the mass action law. Thus, the vapor phase of the RPM remains conducting at all states. The existence of a conducting vapor is confirmed by MC simulations of the RPM [209, 258, 259] and also by MD simulations of a model for NaCl [51]. This supersedes earlier speculations [180, 181] that in $D = 3$ the RPM vapor may be insulating, as observed in $D = 2$ [261].

Similar conclusions can be drawn from an analysis of the heat capacity of the RPM. In general, the presence of chemical equilibria is known to give maxima of the heat capacity along isobars and isotherms in the density range, where the changes in the species distribution are largest. This is also expected for association equilibria between free ions and pairs. Camp and Patey [258, 259] found such maxima in MC simulations of the isobaric heat capacity along isobars. Weingärtner et al. [68] mentioned that along the isotherms in Fig. 10, one gets *two* maxima in the C_V versus ρ^* plot, located near the inflection points of the low- and high-density branches of the α versus ρ^* curves. This clearly shows a structural transition from dissociated states as $\rho^* \to 0$ to essentially nonconducting dipolar states, followed by a redissociation process toward a conducting ionic melt. Only the maximum of C_V of the high concentration branch interferes with the two-phase regime. However, above T_c both the heat capacities and the electrical conductances show gradual transitions that clearly are not real thermodynamic phase transitions.

Interestingly, there have been repeated attempts to view such gradual structural changes due to chemical equilibria as phase transitions as well [262]. Instructive examples for such an analysis are "living polymers" — for example, in demixing solutions of polymers [263] and in sulfur [264], where the reversible polymerization process has been treated as a second-order phase transition. The experimental evidence for such an interpretation is, however, at best weak [265], and classical association models [266] describe the thermodynamic properties equally well.

B. Hard Spheres of Different Size and Charge

The unrestricted form of the primitive model (UPM) becomes important for more complex fluid systems. Stell argued that symmetry breaking in the UPM may play an important role in determining critical behavior [17]. In spite of this potential utility, the UPM is rarely explored. In MC simulations of the cluster structure in the UPM, Camp and Patey [259] compared results for asymmetrical charges $\lambda_q = z_+/z_- = 1, 2, 4$ at the diameter ratio $\lambda_\sigma = \sigma_-/\sigma_+ = 1$. Again the vapor phase contains, above all, neutral clusters such as trimers for $\lambda_q = 2$ and tetrahedral pentamers for $\lambda_q = 4$, as well as higher clusters. At $\lambda_q = 4$ asymmetry effects not covered by simple theories seem to play a role.

There have been few attempts to generalize mean-field theories to the unrestricted case. Netz and Orland [227] applied their field-theoretical model to the UPM. Because such lattice theories yield quite different critical properties from those of continuum theories, comparison of their results with other data is difficult. Outhwaite and coworkers [204–206] considered a modification of their PB approach to treat the UPM. Their theory was applied to a few conditions of moderate charge and size asymmetry.

A more comprehensive analysis is available for MSA theory, which provides an analytical solution [267, 268]. Gonzalez-Tovar [269] investigated in detail phase equilibria for the UPM at size ratios $0.1 < \lambda_\sigma < 1$ and charge ratios $0.01 < \lambda_q < 100$. It seems that anisotropy effects that violate simple corresponding state arguments play an essential role at large size and charge ratios. Surprisingly, for $\lambda_\sigma = 1$ the RPM results are recovered, irrespective of the charge anisotropy λ_q. This result can be formally proved, but seems to be a peculiarity of the MSA, not shared with other approaches [206, 227]. Of course, the known deficits of linear theories also apply to the UPM. To our knowledge, no pairing theories have been examined to correct for the effects of linearization at low T^*.

In passing, we note that UPM-based models are necessary for treating salt mixtures, *e.g.* with three ionic species. Again, there have been only a few studies in this direction which were mostly based on MSA-based theories [270, 271]. These clearly indicate the possibility of liquid–liquid phase equilibria. By judicious choice of the parameters, one can generate situations in which this demixing interferes with the liquid–gas critical lines. In molten salt mixtures such liquid–liquid phase equilibria are known for a long time, and some consolute points are located in a temperature range around 500 K [272], where fairly precise experiments might be possible. However, this path has not yet been followed up in studies of critical phenomena. We note that such models for salt mixtures may apply to solutions of charged macroions which usually contain a supporting electrolyte.

C. Beyond Primitive Models

The simplest salt–solvent model comprises charged spheres mixed with neutral spheres. Intuitively, one expects wide liquid–liquid miscibility gaps, owing to the very different nature of the constituents. In fact, MSA calculations indicate type III phase behavior with a wide liquid–liquid miscibility gap in the solvent-rich regime that breaks the L–G critical curve [149, 230–232].

More realistic approaches should, of course, comprise solvent models that give rise to electrostatic interactions. Shelley and Patey [273] used grand canonical MC simulations to investigate the demixing transition in model ionic solutions where the solvent is explicitly included. Charged hard-sphere ions in neutral, dipolar, and quadrupolar hard-sphere solvents were considered. For all model solvents the demixing transition is in the same general region of the phase diagram and is roughly described by the liquid–vapor equilibrium in the RPM. Of course, in more detail, the precise location of the critical point and the width of the unstable region depends upon the nature of the solvent.

Mixtures of equisized charged spheres were also treated by the MSA. Such a system is then uniquely characterized by the ratio of the critical temperatures of the pure components. Harvey [235] found that a continuous critical curve from the dipolar solvent to the molten salt is maintained until the critical temperature of the ionic component exceeds that of the dipolar component by a factor of about 3.6. This ratio is much higher than theoretically predicted for nonionic model fluids. We recall that for NaCl the critical line is still continuous at a critical temperature ratio of about 5. Thus, the MSA of the charged-hard-sphere–dipolar-hard-sphere system captures, at least in part, some unusual features of real salt–water systems with regard to their critical curves.

D. Inhomogeneous Fluids and Fluctuations

1. Charge and Density Fluctuations

We turn now to attempts that aim at an understanding of density and charge fluctuations in ionic fluids. It is fair to say that the extension of conventional electrolyte theories to allow for a description of fluctuations forms one of the major recent achievements in theories of ionic criticality.

Let us first recall that standard DH theory presumes constant ion density, so that the pair correlation function cannot say anything about *density* fluctuations. In contrast, simple DH theory describes *charge* fluctuations via the well-known screening decay as $\exp(-\Gamma_D r)$. Note, however, that this result does not satisfy a rigorous condition for the second moment of the charge–charge correlation function first derived by Stillinger and Lovett (SL) [39]:

$$\rho \int h_{zz}(r)4\pi r^2 dr = -6/\Gamma_D^2. \qquad (24)$$

Also, the decay of $h_{zz}(r)$ is monotonous, ignoring the need for an oscillatory decay at high ion densities, anticipated from various other treatments [274, 275]. Thus, also with respect to charge correlations, DH theory is quite insufficient.

There are several possibilities to remedy these defects. The most successful one is the generalized DH theory of Lee and Fisher [238, 276]. We recall that GDH theory is the solution of DH theory for inhomogeneous systems. If applied to density–density correlations, it yields an exponentially decaying density–density correlation function $h_{\rho\rho}(r)$. At low ion densities, the correlation length ξ is given by

$$\xi(\rho^*, T^*) = \left(\frac{b}{48\Gamma_D}\right)^{1/2} \left[1 + \Gamma_D b/8 + \cdots\right] \propto (T^*\rho^*)^{-1/4}, \qquad (25)$$

where b is the Bjerrum length defined earlier. Thus, in the low-density limit this expression is universal (i.e., independent from the diameter σ) and diverges as $(T\rho)^{-1/4}$. The correctness of this universal limiting law was later confirmed by explicit consideration of the cluster expansion of $h_{\rho\rho}(r)$ [277–279].

The results are in striking contrast with the result for the screening length, $\xi_z \cong 1/\Gamma$ with $\Gamma \to \Gamma_D$, which, according to Eq. (14), behaves as $(T/\rho)^{1/2}$. Thus, at low ion densities the correlation length ξ and ξ_z of the charge and density fluctuations are interrelated by

$$\xi \cong (b\xi_z/48)^{1/2} + \cdots, \qquad (26)$$

It is obvious that, as $\rho^* \to 0$, the density fluctuations decay on a shorter scale than the charge fluctuations.

We note that, in contrast to the simple DH expression, the GDH result for the charge–charge correlation function [276, 278] $h_{zz}(r)$ satisfies the SL second-moment condition (24). When ion pairs are introduced, the SL second-moment condition is, however, only satisfied up to order ρ^2.

Furthermore, at low ion densities, GDH theory predicts an exponentially decaying charge–charge correlation function $h_{zz}(r)$, but this transforms into an oscillating behavior at high ion densities. Specifically, charge oscillations arise at about $x = \Gamma_D\sigma > 1.178$ with x evaluated at the total ion density. This is quite near the value of x found at the critical point. Thus, charge–density waves may compete with the density fluctuations in driving criticality.

As already stated, it is also possible to calculate the density fluctuations by Landau–Ginzburg theory on the basis of traditional DH theory and its extensions, provided that ways are found to work out the square gradient term. Actually, it turned out that such a procedure depends largely on the local density approximation applied in these calculations [239–242, 280]. Agreement with GDH theory concerning the low-density limit was obtained with a quadratic HNC approximation termed AHNC [242]. This approximation implies that $c_{\alpha\beta}(r) = (h_{DH}(r))^2$, where $h_{DH}(r)$ is the total correlation function of DH theory. However, if ion pairing is explicitly accounted for in this approach, the divergence of the correlation length is apparently suppressed [242]. A rationale may be that the depletion of free ions shifts the divergence to very low ion densities, not covered by the calculations.

Attempts have also been made to treat classical fluctuations by MSA-based theories. For the normal MSA the critical fluctuations are bounded. This implies that the coefficient c_0 in the Landau-Ginzburg expansion (21) vanishes, if calculated via the MSA direct correlation function. Leote de Carvalho and Evans [281] used instead the generalized mean spherical approximation of Høye, Lebowitz, and Stell [195] which remedies some deficits of the simple MSA by adding to $c_{\alpha\beta}(r)$ a term that generates thermodynamic consistency. Note the different meanings of the term "generalized" in GMSA and GDH theory. If one works out the low-density behavior of GMSA theory, one obtains however a completely incorrect dependence on T^* and ρ^* which still involves the substance-specific diameter σ [238, 282].

2. The Range of Validity of Mean-Field Theories

On the basis of the theories described in the preceding section, one can work out the Ginzburg criterion for the range of validity of classical theory. To compare with nonionic fluids, we first consider the Ginzburg number obtained from simple van der Waals theory which yields the coefficients $a_0 = 6$ and $u_0 = 9$ of the Landau–Ginzburg expansion. If one sets in Eq. (23) $v_0 \cong \frac{4}{3}\pi\xi_0^3$, one finds a Ginzburg number of the order of $N_{Gi} \cong 0.01$, in surprisingly good agreement with data for nonionic fluids [45].

There have been several attempts to treat the RPM on an analogous basis. To this end, Leote de Carvalho and Evans [281] used the GMSA, Lee and Fisher [283] used the GDH, and Weiss and Schröer [239, 280, 284] examined several DH-based models that approximate the direct correlation function or the pair correlation function. In some cases the results depended significantly on details of the approximations. In total, none of these studies, whatever theory used, gave evidence that N_{Gi} may be significantly smaller than observed for simple nonionic fluids. Rather the opposite seems to be the case. From this perspective, the experimental results for some ionic systems remain a mystery.

3. Interfacial Properties

Once the free energy of an inhomogeneous system is given, one can calculate by standard methods the properties of the interface—for example, the inter-facial tension or the density profile perpendicular the interface [285]. Weiss and Schröer compared the various approximations within square-gradient theory discussed earlier in Section IV.F for studying the interfacial properties for pure DH and FL theory [241, 242]. In theories based on local density approximations the interfacial thickness and the interfacial tension were found to differ by up to a factor of four in the various approximations. This contrasts with nonionic fluids, where the density profiles and interfacial tensions are comparatively insensitive to the local density approximation [286]. The quadratic HNC theory (AHNC), which, as GDH theory, satisfies the correct limiting law for the density fluctuations at low T^*, was judged to give acceptable results [242].

Interfacial properties were also determined via a modified MSA approach [243] and via the GMSA [287]. Despite all differences regarding the coexistence curve, the few reduced surface tensions available from these MSA-based approaches agree quite well with those obtained by DH theory. It is not known how the MSA would react toward the various local density approximations.

E. Summary

Most work has dealt with the RPM as a generic model for ionic criticality. MC data suggest that the replacement of the solvent's dielectric continuum by discrete solvent molecules does not change the principal topology of the phase diagram. This ensures that the simple RPM covers the major features of real ionic fluids, at least in cases where Coulombic interactions prevail.

MC simulations have provided information on the major properties of the RPM fluid, and more recent simulations agree quite well on the location of the critical point. However, it is fair to say that such simulations provide little insight into the mechanism of the phase transition. Therefore, statistical–mechanical theories will remain an important tool for understanding ionic criticality. Any such theory must start with assumptions concerning the contributions to the free energy, and it is necessarily based on approximations in the evaluation of the EOS. There is, however, in principle, no need to speculate about the net inaccuracies in these approaches, because these questions can be answered by comparison with "exact" MC data.

At the present stage, only MSA- and DH-based theories have been widely applied. From these theories, and their comparison with MC data, there is now a quite consistent picture of the major mechanisms driving these ionic phase transitions, although there are still difficulties to predict the critical point on a quantitative level. The results point toward the crucial importance of ion pairing and of interactions between ion pairs and the remainder of the ionic fluid. Deficits in predicting the critical point quantitatively seem to be closely connected to a substantial overestimate of ionicity in almost all pairing theories.

An important prerequisite for any theory of potential use for describing critical phenomena is the proper description of fluctuations. In their original form, both DH and MSA theories are not suited for this purpose. One of the major achievements is the development of various approaches, and in particular the GDH theory, which enable the description of fluctuations at the mean-field level. In fact, GDH theory provides the only description of ionic correlation functions that satisfies the second-moment condition, exhibits the correct low-density behavior, and predicts the anticipated charge oscillations at high ion densities.

VI. THEORIES OF CRITICAL BEHAVIOR

A. Critical Phenomena and Range of Interactions

We turn now to theories of ionic criticality that encompass nonclassical phenomena. Mean-field-like criticality of ionic fluids was debated in 1972 [30]; and according to a remark by Friedman in this discussion [69], this subject seems to have attracted attention in 1963. Arguments in favor of a mean-field criticality of ionic systems, at least in part, seem to go back to the work of Kac et al. [288], who showed in 1962 that in $D = 1$ classical van der Waals behavior is obtained for a potential of the form $\phi(r) \propto k \exp(-kr)$ in the limit $k \to 0$, where interactions become infinitely weak and of infinite range. One should, however, appreciate that ionic fluids with attractive and repulsive Coulombic interactions have little in common with the simple Kac fluid.

A different set of theories focuses on potentials of the general form $\phi(r) \propto r^{-p} = r^{-(D+\sigma)}$, where σ is any number (not to be confused with the diameter used earlier). We consider here results for $D = 3$, but note that for systems of higher dimensionality $(D > 4)$, mean-field theory becomes exact for all σ (note that RG analysis shows that potentials with $\sigma > 2 - \eta = 1.97$ fall into the Ising universality class [21]). For smaller σ, departures from Ising behavior set in, with critical exponents depending on σ [21]. For $0 \leq \sigma \leq 1.5$ one finds generalized mean-field behavior with mean-field exponents α, β, γ, and δ for the thermodynamic properties [289]. In the same range, ν and η still depend on σ with $\nu = 1/\sigma$ and $\eta = 2 - \sigma$ [290]. Only for $\sigma = 2$ the latter exponents correspond to the values obtained by classical mean-field theory. While there were speculations that the bare Coulombic potential $(\sigma = -2)$ exhibits classical criticality, no decisive theoretical results are available. For $\sigma \leq 0$, the behavior sensitively depends on the system's size, and the existence of a thermodynamic limit cannot be taken for granted. Thus, from this perspective, claims for mean-field behavior of long-range systems have also no rigorous basis.

B. Lattice Models

For nonionic systems the lattice gas has provided essential insight into the criticality of fluids. The striking feature of results discussed now for the lattice analogue of *ionic* fluids is that this Coulomb gas possesses obviously quite different properties from those of the Coulomb fluid.

The generic case of a 3D-Coulomb lattice gas is certainly the lattice analogue of the RPM, where each charge just occupies one site. In this case the ratio ζ of the ion size to the lattice unit cell dimension is $\zeta = 1$. Dickman and Stell [291] were able to solve this model analytically. The striking result is tricritical behavior, in clear contrast to the ordinary critical point of the continuum RPM. Dickman and Stell observed a Néel line of second-order phase transitions between a disordered and a antiferroelectric (termed antiferromagnetic) phase terminating at a tricritical point, where it meets a line of first-order coexistence. In $D = 3$ a tricritical point gives rise to mean-field behavior. MC simulations performed by Panagiotopoulos and Kumar [292] confirm this topology of the phase diagram. According to their data the tricritical point is located at $T^* \cong 0.15$ and $\rho^* \cong 0.48$.

The transition to the continuum fluid may be mimicked by a discretization of the model choosing $\zeta > 1$. To this end, Panagiotopoulos and Kumar [292] performed simulations for several integer ratios $1 \leq \zeta \leq 5$. For $\zeta = 2$ the tricritical point is shifted to very high density and was not exactly located. The absence of a liquid–vapor transition for $\zeta = 1$ and 2 appears to follow from solidification, before a liquid is formed. For $\zeta \geq 3$, ordinary liquid–vapor critical points were observed which were consistent with Ising-like behavior. Obviously, for finely discretisized lattice models the behavior approaches that of the continuum RPM. Already at $\zeta = 4$ the critical parameters of the lattice and continuum RPM agree closely. From the computational point of view, the exploitation of these discretization effects may open many possibilities for methodological improvements of simulations [292]. From the fundamental point of view these discretization effects need to be explored in detail.

Another interesting route may result from considering lattice models in $D = 2$. In the context of RG theory, this may allow an RG ($\epsilon = 2 - D$) expansion. In $D = 2$, the Coulomb gas undergoes a so-called Kosterlitz–Thouless (KT) metal-insulator transition of infinite order and characterized by a diverging Debye screening length [261]. This transition line, which terminates at $\rho^* = 0$ and $T^* = 1/4$, intersects the liquid–gas coexistence curve at the critical point. In the low-temperature insulating state, all ions are associated into dipolar pairs, while in the high-temperature state there exists a fraction of free charges. Note that this tricritical scenario is completely different from that of the $3D$ Coulomb gas.

As the density increases, the validity of KT theory becomes, however, more and more questionable. There are conflicting views about the fate of the KT transition. It was suggested that the KT transition is replaced by some discontinuous first-order transition or by a first-order coexistence curve between an insulating vapor and a conducting fluid-like phase [293]. Minnhagen and Wallin [294, 295] found that the KT transition terminates in a critical end point. In contrast, DH theory predicts in $D = 2$ the KT line to terminate in a tricritical point, after which the insulating vapor phase coexists with a conducting liquid phase [296]. The latter conclusion was essentially confirmed by a sine–Gordon field theoretical study of the Coulomb gas by Diehl et al. [228].

Returning to $3D$ lattice models, one may note that sine–Gordon field theory of the Coulomb gas should enable an RG $(\epsilon = 4 - D)$ expansion [15], but this path has obviously not yet followed up. An attempt to establish the universality class of the RPM by a sine–Gordon-based field theory was made by Khodolenko and Beyerlein [105]. However, these authors did not present a scheme for calculating the critical exponents. Rather they argued that the grand partition function can be mapped onto that of the spherical model of Kac and Berlin [106, 297] which predicts a parabolic coexistence curve, i.e. $\beta = 1/2$. This analysis was severely criticized by Fisher [298]. Actually, the spherical model has some unpleasant thermodynamic features, never observed in real fluids. In particular, it is associated with a divergence of the compressibility K_T as the coexistence curve (rather than the spinodal line) is approached. By a determination of the exponent γ, this possibility could also be ruled out experimentally [95, 97].

Fisher [15] has discussed various other field-theoretical scenarios that could be profitably employed for establishing the universality class of the RPM. In particular, he has advocated the use of a four-state lattice model with lattice sites occupied by cations, anions, and neutral pairs, with the number of pairs controlled by the mass action law. Again this path has not been followed up in detail, and only a highly speculative phase diagram has been given [15].

In the spirit of Fisher's idea, Moreira et al. [229] examined how the introduction of weak electrostatic interactions may affect the critical behavior of an Ising-like fluid. To do so, the authors considered an ordinary fluid with short-range interactions and an Ising critical point at temperature T_c^0. They then examined how the crossover temperature changes, when an electrostatic interaction is switched on, so that the particles become positively and negatively charged. If the strength of the Coulombic interaction is characterized by $e^2/\varepsilon_S\sigma$, one may define the ionicity by the ratio of $e^2/\varepsilon_S\sigma$ to $k_B T_c^0$. The ionic model was treated by sine–Gordon field theory, yielding a Hamiltonian with coefficients depending on the ionicity. The results indicate that, in general, when increasing the ionicity, the crossover temperature decreases only slightly, and far too little to describe the experimental observations.

C. Monte Carlo Simulations of Fluid Models

MC simulations can reflect the nonclassical critical fluctuations if the simulation box is sufficiently large or if special techniques are applied to analyze the fluctuations. Simulations for simple nonionic models such as the square-well fluid (SCF) [52] show that there is indeed a good chance to study details of criticality. As noted, MC simulations have also been profitably exploited for studying the lattice analogue of the RPM [292]. For the continuum RPM fluid the few reported results are still conflicting.

Orkoulas and Panagiotopulos [52] performed a careful analysis of the RPM coexistence curve. In the absence of any certainty about the universality class of the RPM, they presumed Ising-like behavior. They then showed that their data are consistent with this hypothesis, but did not exclude other possibilities. Similar conclusions were reached by Caillol et al. [247]. In a study of the isochoric heat capacity, Valleau and Torrie [299] obtained opposite results. We recall that the Ising model predicts a divergence of the isochoric heat capacity which is not present in the mean-field case. Valleau and Torrie found no evidence for such a divergence. In view of the extraordinary difficulty of such studies and the limited range of system sizes, it is easy to raise objections against this observation. However, the careful analysis seems to ensure that the result is reliable. It may therefore be some time before more decisive MC simulations for the RPM provide deeper insight.

D.   Analytical Theories of Fluids

1.   The Restricted Primitive Model

From the view point of statistical mechanical theory there are little more than plausible arguments about the universality class of primitive models. The MSA is analytical and thus it yields classical exponents. Therefore, applications to calculate critical exponents or to prove the apparent mean-field criticality of ionic systems [230] are not meaningful [300]. By construction, other integral equation theories also are also not suitable for determining critical exponents [301]. The ultimate goal of an RG analysis of the RPM or UPM has not yet been reached.

One of the few attempts to tackle the problem of ionic criticality more quantitatively was made by Hafskjold and Stell in 1982 [36], and was later taken up by Høye and Stell [17, 302, 303]. Based on a comparative analysis of the correlation functions for nonionic and ionic fluids, these authors asserted that the critical point of the RPM is Ising-like. To this end, they argued that the density–density correlation function $h_{\rho\rho}(r)$ and the associated direct correlation function $c_{\rho\rho}(r)$ obey essentially the same OZ equation and closure as that of a single-component, nonionic fluid. It was assumed that this analogy suffices to ensure that the critical exponents are Ising-like.

A crucial assumption for such a scenario is that the charge–charge correlation function $h_{zz}(r)$ decays as $\exp(-\Gamma r)$, where the inverse charge–charge correlation length $\Gamma = 1/\xi_z$ does *not* vanish at the critical point. At low ion densities, where $\Gamma \to \Gamma_D$, this has meanwhile been proved for DH theory. Then, at the critical point, only the density fluctuations become infinite, and the resulting behavior should be the same as that of uncharged fluids. Fisher [15] and Stell [17] discussed the reliability of those arguments, but did not agree on firm conclusions. There are arguments for Ising-like criticality in the RPM, even if one were to assume $\Gamma \to 0$ [18]. Fisher [15] discussed various other subtle problems regarding Stell's analysis.

2. The Unrestricted Primitive Model

In the asymmetric case, the high symmetry of the RPM is broken and the correlation functions $h_{\rho\rho}(r)$ and $h_{zz}(r)$ are no longer given by the simple sum and difference combinations introduced in Section IV.F. Stell [17] showed that when one loses the symmetry of the RPM, a strong coupling between charge–charge and density–density correlations ensues. Stell argued that these differences in symmetry between the UPM and RPM may cause both models to fall in different universality classes, with the UPM possibly exhibiting mean-field behavior.

In real fluids such a symmetry breaking can result from several effects such as different charges or sizes of the cation and anion. It can also be broken by any disparity between anion–solvent and cation–solvent interactions. Such a disparity is certainly present in all systems studied experimentally. It would be ironic [17] that the RPM that has been used as the generic case of Coulombic phase separation suppresses the effects causing mean-field-like criticality. On the other hand, the only report of pure mean-field criticality concerns Pitzer's system comprising the symmetric $HexEt_3N^+$ and $HexEt_3B^-$ ions. Among all systems studied experimentally, one would expect just this system to exhibit high symmetry, although one cannot exclude symmetry breaking by subtle interactions of the cations and anions with the solvent diphenylether or by the nonspherical shape of the ions.

3. The Role of r^{-4}-Dependent Interactions

In the sense of RG theory, bare interactions with $\phi(r) \propto r^{-4}$ represent long-range interactions [289]. One source for r^{-4}-dependent interactions is the attractive interaction of ions with rotating dipoles in solvent molecules. Høye and Stell [304] argued that theories for the ion–solvent term can also be profitably employed for modeling ion–ion pair interactions.

Another option evolves from the cavity terms already mentioned in Section IVA. We recall that these cavity interactions arise from the assumption that in the RPM the dielectric continuum penetrates the ions, so that the dielectric constant ε_{in} inside the ions is equal to ε_S. To counterbalance this effect, a cavity term is introduced. The standard situation in electrolyte solutions is certainly that $\varepsilon_S > \varepsilon_{in}$, which implies that the cavity interaction is repulsive. $\varepsilon_{in} > \varepsilon_S$ would imply an attractive interaction.

If the r^{-4}-dependent *attractive* interactions were unscreened, they would indeed give rise to mean-field criticality. On the other hand, an unscreened *repulsive* interaction would suppress the development of a critical point, giving rise instead to a charge–density wave instability in the neighborhood of where criticality would take place in the absence of the r^{-4} cavity term. The latter scenario was developed some time ago by Nabutovskii et al. [305] using a Ginzburg-type analysis.

Again the issue of screening is of crucial interest, because one expects the bare r^{-4}-dependent interactions to be screened by the free ions in the same way as the bare Coulombic potential is screened. The problem of screening of the r^{-4} terms has been treated from various aspects by Fisher and coworkers [170, 171] and by Høye and Stell [169, 303]. All work indicates that at nonzero ion density these are screened to shorter range by the factor $\exp(-2\Gamma_D r)$.

Specifically with regard to cavity interactions, Fisher and coworkers [170] found that, within DH theory, representing the ions as spheres with $\varepsilon_{in} \neq \varepsilon_S$ does not change the critical parameters. Thus, they concluded that explanations other than cavity interactions may be looked for to explain the peculiar critical behavior of ionic fluids. In contrast, Høye and Stell [169] argued that one may or may not have charge–density waves, depending on the relative magnitude and range of the cavity term in relation to other terms. Similar arguments were presented with respect to *attractive* ion–dipole interactions. Høye and Stell [18, 303] concluded that the presence of a screened charge–dipole term will indeed perturb the Ising-like thermodynamic behavior, thus leading to crossover scenarios. These are, however, expected to be strongly system-dependent.

While there is thus much spirited theoretical discussion, there is no direct experimental evidence for such effects, except that this scenario might be a rationale for the observed [108, 109] sharp crossover. Balevicius and Fuess [306] have recently attributed some puzzling visual observations of salt precipitation and turbidity phenomena in ternary aqueous systems to the presence of charge–density waves, but this seems to be pure speculation.

E.   Crossover Theories and Tricriticality

We recall that comparatively sharp and even nonmonotonous crossover from Ising to mean-field behavior has been deduced from experiments for a diversity of ionic systems. We note that this unusually sharp crossover is a striking feature of some other complex systems as well; we quote, for example, solutions of polymers in low-molecular-weight solvents [307], polymer blends [308–311], and microemulsion systems [312]. Apart from the fact that application of the Ginzburg criterion to ionic fluids yields no particularly small   nonclassical regime, we recall that Ginzburg's theory does not say anything about the form of the crossover function.

Extension of the classical Landau–Ginzburg expansion to incorporate nonclassical critical fluctuations and to yield detailed crossover functions were first presented by Nicoll and coworkers [313, 314] and later extended by Chen et al. [315, 316]. These extensions match Ginzburg theory to RG theory, and thus interpolate between the lower-order terms of the Wegner expansion at $\tilde{T} \ll N_{Gi}$ and mean-field behavior at $\tilde{T} \gg N_{Gi}$.

As an essential feature of these theories, crossover behavior is governed by two physical parameters [317]: (1) a scaled coupling constant $\bar{u}$ which reflects the strength and range of the intermolecular forces as represented by ξ_0 and (2) a cutoff wave number Λ which is assumed to be inversely proportional to a structural length ξ'. When $\xi' = \xi_0$, one has only one length scale, and one recovers the Ginzburg number with $N_{Gi} \propto (\bar{u}\Lambda)^2$.

This crossover theory has been repeatedly tested with regard to MC simulations of the 3D lattice gas with variable interaction range. For example, a recently developed MC algorithm [318] allows the ratio of $\tilde{T}/N_{Gi}$ to be varied over eight orders of magnitude to cover the full crossover region [319]. The crossover theory gives an excellent representation of these data [320].

For more complex fluids, one expects $\xi' \neq \xi_0$. Then, mean-field behavior can result from two different processes. First, the long-range nature of the intermolecular forces may cause $\bar{u}$ to be small, while Λ is not small. Second, ξ' may be large or even diverging. Then, Λ and N_{Gi} will be small, while $\bar{u}$ is not necessarily small. This case is expected to give a sharp or even nonmonotonous crossover, because a second length scale is present.

At first, one would tend to reconsider conventional crossover due to mean-field criticality associated with long-range interactions in terms of the refined theories. Conventional crossover conforms to the first case mentioned—that is, small $\bar{u}$ with the correlation length of the critical fluctuations to be larger than ξ_0. However, in the latter case one expects smooth crossover with slowly and monotonously varying critical exponents, as observed in nonionic fluids. Thus, the sharp and nonmonotonous behavior cannot be reconciled with one length scale only.

Because, from these arguments, ordinary crossover associated with long-range interactions may not be responsible for an apparent classical behavior, there is the pressing question, What else may cause it? A possible scenario is the crossover to some real or virtual tricritical point. A well-understood example of the latter type is crossover to theta-point tricriticality in polymer solutions [320]. Here, one can identify ξ' with the radius of gyration. Then, ξ' increases with increasing molecular weight, giving rise to a competing order parameter that causes crossover to a tricritical theta point at infinite molecular weight.

Turning to ionic systems, we have already encountered several scenarios that suggest tricriticality: First, tricriticality is present in the $2D$ RPM and in the $3D$ Coulomb gas, although it is of different origin in both cases. If such a scenario is applicable to the $3D$ fluid as well, one has to search for a hidden phase transition that corresponds to the Kosterlitz transition in the $2D$ RPM or the Néel line in the $3D$ Coulomb gas. For electrolyte solutions a natural possibility would be an insulator–conductor transition. As seen earlier, some change from nonconducting to conducting states is certainly present near the critical regime of real systems and the RPM model fluid; but of course, this is no thermodynamic phase transition. Similar arguments are applicable to solutions of sodium in ammonia, where the presence of a nonmetal–metal transition meeting the coexistence curve near the critical point has been debated for example, on the grounds of a rapid change or even jump in the heat capacity in the supercritical regime [321].

Second, there is a line of charge-ordering in the $T^*-\rho^*$ plane, where the charge-charge correlation function begins to oscillate. This line, as established from GDH theory, passes close to the critical point and may generate a virtual tricritical state. A charge–density wave scenario also arises from r-dependent cavity interactions.

Third, in real ionic solutions, solvophobic and Coulombic interactions may define different length scales. This case is, of course, not covered by the RPM and similar continuum models. Anisimov et al. [322] have argued that such a mechanism may be responsible for the observed shift of the crossover temperature closer to T_c found in solutions of a picrate in a homologous series of alcohols.

Fourth, a crossover scenario is comparatively easily accounted for ternary systems, where three phases in equilibrium certainly form a more natural scenario than in two-component systems or one-component systems such as the RPM. In fact, the most clear evidence for crossover with $\tilde{T}_\times \to 0$ has been found in the ternary system water $+3$-MP $+$ NaBr [165]. In the latter case, from small-angle X-ray scattering, there is some vague indication for clusters that could serve to establish a second length scale [323].

Finally, we recall that in high-temperature aqueous solutions of NaCl near the L–G critical line, crossover has also been observed. Again, it has been concluded [152] that the critical locus may be affected by a virtual tricritical point.

VII. CONCLUSIONS

About 30 years after Buback and Franck's pioneering study on the critical point of molten NH_4Cl and 10 years after Singh and Pitzer's report on a mean-field nature of the liquid–liquid critical point in an electrolyte solution, several precise studies have substantiated Pitzer's conjecture that ionic criticality may be different from that of nonionic fluids.

From a global assessment of these results, it seems inescapable to conclude that mean-field behavior does not remain valid asymptotically close to the critical point. Rather, ionic systems seem to show Ising-to-mean-field crossover. Such a crossover has been a recurring result observed near liquid–liquid consolute points in "Coulombic" electrolyte solutions, in ternary aqueous electrolyte solutions containing an organic cosolvent, and in binary aqueous solutions of NaCl near the liquid–vapor critical line.

On the other hand, it is by no means straightforward to deduce from these experimental results that a generic ionic liquid would show asymptotic Ising behavior as well. In real systems there is certainly an interplay between Coulombic and specific non-Coulomic interactions. It has become clear that even small changes in the solvent properties along homologous solvent series can change this interplay in such a way that the location and extension of the crossover regime is largely affected. While there is no evidence for anything but normal Ising behavior in the asymptotic regime of real ionic fluids, one may speculate from results for homologous solvent series that in the limit of a noninteracting solvent the asymptotic critical behavior would be mean-field-like. In the light of the present knowledge, such a mean-field-like criticality would, however, come as a surprise to theoreticians.

In view of these observations, one would like to establish the Ising-like nature of the critical point by an RG treatment. Unfortunately, lattice models, as successfully applied to describe the criticality of nonionic fluids, may be of little help in this regard, because predictions for the Coulomb gas have proved to be surprisingly different from those for the continuum RPM. Discretization effects—and, more generally, the relevance of the results of lattice models with respect to the fluid—still need to be explored in detail. On the other hand, an RG treatment of the RPM or UPM is still lacking and, as Fisher [278] notes, the way ahead remains misty.

A major ingredient for an RG treatment is a simple and transparent characterization of the molecular forces driving phase separation. This situation calls for mean-field theories of the ionic phase transition. The past decade has indeed seen the development of several approximate mean-field theories that seem to provide a reasonable, albeit not quantitative, picture of the properties of the RPM. Thus, the major forces driving phase separation seem now to be identified. Moreover, the development of a proper description of fluctuations by GDH theory has gone some way to establish a suitable starting point for RG analysis. Needless to say, these developments are also of prime importance in the more general context of electrolyte theory.

Actually, MC simulations should reflect the nonclassical critical fluctuations as well, thus allowing us to identify the critical exponents and the universality class. More recent work of this type has not given a final word concerning the universality class. In contrast, one may now have confidence in results from MC simulations at the mean-field level—for example, the topology of the phase diagram, the approximate location of the critical point, or the ion distribution near criticality.

Even if the critical point of an ionic fluid were mean-field-like, there is no certainty that this mean-field behavior is related to the long-range nature of the interionic forces. By analogy with the behavior of more complex fluids, crossover may be controlled by the approach toward a real or virtual tricritical point which in $D = 3$ is mean-field like. One crucial argument in favor of such an interpretation is that the observed crossover is comparatively sharp or even nonmonotonous. Several attempts to calculate the Ginzburg number for RPM-based models failed to explain crossover with a particularly small Ising regime. Sharp or nonmonotonous crossover suggests the presence of a second length scale in the fluid.

In fact, there are various scenarios for a real or incipient tricritical point in ionic fluids. In contrast to the case of polymer tricriticality, which is well-established from theory and experiment, for ionic fluids these scenarios are still speculative and call for more detailed theoretical and experimental investigations. In particular, it remains to identify the physical origin of the mesoscopic, second length scales.

In conclusion, we see that neither the theoretical nor the experimental understanding of the criticality of ionic fluids is complete. On the basis of the theoretical and experimental developments achieved during the past decade, one can reasonably expect that rapid further progress is imminent. Thus, the study of ionic criticality will remain a field of intense research in the foreseeable future.

Acknowledgment

We would like to thank M. A. Anisimov, M. E. Fisher, H. L. Friedman, J. M. H. Levelt Sengers, J. V. Sengers, and G. Stell for many stimulating discussions, and for making preprints of their work available prior to publication. Over the years, we have benefited greatly from the contributions of many coworkers, especially of Drs. S. Kāshammer, M. Kleemeier, V. C. Weiss and S. Wiegand.

References

1. J. E. Verschaffelt, Proc. *Kon. Acad. Sci. Amsterdam* **2**, 5777 (1900).

2. P. Walden and M. Centnerszwer, *Z. Phys. Chem.* **42**, 432 (1903).

3. H. L. Friedman, *J. Phys. Chem.* **66**, 1595 (1962).

4. K. S. Pitzer, *Acc. Chem. Res.* **23**, 333 (1990).

5. H. Weingärtner, M. Kleemeier, S. Wiegand and W. Schröer, *J. Stat. Phys.* **78**, 169 (1995).

6. S. C. Greer and M. R. Moldover, *Annu. Rev. Phys. Chem.* **32**, 233 (1981).

7. J. V. Sengers and J. M. H. Levelt Sengers, *Annu. Rev. Phys. Chem.* **37**, 189 (1986).

8. A. Kumar, H. R. Krishnamurthy, and E. S. R. Gopal, *Phys. Rep.* **98**, 57 (1983).

9. See, for example, J. S. Rowlinson and F. L. Swinton, *Liquids and Liquid Mixtures*, Butterworth, London, 1982.

10. See, for example, H. E. Stanley, *Introduction to Phase Transitions and Critical Phenomena*, Oxford University Press, Oxford, 1971.

11. See, for example, P. Pfeuty and G. Tolouse, *Introduction to the Renormalization Group and to Critical Phenomena*, Wiley, London, 1977.

12. See, for example, C. Domb, *The Critical Point. A Historical Introduction to the Modern Theory of Critical Phenomena*, Taylor and Francis, London, 1996.

13. K. S. Pitzer, *J. Phys. Chem.* **99**, 13070 (1995).

14. J. M. H. Levelt Sengers and J. A. Given, *Mol. Phys.* **80**, 899 (1993).

15. M. E. Fisher, *J. Stat. Phys.* **75**, 1 (1994).

16. M. E. Fisher, *J. Phys. Condens. Matter* **8**, 9103 (1996).

17. G. Stell, *J. Stat. Phys.* **78**, 197 (1995).

18. G. Stell, *J. Phys. Condens. Matter* **8**, 9329 (1996).

19. See, for example, R. Wilson, *Rev. Mod. Physics* **55**, 583 (1983) and references cited therein.

20. See, for example, M. E. Fisher, *Rev. Mod. Phys.* **70**, 653 (1998).

21. R. F. Kayser and H. J. Raveche, *Phys. Rev. A* **29**, 1013 (1984).

22. M. E. Fisher, *J. Math. Phys.* **5**, 944 (1964).

23. R. Guida and J. Zinn-Justin, *J. Phys. A: Math. Gen.* **31**, 8103 (1998).

24. See, for example, C. Domb and M. S. Green (eds.), *Phase Transitions and Critical Phenomena*, Vol. 3, Academic Press, London, 1974.

25. See, for example, K. Binder, in: *Phase Transitions and Critical Phenomena*, Vol. 5b, edited by C. Domb and M. S. Green, Academic Press, New York, 1976, p. 1.

26. K. Binder, H. W. J. Böte, E. Luijten, and J. R. Hering, *J. Phys. A* **28**, 6289 (1995).

27. S. Jüngst, B. Knuth, and F. Hensel, *Phys. Rev. Lett.* **55**, 2160 (1985).

28. F. Hensel, B. Knuth, S. Jüngst, H. Uchtmann, and M. Yao, *Physica* **139/140**, 90 (1986).

29. F. Hensel, *J. Phys. Condens. Matter* **2**, SA33 (1990).

30. See, for example, discussion remarks in: *J. Solution Chem.* **2**, pp. 353–355 (1972).

31. J. G. Kirkwood, *J. Chem. Phys.* **2**, 351 (1934).

32. See, for example, H. Falkenhagen and W. Ebeling, in: *Ionic Interactions*, Vol. 1, edited by S. Petrucci, Academic Press, New York, 1971.

33. M. Blume, V. J. Emery, and R. B. Griffiths, *Phys. Rev. A* **4**, 1071 (1971).

34. M. Buback and E. U. Franck, *Ber. Bunsenges. Phys. Chem.* **76**, 350 (1972).

35. R. R. Singh and K. S. Pitzer, *J. Chem. Phys.* **92**, 6775 (1990).

36. B. Hafskjold and G. Stell, in: *The Liquid State of Matter*, edited by E. W. Montroll and J. L. Lebowitz, North-Holland, Amsterdam, 1982.

37. H. L. Friedman and B. Larsen, *J. Chem. Phys.* **70**, 92 (1979).

38. D. A. McQuarry, *J. Phys. Chem.* **66**, 1508 (1962).

39. F. H. Stillinger and R. Lovett, *J. Chem. Phys.* **48**, 3858 (1968); **49**, 1991 (1968).

40. P. N. Vorontsov-Veliaminov, *Teplofiz. Vys. Temp.* **8**, 177 (1970).

41. B. P. Chasovskikh and P. N. Vorontsov-Veliamov, *High Temp. (USSR)* **14**, 174 (1976).

42. G. Stell, K. C. Wu and B. Larsen, *Phys. Rev. Lett.* **37**, 1369 (1976).
43. F. J. Wegner, *Phys. Rev. B* **5**, 4529 (1972); **6**, 1891 (1972).
44. R. R. Singh and K. S. Pitzer, *J. Chem. Phys.* **90**, 5742 (1989).
45. M. A. Anisimov, S. B. Kiselev, J. V. Sengers, and S. Tang, *Physica A* **188**, 487 (1992).
46. P. Chieux and M. Sienko, *J. Chem. Phys.* **53**, 566 (1970).
47. V. L. Ginzburg, *Sov. Phys. Solid* **2**, 1824 (1962).
48. A. D. Kirshenbaum, J. A. Cahill, P. J. McGonigal, and A. V. Grosse, *J. Inorg. Nucl. Chem.* **24**, 1287 (1962).
49. K. S. Pitzer, *Chem. Phys. Lett.* **105**, 484 (1984).
50. K. S. Pitzer, *J. Phys. Chem.* **88**, 2689 (1984).
51. Y. Guissani and B. Guillot, *J. Chem. Phys.* **101**, 490 (1994).
52. G. Orkoulas and A. Z. Panagiotopoulos, *J. Chem. Phys.* **110**, 1581 (1999).
53. J. M. Caillol, D. Levesque, and J. J. Weis, *J. Chem. Phys.* **105**, 1565 (1997).
54. A. Z. Panagiotopoulos, *Mol. Phys.* **61**, 813 (1987).
55. M. Buback and E. U. Franck, *Ber. Bunsenges. Phys. Chem.* **77**, 1074 (1973).
56. B. Guillot and Y. Guissani, *Mol. Phys.* **87**, 37 (1996).
57. B. Guillot and Y. Guissani, in: *Steam, Water and Hydrothermal Systems: Physics and Chemistry at the Needs of Industry*, edited by P. Tremaine, P. G.Hill, D. E. Irish, and P. V. Balakrishnan, NRC Press, Ottawa (2000).
58. M. A. Anisimov, E. E. Goredetskii, V. D. Kulikov, and J. V. Sengers, *Phys. Rev. E* **51**, 1199 (1995).
59. Discussion remark by F. H. Stillinger, *J. Solution Chem.* **2**, 354 (1972).
60. J. W. Johnson and D. Cubiciotti, *J. Phys. Chem.* **68**, 2235 (1964).
61. J. M. H. Levelt Sengers and S. C. Greer, *Int. J. Heat Mass Transfer* **15**, 1865 (1972).
62. W. B. Holzapfel and E. U. Franck, *Ber. Bunsenges. Phys. Chem.* **70**, 1105 (1966).
63. I. G. Dillon, P. A. Nelson, and B. S. Swanson, *J. Chem. Phys.* **44**, 4229 (1966).
64. A. C. Maggs and N. W. Ashcroft, *Phys. Rev. Lett.* **59**, 113 (1987).
65. R. E. Goldstein, A. Parola, and A. P. Smith, *J. Chem. Phys.* **91**, 1843 (1989).
66. R. E. Goldstein and A. Parola, *Acc. Chem Res.* **22**, 77 (1989).
67. R. E. Goldstein and N. W. Ashcroft, *Phys. Rev. Lett.* **55**, 2164 (1985).
68. H. Weingärtner, V. C. Weiss, and W. Schröer, *J. Chem. Phys.*, **113**, 762 (2000).
69. Discussion remark by H. L. Friedman, *J. Solution Chem.* **2**, 354 (1972).
70. J. E. Gordon, *J. Am. Chem. Soc.* **87**, 4347 (1965).
71. R. R. Singh and K. S. Pitzer, *J. Am. Chem. Soc.* **110**, 8723 (1988).
72. H. Weingärtner, T. Merkel, U. Maurer, J.-P. Conzen, H. Glassbrenner, and S. Käshammer, *Ber. Bunsenges. Phys. Chem.* **95**, 1579 (1991).
73. V. M. Valyashko, M. A. Urusova, and K. G. Kravchuk, *Dokl. Akad. Nauk SSSR* **272**, 390 (1983).
74. W. L. Marshall, J. S. Gill, and C. H. Secoy, *J. Am. Chem. Soc.* **76**, 4279 (1974).
75. A. Mugnier de Tobriand and M. Lucas, *J. Inorg. Nucl. Chem.* **41**, 1214 (1979).
76. H. Weingärtner, *Ber. Bunsenges. Phys. Chem.* **93**, 1058 (1989).
77. M. L. Japas and J. M. H. Levelt Sengers, *J. Phys. Chem.* **94**, 5361 (1990).
78. E. Steinle and H. Weingärtner, *J. Phys. Chem.* **96**, 2407 (1992).

79. M. Kleemeier, W. Schröer and H. Weingärtner, *J. Mol Liq.* **73/74**, 501 (1997).

80. H. Weingärtner, *J. Chem. Thermodyn.* **29**, 1409 (1997).

81. H. Xu, H. L. Friedman, and F. O. Raineri, *J. Solution Chem.* **20**, 739 (1991).

82. H. Weingärtner, D. Klante, and G. M. Schneider, *J. Solution Chem.* **28**, 435 (1999).

83. See, for example, D. Eagland, in: *Water. A Comprehensive Treatise*, Vol. 4, edited by F. Francks, Plenum, New York, 1975, p. 305.

84. See, for example, K.D. Collins and M. W. Washabaugh, *Q. Rev. Biophys.* **18**, 323 (1985).

85. H. Weingärtner, T. Merkel, S. Käshammer, W. Schröer, and S. Wiegand, *Ber. Bunsenges. Phys. Chem.* **97**, 1970 (1993).

86. C. Bagnuls and C. Bervillier, *Phys. Rev. B* **32**, 7209 (1985).

87. M. Kleemeier, S. Wiegand, W. Schröer, and H. Weingärtner, *J. Chem.Phys.* **110**, 3085 (1999).

88. See, for example, P.-G. de Gennes, *Scaling Concepts in Polymer Physics*, Cornell University Press, Ithaca, New York, 1979.

89. M. Bonetti, A. Oleinikova and C. Bervillier, *J. Phys. Chem. B* **101**, 2164 (1997).

90. T. Narayanan and A. Kumar, *Phys. Rep.* **249**, 135 (1994).

91. S. Wiegand, M. Kleemeier, J. M. Schröder, W. Schröer, and H. Weingärtner, *Int. J. Thermophys.* **15**, 1045 (1994).

92. W. Schröer, M. Kleemeier, M. Plikat, V. Weiss, and S. Wiegand, *J. Phys. Condens. Matter* **8**, 9321 (1996).

93. M. Kleemeier, Ph.D. Thesis, University of Bremen, 2000.

94. M. C. P. de Lima, D. R. Schreiber, and K. S. Pitzer, *J. Phys. Chem.* **89**, 1854 (1985).

95. K. C. Zhang, M. E. Briggs, R. W. Gammon, and J. M. H. Levelt Sengers, *J. Chem. Phys.* **97**, 8692 (1992).

96. S. Wiegand, M. E. Briggs, J. M. H. Levelt Sengers, M. Kleemeier, and W. Schröer, *J. Chem. Phys.* **109**, 9038 (1998).

97. H. Weingärtner, S. Wiegand, and W. Schröer, *J. Chem. Phys.* **96**, 848 (1991).

98. See, for example, M. J. Sienko (ed.), *Metal–Ammonia Solutions*, Benjamin, New York, 1964.

99. B. K. Das and S. C. Greer, *J. Chem. Phys.* **74**, 3630 (1981).

100. M. Ley-Khoo and M. S. Green, *Phys. Rev. A* **23**, 2650 (1981).

101. J. V. Sengers, D. Bedeaux, P. Mazur, and S. C. Greer, *Physica A* **104**, 573 (1980).

102. B. Widom, *Proc. Robert A. Welch Found. Conf. Chem. Res.* **16**, 161 (1972).

103. See, for example, K. Kawasaki, in: *Phase Transitions and Critical Phenomena*, Vol. 5A, edited by C. Domb and J. L. Lebowitz, Academic Press, New York, 1983.

104. A. L. Khodolenko and A. L. Beyerlein, *J. Chem.Phys.* **93**, 8403 (1990).

105. A. L. Khodolenko and A. L. Beyerlein, *Phys. Lett.* **132**, 347 (1988).

106. See, for example, G. S. Joyce, in: *Phase Transitions and Critical Phenomena*, Vol. 2, edited by C. Domb and M. S. Green, Academic Press, New York, 1972, Chapter 10.

107. W. Schröer, S. Wiegand, and M. Kleemeier, unpublished data.

108. T. Narayanan and K. S. Pitzer, *J. Chem. Phys.* **102**, 8118 (1995).

109. T. Narayanan and K. S. Pitzer, *Phys. Rev. Lett.* **73**, 3002 (1994).

110. T. Narayanan and K. S. Pitzer, *J. Phys. Chem.* **98**, 9170 (1994).

111. W. Schröer and M. Kleemeier, unpublished data.

112. F. Leclercq, P. Damay, and P. Chieux, *Z. Phys. Chem.* **156**, 183 (1988).
113. P. Chieux, J.-F. Jal, L. Hily, J. Dupuy, F. Leclercq, and P. Damay, *J. Phys. IV* **C5**, 3 (1991) and references cited therein.
114. P. Chieux, P. Damay, J. Dupuy, and J. F. Jal, *J. Phys. Chem.* **84**, 1211 (1980).
115. J. F. Jal, J. Dupuy, J. P. Dupin, and P. Chieux, *J. Phys. C: Solid State Phys.* **18**, 1347 (1985).
116. W. Schröer, S. Wiegand and H. Weingärtner, *Ber. Bunsenges. Phys. Chem.* **97**, 975 (1993).
117. M. Bonetti and P. Calmettes, *Int. J. Thermophys.* **19**, 1555 (1998).
118. M. Bonetti, C. Bagnuls, and C. Bervillier, *J. Chem. Phys.* **107**, 550 (1997).
119. See, for example, G. Hohenberg and P. I. Halperin, *Rev. Mod. Phys.* **49**, 435 (1977).
120. See, for example, J. V. Sengers, in: *Supercritical Fluids: Fundamentals for Application*, edited by E Kiran and J. M. H. Levelt Sengers, Kluwer, Dordrecht, 1994.
121. R. F. Berg and M. R. Moldover, *J. Chem. Phys*, **93**, 1926 (1990).
122. M. Kleemeier, S. Wiegand, T. Derr, V. Weiss, W. Schröer, and H. Weingärtner, *Ber. Bunsenges. Phys. Chem.* **100**, 27 (1996).
123. J. Douglas, *Macromol.* **25**, 1468 (1992).
124. J. K. Battacharjee, R. A. Ferrell, R. S. Basu, and J. V. Sengers, *Phys. Rev. A* **24**, 1469 (1981).
125. S. Wiegand, R. F. Berg, and J. M. H. Levelt Sengers, *J. Chem. Phys.* **109**, 4533 (1998).
126. A. Oleinikova and M. Bonetti, *J. Chem. Phys.* **104**, 3111 (1996).
127. M. E. Fisher and J. S. Langer, *Phys. Rev. Lett.* **20**, 665 (1968).
128. D. Jasnow, W. I. Goldburg, and J. S. Semura, *Phys. Rev. A* **9**, 355 (1974).
129. T. Narayanan, A. Kumar, and E. S. R. Gopal, *Phys. Lett. A* **144**, 371 (1990).
130. C. H. Shaw and W. I. Goldburg, *J. Chem. Phys.* **65**, 4906 (1976).
131. J. Ramakrishnan, N. Nagarajan, A. Kumar, E. S. R. Gopal, P. Chandrasekhar, and G. Ananthakrishna, *J. Chem. Phys.* **68**, 4098 (1978).
132. E. M. Andersen and S. C. Greer, *Phys. Rev. A* **30**, 3129 (1984).
133. A. Oleinikova and M. Bonetti, *Phys. Rev. Lett.* **83**, 2985 (1999).
134. C. L. Caylor, B. M. Law, P. Senanyake, V. L. Kuzmin, V. P. Romanov, and S. Wiegand, *Phys. Rev. E* **56**, 4441 (1997).
135. V. L. Kuzmin and V. P. Romanov, *Phys. Rev. E* **49**, 2049 (1994).
136. J. W. Schmidt, *Phys. Rev. A* **38**, 567 (1988).
137. D. R. Schreiber, M. C. P. de Lima, and K. S. Pitzer, *J. Phys. Chem.* **91**, 4087 (1987).
138. H. Weingärtner and W. Schröer, in: *Steam, Water and Hydrothermal Systems: Physics and Chemistry at the Needs of Industry*, edited by P. Tremaine, P. G. Hill, D. E. Irish, and P. V. Balakrishnan, NRC Press, Ottawa (2000).
139. H. Weingärtner, H. Nadolny, and S. Käshammer, *J. Phys. Chem. B* **103**, 4738 (1999).
140. N. Bjerrum, *Kgl. Danske Vidensk. Selsk. Mat.-Fys. Medd.* **7**, 1 (1926).
141. K. S. Pitzer and D. R. Schreiber, *Mol. Phys.* **60**, 1067 (1987).
142. See, for example, K. S. Pitzer, *J. Chem. Thermodyn.* **21**, 1 (1989).
143. K. S. Pitzer, J. L. Bischoff, and R. Rosenbauer, *Chem. Phys. Lett.* **134**, 60 (1987).
144. K. S. Pitzer, *J. Phys. Chem.* **90**, 1502 (1986).
145. R. L. Scott and P. H. van Konynenburg, *Disc. Faraday Soc.* **49**, 87 (1970).
146. P. H. van Konynenburg and R. L. Scott, *Philos. Trans. A* **298**, 495 (1980).

147. X. S. Chen and F. Forstmann, *J. Chem. Phys.* **97**, 3696 (1992); *Mol. Phys.* **76**, 1203 (1992).
148. C. P. Ursenbach and G. N. Patey, *J. Chem. Phys.* **100**, 3827 (1994).
149. Y. Q. Zhou and G. Stell, *J. Chem. Phys.* **102**, 5796 (1995).
150. V. M. Valyashko, *Pure Appl. Chem.* **67**, 569 (1995).
151. W. L. Marshall, *J. Chem. Soc. Faraday Trans.* **86**, 1807 (1990).
152. A. A. Povodyrev, M. A. Anisimov, J. V. Sengers, W. L. Marshall, and J. M. H. Levelt Sengers, *Int. J. Thermophys.* **20**, 1529 (1999).
153. J. L. Bischoff and R. Rosenbauer, *Geochim. Cosmochim. Acta* **52**, 2121 (1988).
154. A. H. Harvey and J. M. H. Levelt Sengers, *Chem. Phys. Lett.* **156**, 415 (1989).
155. K. S. Pitzer and J. C. Tanger IV, *Chem. Phys. Lett.* **156**, 418 (1989).
156. See, for example, J. M. H. Levelt Sengers, in: *Supercritical Fluid Technology: Reviews in Modern Theory and Applications*, edited by T. J. Bruno and J. F. Ely, CRC Press, Boca Raton, FL, 1991, Chapter 1, p.1.
157. M. A. Anisimov, M. M. Bochkov, S. B. Kiselev, and A. A. Povodyrov, in: *Properties of Water and Steam*, edited by M. Pichal and O. Sifner, Hemisphere, New York, 1990, p. 189.
158. I. M. Abdulgatov, V. I. Dvorianchikov, and I. M. Abdulrakhmanov, in: *Properties of Water and Steam*, edited by M. Pichal and O. Sifner, Hemisphere, New York, 1990, p. 203.
159. M. E. Fisher, *Phys. Rev.* **176**, 257 (1968).
160. B. M. J. Ali and A. Kumar, *J. Chem. Phys.* **107**, 8020 (1998).
161. B. M. J. Ali, J. Jacob and A. Kumar, *Pure Appl. Chem.* **70**, 591 (1998).
162. L. A. Bulavin and A. V. Oleinikova, *Rep. Bolgoliubov Inst. Ukr. Acad. Sci.* 1 (1994).
163. L. A. Bulavin, A. V. Oleinikova, and A. V. Petrovitskij, *Int. J. Thermophys.* **17**, 137 (1996).
164. A. Oleinikova, L. Bulavin, and V. Pipich, *Chem. Phys. Lett.* **278**, 121 (1997).
165. M. A. Anisimov, J. Jacob, A. Kumar, V. Agayan and J. V. Sengers, *Phys. Rev. Lett.* in press.
166. A. J. Liu and M. E. Fisher, *Physica A* **156**, 35 (1989).
167. S. A. Adelman, *J. Chem. Phys.* **64**, 724 (1976).
168. P. S. Ramanathan and H. L. Friedman, *J. Chem. Phys.* **54**, 1086 (1971).
169. J. S. Høye and G. Stell, *J. Chem. Phys.* **102**, 2841 (1995).
170. X. Li, Y. Levin, and M. E. Fisher, *Europhys. Lett.* **26**, 683 (1994).
171. M. E. Fisher, Y. Levin, and X. Li, *J. Chem. Phys.* **101**, 2273 (1994).
172. M. P. Tosi and F. G. Fumi, *J. Phys. Chem. Solids* **25**, 45 (1964).
173. H. L. Friedman, *Annu. Rev. Phys. Chem.* **32**, 179 (1979).
174. F. Vericat and L. Blum, *J. Stat. Phys.* **22**, 593 (1980).
175. J. S. Høye and E. Lomba, *J. Chem. Phys.* **88**, 5790 (1988).
176. See, for example, M. P. Allen and D. J. Tildesley, *Computer Simulation of Liquids*, Clarendon Press, Oxford, 1987.
177. I. Prigogine and S. A. Rice, *Advances in Chemical Physics*, Vol. 105, 1999.
178. G. M. Torrie and J. P. Valleau, *J. Comput. Phys.* **23**, 187 (1977).
179. A. Z. Panagiotopoulos, *Fluid Phase Equil.* **76**, 97 (1992).
180. J. M. Caillol, *J. Chem. Phys.* **100**, 2161 (1994).
181. J. M. Caillol, *J. Phys. Condens. Matter* **A171** 6 (1994).
182. A. M. Ferrenberg and R. H. Swendson, *Phys. Rev. Lett.* **61**, 2635 (1988).

183. J. L. Cardy (ed.), *Finite-Size Scaling*, North-Holland, Amsterdam, 1988.

184. K. Binder, *Rep. Prog. Phys.* **60**, 487 (1997).

185. A. D. Bruce and N. B. Wilding, *Phys. Rev. Lett.* **68**, 193 (1992).

186. J. A. Barker and D. Henderson, *Rev. Mod. Phys.* **48**, 587 (1976).

187. J.-P. Hansen and I. R. McDonald, *Theory of Simple Liquids*, Academic Press, New York, 1986.

188. C. Cacammo, *Phys. Rep.* **274**, 1 (1996).

189. E. Waisman and J. L. Lebowitz, *J. Chem. Phys.* **56**, 3086 (1972); **56**, 3093 (1972).

190. J.-P. Hansen and I. R. McDonald, *Phys. Rev. A* **11**, 2111 (1975).

191. G. M. Abernethy and M. J. Gillan, *Mol. Phys.* **39**, 839 (1980).

192. M. Kinoshita and M. Harada, *Mol. Phys.* **65**, 599 (1988).

193. L. Belloni, *J. Chem. Phys.* **88**, 5143 (1988); **98**, 8080 (1993).

194. J. S. Høye, E. Lomba, and G. Stell, *Mol. Phys.* **75**, 1217 (1992).

195. J. S. Høye, J. L. Lebowitz, and G. Stell, *J. Chem. Phys.* **61**, 3253 (1974).

196. P. W. Debye and E. Hückel, *Z. Physik* **24**, 185 (1923).

197. See, for example, D. A. McQuarry, *Statistical Mechanics*, Harper and Row, New York, 1976.

198. M. E. Fisher and Y. Levin, *Phys. Rev. Lett.* **71**, 3826 (1993).

199. Y. Levin and M. E. Fisher, *Physica A* **225**, 164 (1995).

200. V. McGahay and M. Tomozawa, *J. Chem. Phys.* **97**, 2609 (1992).

201. G. A. Mansoori, N. F. Carnahan, K. E. Starling and T. W. Leland, Jr., *J. Chem. Phys.* **54**, 1523 (1971).

202. C. W. Outhwaite, *Chem. Phys. Lett.* **53**, 599 (1978).

203. C. W. Outhwaite, M. Molero, and L. B. Bhuiyan, *Trans. Faraday Soc.* **87**, 3227 (1991).

204. L. B. Bhuiyan, C. W. Outhwaite, M. Molero, and E. Gonzalez-Tovar, *J. Chem. Phys.* **100**, 8301 (1994).

205. C. W. Outhwaite, *J. Chem. Phys.* **100**, 8301 (1994).

206. A. K. Sabir, L. B. Bhuiyan, and C. W. Outhwaite, *Mol. Phys.* **93**, 405 (1998).

207. M. J. Gillan, *Mol. Phys.* **49**, 421 (1983).

208. J. M. Caillol and J. J. Weiss, *J. Chem. Phys.* **102**, 7610 (1995).

209. J. M. Caillol, *J. Chem. Phys.* **102**, 5471 (1995).

210. W. Ebeling, *Z. Phys. Chem. (Leipzig)* **247**, 340 (1971).

211. W. Ebeling, *Z. Phys. Chem. (Leipzig)* **238**, 400 (1968).

212. W. Ebeling and M. Grigo, *Ann. Phys. (Leipzig)* **37**, 21 (1980).

213. D. M. Zuckerman, M. E. Fisher, and B. P. Lee, *Phys. Rev. E* **56**, 6569 (1997).

214. Y. Zhou and G. Stell, *J. Chem. Phys.* **96**, 1504 (1992); **96**, 1507 (1992).

215. Y. Q. Zhou, S. Yeh, and G. Stell, *J. Chem. Phys.* **102**, 5785 (1995).

216. R. M. Fuoss, *J. Am. Chem. Soc.* **80**, 5059 (1958).

217. See, for example, M. C. Justice and J. C. Justice, *J. Solution Chem.* **5**, 543 (1976); **6**, 819 (1977).

218. S. Yeh, Y. Q. Zhou, and G. Stell, *J. Phys. Chem.* **100**, 1415 (1996).

219. A. M. Sukhotin, *Russ. J. Phys. Chem.* **34**, 19 (1960); see also Ref. 37.

220. E. A. S. Cavell and P. C. Knight, *Z. Phys. Chem. Neue Folge* **57**, 331 (1968).

221. V. C. Weiss and W. Schröer, *J. Chem. Phys.* **108**, 7747 (1998).

222. P. Minnhagen, *Rev. Mod. Phys.* **59**, 1001 (1987).

223. S. F. Edwards, *Philos. Mag.* **4**, 1171 (1959).

224. Y. Saito, *Prog. Theor. Phys.* **62**, 927 (1979).

225. C. Deutsch, Y. Furutani, and M. M. Gombert, *Phys. Rep.* **69**, 85 (1981).

226. A. L. Khodolenko and A. L. Beyerlein, *Phys. Rev. A* **34**, 3309 (1986).

227. R. R. Netz and H. Orland, *Europhys. Lett.* **42**, 419 (1998).

228. A. Diehl, M. C. Barbosa, and Y. Levin, *Phys. Rev. E* **56**, 619 (1997).

229. A. G. Moreira, M. M. Telo da Gama, and M. E. Fisher, *J. Chem. Phys.* **110**, 10058 (1999).

230. C. Caccamo and A. Giacoppo, *Phys. Rev. A* **42**, 6285 (1990).

231. A. H. Harvey and J. M. H. Levelt Sengers, *Phys. Rev. A* **46**, 1148 (1992).

232. P. U. Kencare, C. K. Hall, and C. Caccamo, *J. Chem. Phys.* **103**, 8089 (1995).

233. L. Blum and D. Q. Wei, *J. Chem. Phys.* **87**, 555 (1987).

234. L. Blum, F. Vericat and W. R. Fawcett, *J. Chem. Phys.* **96**, 3039 (1992).

235. A. H. Harvey, *J. Chem. Phys.* **95**, 479 (1991).

236. See, for example, L. D. Landau and E. M. Lifshitz, *Statistical Physics*, Pergamon, New York, 1958.

237. See, for example, B. Widom and J. S. Rowlinson, *Translation of J. D. van der Waals: The Thermodynamic Theory of Capillarity Under the Hypothesis of a Continuous Variation of Density, J. Stat. Phys.* **20**, 197 (1979).

238. B. P. Lee and M. E. Fisher, *Phys. Rev. Lett.* **76**, 2906 (1996).

239. W. Schröer and V. C. Weiss, *J. Chem. Phys.* **109**, 8504 (1998).

240. W. Schröer and V. C. Weiss, *J. Chem. Phys.* **110**, 4687 (1999).

241. V. C. Weiss and W. Schröer, *J. Phys. Condens. Matter* **10**, L705 (1998).

242. V. C. Weiss and W. Schröer, *J. Phys. Condens. Matter* **12**, 2637 (2000).

243. B. Groh, R. Evans, and S. Dietrich, *Phys. Rev. E* **57**, 6944 (1998).

244. P. C. Albright, Z. Y. Chen, and J. V. Sengers, *Phys. Rev. B* **36**, 877 (1987).

245. M. Yu. Belyakov, and S. B. Kiselev, *Physica A* **190**, 75 (1992).

246. J. P. Valleau, *J. Chem. Phys.* **95**, 584 (1991).

247. J. M. Caillol, D. Levesque, and J. J. Weis, *Phys. Rev. Lett.* **77**, 4039 (1996).

248. Unpublished work of J. Valleau quoted by M. E. Fisher in Ref. 16.

249. A. K. Wyczalkowska, M. A. Anisimov, and J. V. Sengers, *Fluid Phase Equilibria* **158–160**, 523 (1998).

250. J. C. Shelley and G. N. Patey, *J. Chem. Phys.* **103**, 8299 (1995).

251. J. J. Weiss and D. Levesque, *Phys. Rev. Lett.* **71**, 2729 (1993).

252. B. Smit, K. Esselink and D. Frenkel, *Mol. Phys.* **87**, 159 (1996).

253. L. Onsager, *J. Phys. Chem.* **43**, 189 (1939).

254. H. Totsuji, *Phys. Rev. A* **24**, 1077 (1981).

255. M. J. Gillan, *Mol. Phys.* **41**, 75 (1980).

256. A. Tani and D. Hendersen, *J. Chem. Phys.* **79**, 2390 (1983).

257. D. Laria, H. R. Corti, and R. Fernandez-Prini, *J. Chem. Soc. Faraday Trans.* **86**, 1051 (1990).

258. P. J. Camp and G. N. Patey, *Phys. Rev. E* **60**, 1063 (1999).

259. P. J. Camp and G. J. Patey, *J. Chem. Phys.* **111**, 9000 (1999).
260. R. M. Fuoss and C. A. Kraus, *J. Am. Chem. Soc.* **55**, 2378 (1933).
261. J. M. Kosterlitz, and D. J. Thouless, *J. Phys. C* **6**, 1181 (1973).
262. See, for example, S. Greer, *Int. J. Thermophys.* **9**, 761 (1988).
263. S. J. Kennedy and J. C. Wheeler, *J. Chem. Phys.* **78**, 953 (1983).
264. J. C. Wheeler and P. C. Pfeuty, *J. Chem. Phys.* **74**, 6415 (1981).
265. S. C. Greer, *J. Phys. Chem. B* **102**, 5413 (1998).
266. J. Dudowitz, K. F. Freed, and J. Douglas, *J. Chem. Phys.* **111**, 7116 (1999).
267. L. Blum, *Mol. Phys.* **30**, 1529 (1975).
268. L. Blum and J. S. Høye, *J. Phys. Chem.* **81**, 1311 (1977).
269. E. Gonzalez-Tovar, *Mol. Phys.* **97**, 1203 (1999).
270. C. Caccamo and G. Malescio, *J. Phys. Chem.* **90**, 1091 (1989).
271. P. U. Kencare, C. K. Hall, and C. Caccamo, *J. Chem. Phys.* **103**, 8111 (1995).
272. See, for example, J. Kendall, E. D. Crittenden, and K. H. Miller, *J. Am. Chem. Soc.* **45**, 963 (1923).
273. J. C. Shelley and G. N. Patey, *J. Chem. Phys.* **110**, 1633 (1999).
274. J. G. Kirkwood, *Chem. Rev.* **19**, 275 (1936).
275. See, for example, C. W. Outhwaite, in: *Statistical Mechanics, A Specialist Periodical Report*, Vol. 2, edited by K. Singer, The Chemical Society, London, 1978, p. 188.
276. B. P. Lee and M. E. Fisher, *Europhys. Lett.* **39**, 611 (1997).
277. M. E. Fisher and S. Bekiranov, *Physica A* **263**, 466 (1999).
278. M. E. Fisher, in: *New Approaches to Problems in Liquid State Theory*, edited by C. Caccamo, J.-P. Hansen and G. Stell, NATO ASI Series C, Kluwer, Dordrecht, 1999, p. 71.
279. S. Bekiranov and M. E. Fisher, *Phys. Rev. Lett.* **81**, 5836 (1998); *Phys. Rev. E* **59**, 492 (1999).
280. W. Schröer and V. C. Weiss, *J. Chem. Phys.* **106**, 7458 (1997).
281. R. J. F. Leote de Carvalho and R. Evans, *Mol. Phys.* **83**, 619 (1994); *J. Phys. Condens. Matter* **7**, L575 (1995).
282. J. Ennis, R. Kjellander and D. J. Mitchell, *J. Chem. Phys.* **102**, 975 (1995).
283. M. E. Fisher and B. P. Lee, *Phys. Rev. Lett.* **77**, 3561 (1996).
284. V. C. Weiss and W. Schröer, *J. Chem. Phys.* **106**, 1930 (1997).
285. See, for example, H. T. Davis, *The Statistical Mechanics of Phases, Interfaces and Thin Films*, VCH, New York, 1995.
286. B. F. McCoy and H. T. Davis, *Phys. Rev. A* **20**, 1201 (1979).
287. M. M. Telo da Gama, R. Evans and T. J. Sluckin, *Mol. Phys.* **41**, 1355 (1980).
288. M. Kac, G. E. Uhlenbeck and P. C. Hemmer, *J. Math. Phys.* **4**, 216 (1963); see also P. C. Hemmer and J. L. Lebowitz, in: *Phase Transitions and Critical Phenomena*, Vol. 5b, edited by C. Domb and M. S. Green, Academic Press, New York, 1976.
289. G. Stell, *Phys. Rev. B* **1**, 2265 (1970).
290. M. E. Fisher, S.-K. Ma and B. G. Nickel, *Phys. Rev. Lett.* **29**, 917 (1972).
291. Unpublished work cited by G. Stell in: *New Approaches to Problems in Liquid State Theory*, edited by C. Caccamo, J.-P. Hansen, and G. Stell, NATO ASI Series C, Kluwer, Dordrecht, 1999, pp. 71.
292. A. Z. Panagiotopoulos and S. K. Kumar, *Phys. Rev. Lett.* **83**, 2981(1999).

66   HERMANN WEINGÄRTNER AND WOLFFRAM SCHRÖER

293. G. Orkoulas and A.Z. Panagiotopoulos, *J. Chem. Phys.* **104**, 7205 (1996).

294. P. Minnhagen, *Phys. Rev. Lett.* **54**, 2351 (1985).

295. P. Minnhagen and M. Wallin, *Phys. Rev. B* **36**, 5620 (1987); **40**, 5109 (1989).

296. Y. Levin, X. Li and M. E. Fisher, *Phys. Rev. Lett.* **73**, 2716 (1993).

297. G. Baker, *Quantitative Theory of Critical Phenomena*, Academic Press, New York, 1990.

298. M. E. Fisher, *J. Chem. Phys.* **96**, 3352 (1992).

299. J. Valleau and G. Torrie, *J. Chem. Phys.* **108**, 5169 (1998).

300. A. H. Harvey, C. Cacammo, and J. M. H. Levelt Sengers, *Phys. Rev. E* **46**, 1148 (1992).

301. M. E. Fisher and S. Fishman, *Phys. Rev. Lett.* **47**, 421 (1981).

302. J. S. Høye and G. Stell, *J. Phys. Chem.* **94**, 7899 (1990).

303. G. Stell, *Phys. Rev. A* **45**, 7628 (1992).

304. J. S. Høye and G. Stell, *J. Mol. Liq.* **73**, 453 (1997).

305. V. M. Nabutovskii, N. A. Nemov, and Yu. G. Peisakhovich, *Phys. Lett. A* **79**, 98 (1980); *Mol. Phys.* **54**, 979 (1980).

306. V. Balevicius and H. Fuess, *Phys. Chem.–Chem. Phys.* **1**, 1507 (1999).

307. Y. B. Melnichenko, M. A. Anisimov, A. A. Povodyrev, G. D. Wignall, J. V. Sengers and W. A. van Hook, *Phys. Rev. Lett.* **79**, 5266 (1997).

308. G. Meier, D. Schwahn, K. Mortensen, and S. Janssen, *Europhys. Lett.* **22**, 577 (1993).

309. G. Meier, B. Momper, and E. W. Fischer, *J. Chem. Phys.* **97**, 5884, (1992).

310. D. Schwahn, K. Mortensen, and H. Y. Madeira, *Phys. Rev. Lett.* **58**, 1544 (1987).

311. E. K. Hobbie, L. Read, C. C. Huang, and C. C. Han, *Phys. Rev. E* **54**, 629 (1996).

312. H. Seto, D. Schwahn, M. Nagao, E. Yokoi, S. Komura, M. Imai, and K. Mortensen, *Phys.Rev. E* **54**, 629 (1996).

313. J. F. Nicoll and J. K. Bhattacharjee, *Phys. Rev. B* **23**, 389 (1981).

314. J. F. Nicoll and P. C. Albright, *Phys. Rev. B* **31**, 4576 (1985).

315. Z. Y. Chen, P. C. Albright, and J. V. Sengers, *Phys. Rev. A* **41**, 3161 (1990).

316. Z. Y. Chen, A. Abacci, S. Tang, and J. V. Sengers, *Phys. Rev. A* **42**, 4470 (1990).

317. A. A. Povodyrev, M. A. Anisimov and J. V. Sengers, *Physica A* **264**, 345 (1999).

318. E. Luijten and H. W. J. Blöte, *Int. J. Mod. Phys. C* **6**, 359 (1995).

319. E. Luijten and K. Binder, *Phys. Rev. E* **58**, R4060 (1998).

320. M. A. Anisimov and J. V. Sengers, Critical and Crossover Phenomena in Fluids and Fluid Mixtures, in: *Supercritical Fluids. Fundametals and Applications*, edited by E. Kiran, P. G. Debenedetti and C. J. Peters, Kluwer, Dordrecht, 2000.

321. V. Steinberg, A. Voronel, D. Linsky, and U. Schindewolf, *Phys. Rev. Lett.* **45**, 1338 (1980).

322. M. A. Anisimov, A. A. Povodyrev, V. D. Kulikov, and J. V. Sengers, *Phys. Rev. Lett.* **75**, 3146 (1995); **76**, 4095 (1996).

323. J. Jacob, A. Kumar, S. Asokan, D. Sen, R. Chitra, and S. Mazumder, *Chem. Phys. Lett.* **304**, 180 (1999).

MODE COUPLING THEORY APPROACH TO THE LIQUID-STATE DYNAMICS

BIMAN BAGCHI and SARIKA BHATTACHARYYA

Solid State and Structural Chemistry Unit, Indian Institute of Science, Bangalore, India

CONTENTS

Advances in Chemical Physics, Volume 116, edited by I. Prigogine and Stuart A. Rice.
ISBN 0-471-40541-8 © 2001 John Wiley & Sons, Inc.

I. INTRODUCTION

The dynamic processes in liquids span a very wide range of time scale, beginning from a few tens of femtoseconds for the solvation dynamics of an ion in liquid water to minutes and hours for stress relaxation in glassy liquids. What is even more amazing is that many liquids show an initial response whose rate is many orders of magnitude greater than that of the long time response of the same liquid. There is sometimes an intermediate time regime where dynamics is even slower than the long time response. Bridging the gap between the ultrafast and the ultraslow response exhibited by the same liquid has been an ambitious and challenging task that is still largely unfinished. Recently, however, there has been a notable development that allows one to answer many of the pertinent questions. This new development is the formulation of a very general theoretical framework known as the mode coupling theory (MCT) of liquid-state dynamics. The objective of this chapter is to review some aspects of this theory, with emphasis on liquid-state dynamics. That is, the development pertaining to critical phenomena and glass transition will be discussed, but not in great detail.

Mode coupling theory can mean different things to people working in different areas. The name originated from the theory of dynamical processes near the critical phenomena (in phase transitions) where MCT has been quite successful in explaining a large amount of anomalous behavior [1, 2]. This development took place mostly in the 1960s. More recently, MCT seems to have become synonymous with the microscopic theory of glass transition [3, 4]. However, unlike the theory of critical phenomena, the status of the theory in explaining the myriad of experimental results near the glass transition temperature is unclear and is often controversial. In addition, the scope of the mode coupling theory is much broader than just the application to glass transition—a fact sometimes forgotten these days.

In fact, mode coupling theory can be used to understand dynamical response of a dense liquid away from either the glass transition or the critical point where the divergence of long time scale is absent but the disparity between the initial and the long time scales persists. In many applications of great interest to physical chemistry and chemical physics, we need a theory that can provide a *self-consistent* description of *both* the short and the long time response of the liquid. Such a theory is required to understand solvent effects on such diverse phenomena as vibrational phase relaxation, electron transfer reactions, activated barrier crossing dynamics, and barrier-less chemical reactions, to name a few. In fact, what is often investigated in real experiments is the effects of a large variation of the composition, temperature, and other experimental conditions. This, in turn, leads to a large variation of macroscopic properties such as viscosity and diffusion. However, such variation can have very different effects on the short and the long time dynamics of the liquid. Disregard of this fact can lead to erroneous conclusions.

MCT can be best viewed as a synthesis of two formidable theoretical approaches, namely the renormalized kinetic theory [5–9] and the extended hydrodynamic theory [10]. While the former provides the method to treat both the very short and the very long time responses, it often becomes intractable in the intermediate times. This is best seen in the calculation of the velocity time correlation function of a tagged atom or a molecule. The extended hydrodynamic theory provides the simplicity in terms of the wave-number-dependent hydrodynamic modes. The decay of these modes are expressed in terms of the wavenumber- and frequency-dependent transport coefficients. This hydrodynamic description is often valid from intermediate to long times, although it breaks down both at very short and at very long times, for different reasons. None of these two approaches provides a self-consistent description. The self-consistency enters in the determination of the time correlation functions of the hydrodynamic modes in terms of the transport coefficients (or memory functions); the latter are determined by the well-known Green–Kubo relations that define the transport coefficients in terms of integrations (over time and wavenumber) of these time correlation functions [11].

The relaxation equations for the time correlation functions are derived formally by using the projection operator technique [12]. This relaxation equation has the same structure as a generalized Langevin equation. The mode coupling theory provides microscopic, albeit approximate, expressions for the wavevector- and frequency-dependent memory functions. One important aspect of the mode coupling theory is the intimate relation between the static microscopic structure of the liquid and the transport properties. In fact, even now, realistic calculations using MCT is often not possible because of the nonavailability of the static pair correlation functions for complex intermolecular potential.

At the heart of the mode coupling theory of liquids is the assumption that a separation of time scale exists between different dynamical events. While the time scale separation between the fast collisional events and the slower collective relaxation is explicitly exploited in the formulation of the theory, there is also an underlying assumption of the *separation of length scales* between different relaxation modes. Much of the success of MCT depends on the validity of this separation of length and time scales.

Even with all of its sophistication, the mode coupling theory is still a perturbation theory where dynamics is described in terms of a subset of dynamical variables chosen from the products of hydrodynamic modes. It fails, for example, to describe rare events, such as the activated processes or string-like cooperative motions often found to dictate dynamics in glassy liquids [13–15].

Although a large number of articles have appeared on MCT and even some monographs exist [3, 4, 16], we found no article where a coherent and historical account of the development of MCT is given. Such an account is important for a beginner to appreciate the strengths and the weaknesses of the theoretical framework and the legacy it bears. We have attempted to provide such an account here. We have also discussed some recent improvements of the existing theory made by ourselves. These include the implementation of the full self-consistency between the time (or frequency)-dependent friction and the velocity–time correlation function and a new decomposition of the short time part of the four point time correlation function. We have also discussed applications of the MCT to understand the solvent dynamical effects on several chemical relaxation processes. But we have not discussed applications of MCT to complex systems, like colloids [15, 17] and electrolytes [18], because they are available elsewhere.

II. HYDRODYNAMIC MODES: THE BASIC
DYNAMICAL VARIABLES IN LIQUIDS

The hydrodynamic approach to liquid-state dynamics is based on the assumption that many experimental observables (like the intensity in a light scattering experiment) can be rationalized by considering the dynamics of a few slow variables. The natural choice for the slow variables are the densities of the conserved quantities—that is, the number density, $\rho(\mathbf{r}, t)$, the momentum density, $\mathbf{g}(\mathbf{r}, t)$ and the energy density, $e(\mathbf{r}, t)$. The conservation of number, momentum, and energy are expressed locally by the conservation equations

$$\dot{\rho}(\mathbf{r}, t) + \frac{1}{m}\nabla \cdot \mathbf{g}(\mathbf{r}, t) = 0, \tag{1}$$

$$\dot{\mathbf{g}}(\mathbf{r}, t) + \nabla \cdot \sigma(\mathbf{r}, t) = 0, \tag{2}$$

$$\dot{e}(\mathbf{r}, t) + \nabla \cdot \mathbf{J}^{\mathbf{e}}(\mathbf{r}, t) = 0, \tag{3}$$

where m is the mass of the particles, σ is the momentum current, or stress tensor, and $\mathbf{J}^{\mathbf{e}}$ is the energy current. The above equations apply both to microscopic densities and to locally averaged densities.

The continuity equations are supplemented by the constitutive relations involving the current of number, momentum and energy. The local velocity field $\mathbf{u}(\mathbf{r}, t)$ is defined via the following relation,

$$\mathbf{g}(\mathbf{r}, t) \simeq m\rho(\mathbf{r}, t)\mathbf{u}(\mathbf{r}, t). \tag{4}$$

The stress tensor on a hydrodynamic volume element is obtained from macroscopic consideration and from rotational invariance and is given by [19]

$$\sigma^{\alpha\beta}(\mathbf{r}, t) = \delta_{\alpha\beta}\mathbf{p}(\mathbf{r}, t) - \eta_s\left(\frac{\partial u_\alpha(\mathbf{r}, t)}{\partial r_\beta}\frac{\partial u_\beta(\mathbf{r}, t)}{\partial r_\alpha}\right)$$
$$+ \delta_{\alpha\beta}\left(\tfrac{2}{3}\eta_s - \eta_v\right)\nabla\mathbf{u}(\mathbf{r}, t), \tag{5}$$

where $\mathbf{p}(\mathbf{r}, t)$ is the fluctuating local pressure, η_s is the shear viscosity, and η_v is the bulk viscosity.

The last constitutive relation defines the macroscopic energy current as

$$\mathbf{J}^e(\mathbf{r}, t) = h\mathbf{u}(\mathbf{r}, t) - \lambda\nabla T(\mathbf{r}, t), \tag{6}$$

where $h = (e + P)$ is the equilibrium enthalpy density, $T(\mathbf{r}, t)$ is the local temperature, P is the average pressure, and λ is the thermal conductivity. The second term on the right-hand side is the diffusive part of the energy flux which is assumed to be given by the Fourier's law of heat conduction.

The constitutive relations along with the conservation equations give the basic equations of fluid mechanics, which are a set of five nonlinear partial differential equations involving the seven variables, $\rho, \mathbf{g}, e, P,$ and T. Because five equations [Eqs. (1), (2), (3), (5), and (6)] cannot determine seven quantities, the equations are closed by expressing any two variables of the set (ρ, e, P, T) in terms of the other two remaining variables. This is done by using the assumption of local equilibrium and thermodynamic equations of state.

The density and the temperature can be chosen as independent variables, and the continuity equations are linearized assuming that the fluctuations around the equilibrium value to be small.

Due to the isotropic nature of the liquid, the linearized hydrodynamic equations are easily solved when written in the Fourier (wave number) plane. Thus, the basic equations in fluid mechanics in the wavenumber and Laplace frequency (z) plane are written as

$$-iz\tilde{\rho}_{\mathbf{q}}(z) + i\mathbf{q} \cdot \tilde{\mathbf{g}}_{\mathbf{q}}(z) = \rho_{\mathbf{k}}, \tag{7}$$

$$(-iz + aq^2)\tilde{T}_{\mathbf{q}}(z) + \frac{iT}{\rho^2 c_V}\left(\frac{\partial P}{\partial T}\right)_\rho \mathbf{q} \cdot \tilde{\mathbf{g}}_{\mathbf{q}}(z) = T_{\mathbf{q}}, \tag{8}$$

$$\left(-iz + \frac{\eta_s}{\rho m}q^2 + \frac{\tfrac{1}{3}\eta_s + \eta_v}{\rho m}\mathbf{q}\mathbf{q}.\right)\tilde{\mathbf{g}}_{\mathbf{q}}(z)$$

$$+ \frac{i\mathbf{q}}{m}\left(\frac{\partial P}{\partial \rho}\right)_T \tilde{\rho}_{\mathbf{q}}(z) + \frac{i\mathbf{q}}{m}\left(\frac{\partial P}{\partial T}\right)_\rho \tilde{T}_{\mathbf{q}}(z) = \tilde{\mathbf{g}}_{\mathbf{q}}. \tag{9}$$

Here $\mathbf{q}$ is chosen to be along the z-axis. c_V is the specific heat *per particle* at constant volume and $\rho_{\mathbf{q}}, T_{\mathbf{q}},$ and $\mathbf{g}_{\mathbf{q}}$ are the spatial Fourier components at $t = 0$. $a = \lambda/\rho c_V$. The above set of equations is used to obtain the correlation functions of the hydrodynamic variables.

When Eq. (9) is separated into transverse and longitudinal components, the transverse component of the current gets decoupled from all the longitudinal modes—that is, from the fluctuations in the density, the temperature, and the longitudinal current. On the other hand, it is found from Eqs. (7)–(9) that all the longitudinal modes are coupled together. The coupled longitudinal equations provide the hydrodynamic limiting form of the dynamic structure factor $S(q, \omega)$ which is the spatial and temporal correlation of the density fluctuations in wavenumber and Fourier frequency plane. The spectrum of density fluctuations that can reproduce the Rayleigh–Brillouin spectrum, observed in light scattering experiments [20] is one of the great successes of the classical hydrodynamic theory.

The derivation of $S(q, \omega)$ is available in various textbooks [20–22]. In this chapter, only the final expression is presented and its significance is discussed. The well-known Landau–Placzek formula for the dynamic structure factor is given by

$$S(q, \omega) = \frac{1}{2\pi} S(q) \left[\left(\frac{\gamma - 1}{\gamma} \right) \frac{2 D_T q^2}{\omega^2 + (D_T q^2)^2} \right.$$
$$\left. + \frac{1}{\gamma} \left(\frac{\Gamma q^2}{(\omega + c_s q)^2 + (\Gamma q^2)^2} + \frac{\Gamma q^2}{(\omega + c_s q)^2 + (\Gamma q^2)^2} \right) \right], \quad (10)$$

where $\gamma = c_P / c_V$, c_s is the adiabatic speed of sound,

$$D_T = a/\gamma \quad (11)$$

is the thermal diffusivity, and the acoustic attenuation coefficient Γ is given by

$$\Gamma = \tfrac{1}{2}[(\gamma - 1)D_T + b]. \quad (12)$$

Here $a = \lambda/\rho c_V$ and the kinematic longitudinal viscosity is given by $b = \frac{4}{3}\eta_s + \eta_v/\rho m$.

The spectrum consists of three components. The first term represents an unshifted line called the *Rayleigh line*, which is a Lorentzian with a half-width at half-maximum given by $\Delta \omega_c(q) = D_T q^2$. The next two terms represent a doublet called the *Brillouin doublet*. These are two Lorentzian lines shifted symmetrically from the origin by $\omega = \pm c_s q$, each having half-width at half-maximum, $\Delta \omega_B(q) = \Gamma q^2$.

Physically, the *Brillouin* spectrum arises from the inelastic interaction between a photon and the hydrodynamics modes of the fluid. The doublets can be regarded as the "Stokes" and "anti-Stokes" translational Raman spectrum of the liquid. These lines arise due to the inelastic collision between the photon and the fluid, in which the photon gains or loses energy to the phonons (the propagating sound modes in the fluid) and thus suffer a frequency shift. The width of the band gives the lifetime $(q^2 \Gamma)^{-1}$ of a classical phonon of wavenumber q. The Rayleigh band, on the other hand, represents the scattering of the light by the entropy, or heat fluctuations that are purely diffusive or dissipative modes of the fluid.

It should be noted that the above treatment is based on two assumptions. (1) The fluctuations can be described by the simple linearized hydrodynamic equations and (2) the widths of the Lorentzians are small compared to the shifts; that is, $\gamma D_T q^2 \ll c_s q; \Gamma q^2 \ll c_s q$. This is the case in most of the fluids at small q. For example, in argon at temperature 235 K and mass density 1 g/cm^3, we have $c_s = 6.85 \times 10^4 \text{ cm/sec}$, $D_T = 1.0 \times 10^{-3} \text{cm}^2/\text{sec}$, and $b = 1.6 \times 10^{-3} \text{cm}^2/\text{sec}$. For $\gamma \simeq 1$ and $q = 2.1 \times 10^5 \text{ cm}^{-1}$ (which is typical for light scattering), we have $\gamma D_T q^2/c_s q = 1.5 \times 10^{-3}$ and $\Gamma q^2/c_s q = 1.2 \times 10^{-3}$.

The expression of the transverse current autocorrelation function can also be derived from the linearized hydrodynamic equations. Because it is decoupled from all the longitudinal modes, the derivation is simple and the final expression in wavenumber and Laplace frequency plane can be written as

$$\tilde{C}_t(k, z) = \frac{k_B T}{m(-iz + vk^2)}, \tag{13}$$

where $v = \eta_s/\rho m$ is the kinematic shear viscosity. Similarly, other correlation functions at small wavenumber are obtained from the linearized hydrodynamic equations.

Other than dynamical correlations, transport properties have also been derived using hydrodynamic theory. In hydrodynamics the diffusion of a tagged particle is defined by the Stoke–Einstein relation that is given by the following well-known expression:

$$D = \frac{k_B T}{C\pi \eta_s R}, \tag{14}$$

where k_B is the Boltzmann constant. The above equation is obtained by combining the Einstein equation (which relates the diffusion to the friction) and the Stokes relation (which relates the friction on a spherical molecule to the viscosity of the medium). This Stokes relation is obtained from the Navier–Stokes equation by calculating the frictional force on a sphere with radius R, moving with constant velocity $\mathbf{u}$ in a fluid of shear viscosity η_s [19, 23]. The constant C is obtained from the hydrodynamic boundary condition, which is 6 for "stick" approximation where the fluid velocity everywhere on the surface is considered to be $\mathbf{u}$ and 4 for "slip" approximation where the normal component of the fluid velocity is taken to be equal to the normal component of $\mathbf{u}$.

The relation between friction and viscosity goes beyond the Stokes relation. The Navier–Stokes hydrodynamics has been generalized by Zwanzig and Bixon [23] to include the viscoelastic response of the medium. This generalization provides an elegant expression for the frequency-dependent friction which depends among other things on the frequency-dependent bulk and shear viscosities and sound velocity.

III. SLOW DYNAMICS AT LARGE WAVENUMBERS: DE GENNES NARROWING

In the previous section we have discussed the Rayleigh–Brillouin spectrum, which is observed in light scattering experiments. The wavelengths involved in these scattering experiments are typically about 5000 Å. It is thus possible to calculate the spectral distribution of scattered light from the macroscopic equations of hydrodynamics. The neutron scattering experiments, on the other hand, involves wavelengths of the order of the nearest-neighbor separation in the liquid, thus capturing completely different physics that cannot be explained from the hydrodynamic point of view.

The wavenumbers accessible in inelastic neutron scattering experiments lie typically in between 0.1 and 15 Å^{-1}, *which is the same range as usually studied in molecular dynamics simulations.* There has been extensive study of the density fluctuation spectra by coherent and incoherent inelastic neutron scattering experiments [24] and also by using molecular dynamics simulations [25]. As discussed earlier, the dynamic structure factor $S(q, \omega)$ at reduced wavenumbers $q\sigma \leq 1$ (where σ is the atomic diameter) has a three-peak structure where the two side peaks correspond to propagating sound waves. At shorter wavelengths, the sound waves are strongly damped and the high-frequency structure disappears when $q\sigma \geq 2$, leaving only the Lorentzian-like central peak. The width of the central peak first increases with q, but then shows a *marked decrease at wavenumbers close to* q_m, which corresponds to the wavenumber where the static structure factor has its main peak. At still larger wavenumbers the spectrum broadens again, going over finally to its free-particle limit. This decrease in the width near $q = q_m$ is called "de Gennes narrowing" and cannot be explained from the expression of dynamic structure factor obtained using the linearized hydrodynamic theory [Eq. (10)]. Many years ago, de Gennes observed that for intermediate wavenumbers (that is, for intermediate momentum transfer), correlation effects between neighboring atoms are important [26]. Often this results in a strong narrowing of the density distribution function [26]. This narrowing can also be explained by using a Smoluchowski–Vlasov-type equation to express the number density fluctuations.

At the wavelengths of the order of nearest-neighbor separation, neither the conservation of the momentum nor the energy are relevant constraints to the dynamics. This is because at these wavelengths there is a rapid exchange of momentum and energy between the nearest neighbors, due to binary collisions. On the other hand, *the number density must be conserved at all length scales.* Thus the only slow mode relevant in this regime of wavenumber is the number density. Now a Smoluchowski–Vlasov-type expression can be easily written down for the number density, which involves a mean-field force term due to the cage effect [27, 28]:

$$\dot{\rho}(\mathbf{r}, t) = D\nabla \cdot [\nabla - \beta \mathbf{F}(\mathbf{r}, t)]\rho(\mathbf{r}, t), \tag{15}$$

where $D = k_B T / m\gamma$ is the self-diffusion coefficient of the liquid, γ is the collisional frequency, and $\mathbf{F}$ is the mean field force given by

$$\mathbf{F}(\mathbf{r}, t) = -\nabla V_{\text{eff}}(\mathbf{r}, t). \tag{16}$$

Here $V_{\text{eff}}(\mathbf{r}, t)$ is the effective potential energy that is determined from mean spherical approximation and is written in terms of the two-particle direct correlation function $c(\mathbf{r})$:

$$\beta V_{\text{eff}}(\mathbf{r}, t) = -\int d\mathbf{r}' c(\mathbf{r} - \mathbf{r}')\rho(\mathbf{r}', t). \tag{17}$$

ρ_0 is the equilibrium solution of Eq. (15) and has the following form:

$$\rho_0(\mathbf{r}) = c' \exp\left[\int d\mathbf{r}' c(\mathbf{r} - \mathbf{r}')\rho_0(\mathbf{r}')\right], \tag{18}$$

where c' is a constant determined from the normalization condition.

From Eqs. (15)–(18), the intermediate scattering function $F(k, t)$ can be obtained and has the following form:

$$F(q, t) = S(q)e^{-\frac{D}{S(q)}q^2 t}, \tag{19}$$

where $S(q)$ is the static structure factor. The value of self-diffusion coefficient is about three orders of magnitude smaller than the thermal diffusion coefficient. In addition, for wavenumbers near q_m, the static structure factor is sharply peaked. These two combine to give rise to a marked slowing down of $F(q, t)$. This slowing down in $F(q, t)$ in turn leads to a considerable narrowing of the zero frequency (Rayleigh) peak of the dynamic structure factor, which is the de Gennes narrowing. Underneath all these interpretations lies the fact that in a dense liquid the harsh repulsive part of the intermolecular potential plays an important role. Due to this repulsive potential a cage is created around each molecule which inhibits or hinders the movement of the molecule. This makes the self-diffusion the slowest process; thus the transfer of a molecule becomes the bottleneck of any relaxation process.

Sometimes, this sharp peak in $S(q)$ was used to explore the stability of the liquid [29]. This de Gennes narrowing forms the basis for the current mode coupling theory explanation of the glass transition [30].

IV. EXTENDED HYDRODYNAMICS: DYNAMICS AT INTERMEDIATE LENGTH SCALE

In the preceding sections the validity of hydrodynamics at small q and its breakdown at intermediate q have been discussed. Often in the calculation of the transport coefficients, integration of the time correlation functions over the whole wavenumber space is required. Thus to have a unified description over the whole q plane, the extension of the hydrodynamic theory to intermediate wavenumbers is essential.

This extension of all five hydrodynamic modes (heat, sound, and viscous) to intermediate wavelengths was perhaps first done by de Schepper and Cohen in an elegant paper in 1982 [31]. Their theory was based on a model kinetic equation (generalized Enskog theory) for a hard-sphere fluid. They have shown that at low densities all the five modes are increasingly damped with decreasing wavelength until each ceases to exist after a cutoff wavelength. At high densities the extended heat mode behaves differently. Not only does it soften appreciably for wavelengths of the order of the size of the particles and becomes a diffusion like mode, it also persists until much shorter wavelengths than the other modes. They have also shown that for dense hard-sphere fluids the neutron scattering structure factor can be qualitatively represented for $0 < q\sigma < 15$ as a superposition of the heat and (two) sound modes. One is thus led to a very important conclusion that the extended hydrodynamic modes can be regarded as a seamless extension of the classical hydrodynamic modes used to explain the $S(q, \omega)$ in the light scattering experiments.

There exists another prescription to extend the hydrodynamical modes to intermediate wavenumbers which provides similar results for dense fluids. This was done by Kirkpatrick [10], who replaced the transport coefficients appearing in the generalized hydrodynamics by their wavenumber and frequency-dependent analogs. He used the standard projection operator technique to derive generalized hydrodynamic equations for the equilibrium time correlation functions in a hard-sphere fluid. In the short-time approximation the frequency dependence of the memory kernel vanishes. The final result is a closed set of generalized hydrodynamic equations for the time correlation functions of number density ρ, longitudinal momentum density g_l, temperature T, and transverse momentum density $t_i(i = 1, 2)$. $G_{\alpha\beta}(q, z)(\alpha, \beta = \rho, g_l, T, t_i)$ denotes the normalized time correlation functions in Fourier–Laplace (z) space. The equations of motion are the following:

$$zG_{\rho\beta}(q, z) - \frac{iq}{\sqrt{\beta m S(q)}} G_{l\beta}(q, z) = \delta_{\rho\beta}, \qquad (20)$$

$$zG_{l\beta}(q, z) - \frac{iq}{\sqrt{\beta m S(q)}} G_{\rho\beta}(q, z) - iq\left[\frac{2}{3\beta m}\right]^{1/2}$$

$$\times \left[1 + 2\pi\rho\sigma^3 g(\sigma)\frac{j_1(q, \sigma)}{q\sigma}\right] G_{T\beta}(q, z)$$

$$+ \frac{2}{3t_E}[1 - j_0(q\sigma) + 2j_2(q, \sigma)]G_{l\beta}(q, z) = \delta_{l\beta}, \qquad (21)$$

$$zG_{T\beta}(q, z) - iq\left[\frac{2}{3\beta m}\right]^{1/2}\left[1 + 2\pi\rho\sigma^3 g(\sigma)\frac{j_1(q, \sigma)}{q\sigma}\right] G_{l\beta}(q, z)$$

$$+ \frac{2}{3t_E}[1 - j_0(q\sigma)]G_{T\beta}(q, z) = \delta_{T\beta}, \qquad (22)$$

and $(i = 1, 2)$

$$\left[z + \frac{2}{3t_E}[1 - j_0(q\sigma) - j_2(q\sigma)]\right] G_{t_i\beta}(q, z) = \delta_{t_i\beta}. \qquad (23)$$

Here $j_l(q\sigma)$ is the spherical Bessel function of order l, $g(\sigma)$ is the radial distribution function at contact, and $t_E = \sqrt{\beta m\pi}/4\pi\rho\sigma^2 g(\sigma)$ is the Enskog mean free time between collisions. The transport coefficients in the above expressions are given only by their Enskog values; that is, only collisional contributions are retained. Since it is only in dense fluids that the Enskog values represents the important contributions to transport coefficient, the above expressions are reasonable only for dense hard-sphere fluids. Earlier Alley, Alder, and Yip [32] have done molecular dynamics simulations to determine the wavenumber-dependent transport coefficients that should be used in hard-sphere generalized hydrodynamic equations. They have shown that for intermediate values of q, the wavenumber-dependent transport coefficients are well-approximated by their collisional contributions. This implies that Eqs. (20)–(23) are even more realistic as q and z are increased.

The eigenmodes are obtained by solving the coupled equations (20)–(23). The results are similar to that obtained by de Schepper and Cohen. The extended hydrodynamic also reproduces the softening of the heat mode at intermediate wavenumbers. The softening of the heat mode is shown in Fig. 1.

In addition, the damping of the propagating sound waves at intermediate wavenumber was also recovered. This was interpreted by de Schepper et al. [33] as due to the competition between elasticity and dissipation. There seems to be a trapping of the sound wave on a molecular length scale. The reappearance of the propagating modes at larger q is similar to depinning of sound waves in porous mediums at larger frequencies, where the effective viscosity or damping becomes smaller.

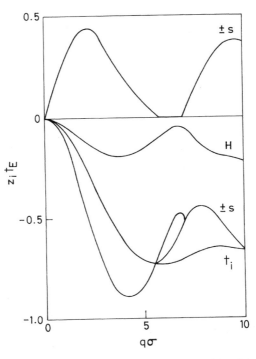

Figure 1. Wavenumber (q) dependence of the eigenvalues (z_i) of the extended hydrodynamic modes. The wavenumber has been scaled by σ^{-1}, where σ is the molecular diameter. The eigenvalues have been scaled by the inverse of the Enskog time (t_E) of the hard-sphere fluid. Here the curve labeled "H" shows the heat mode while the curves labeled "s" shows the sound mode. The graphs in the upper half of the figure are the imaginary part, while the lower section shows the real part of the eigenvalues. Note the softening of the heat mode near $q\sigma = 2\pi$. The reduced density (scaled by σ^{-3}) of the hard-sphere fluid is 0.88. This figure was calculated using the expressions given in Ref. 10.

The complete solution of the heat mode eigenvalue, $z_H(q)$, was found to be well-approximated by the form obtained from the revised Enskog kinetic equation, for not too small q. The later is given by

$$z_H(q) = -\frac{Dq^2}{S(q)} [1 - j_0(q\sigma) + 2j_2(q, \sigma)]^{-1}. \tag{24}$$

From the above expression we can say that the softening of the heat mode happens due to both the peak in $S(q)$ near q_m and also the fact that the heat mode becomes essentially a density mode that decays via self-diffusion.

As discussed by Kirkpatrick [10], this slow mode is important in the theories that include mode coupling effects. Such theories have been used to quantitatively understand the anomalous long-time tails of the stress–stress correlation function and the shear-dependent viscosity [3, 30, 34], observed in computer simulations. As mentioned earlier, a theory of glass transition has also been developed based on the softening of the heat mode.

V. DYNAMICAL CRITICAL PHENOMENA: GENESIS OF THE MODERN MODE COUPLING THEORY

In this section we will not discuss the dynamical critical phenomena in detail because they are beyond the scope of this chapter. Instead, this phenomena will be discussed with a view toward understanding the development of the modern mode coupling theory.

Near the critical point a fluid is known to behave differently, and many anomalies appear in the static and dynamical properties. The important anomalies in the dynamical properties are the critical slowing down of the thermal diffusivity (D_T) in a one-component fluid and the interdiffusion of two species in a binary mixture and also the divergence of the viscosity in a binary mixture.

Underneath all these anomalies lies the fact that near the critical point the correlation length, ξ (which is of the order of few molecular diameter in normal fluids), diverges. This divergence of the correlation length introduces new features in the dynamics which neither the kinetic theory nor the hydrodynamic theory can describe.This is because the kinetic theory is based on the concept of binary collision and thus can not account for the long-range correlation buildup near the critical point. On the other hand, the hydrodynamic theory is based on a macroscopic point of view. While developing the linearized hydrodynamic theory it was considered that the wavelength of fluctuation is large compared to all other relevant length scales appearing in the system. This means that the hydrodynamic theory works in a regime where no microscopic description of the system exists. Instead, the system is described only by the conservation equations (1)–(3) and the macroscopic thermodynamical quantities and transport coefficients. Near the critical point the hydrodynamical regime is no longer defined due to the appearance of this new divergent length scale ξ. The divergence of the correlation length implies that a microscopic description of the system is required even at large length scales.

There have been several treatments to calculate correlation functions and the transport coefficients near the critical point (Fixman [35], Kawasaki [36] and Kadanoff and Swift [37]). All these treatments embody essentially the same physical ideas and contains the genesis of the modern mode coupling theory. Here we discuss the treatment of Kadanoff and Swift [37] because this is physically the most transparent one and seems to have influenced the latter development of the mode coupling theory in a more significant manner.

A. Conservation Equations in the Operator Form

Kadanoff and Swift have considered that the time evolution of the state is described by the Liouville equation. They also wrote down conservation equations for the number, momentum, and energy density similar to the ones given by Eqs. (1)–(3). The only difference was that in the treatment of Kadanoff and Swift the densities and currents are operators. The time derivative of the densities are replaced by the commutator of the respective density operator and the Liouville operator, L (as the Liouville operator governs the time evolution). The suffix "op" in the following equations stands for operator:

$$[L, \rho_{op}(\mathbf{r}, t)] + \frac{1}{m} \nabla \cdot \mathbf{g}_{op}(\mathbf{r}, t) = 0, \tag{25}$$

$$[L, \mathbf{g}_{op}(\mathbf{r}, t)] + \nabla \cdot \sigma_{op}(\mathbf{r}, t) = 0, \tag{26}$$

$$[L, e_{op}(\mathbf{r}, t)] + \nabla \cdot \mathbf{J}_{op}^{e}(\mathbf{r}, t) = 0, \tag{27}$$

The next step is the choice of the basis set of these operators. The choice of the basis set depends on the ensemble. Kadanoff and Swift have considered a grand canonical ensemble with the temperature, chemical potential, and the velocity as the equilibrium parameters. The states that describe a situation where these equilibrium parameters vary slowly are the local equilibrium states. Linear combinations of the densities of the conserved operators acting on the equilibrium state $(|\rangle$, where $L|\rangle = 0)$ provide the desired local equilibrium states. They form the basis set and are given by

$$
\begin{aligned}
|i, \mathbf{q}\rangle &= a_i(\mathbf{q})|\ \rangle \\
\langle i, \mathbf{q}| &= \langle\ |a_i(-\mathbf{q}),
\end{aligned}
\tag{28}
$$

where the $a_i's$ with $i = 1, 2, 3, 4, 5$ are now linear combinations of the density of the conserved operators. The set needs to be properly orthonormalized such that

$$
\langle i, \mathbf{q}\,|\,j, \mathbf{q}'\rangle = \delta_{i,j}(2\pi)^3 \delta(\mathbf{q} - \mathbf{q}').
\tag{29}
$$

Because the liquid is isotropic and translationally invariant, it is easier to work with wavenumbers. Thus, the densities are written in the Fourier plane.

B. Construction of the Eigenfunctions

As discussed in Section II, the five hydrodynamic variables are the number, energy, and momenta densities. Instead of considering the number and energy density, it turns out to be more convenient to consider the entropy density, which can be expressed as a combination of energy and number density. The second variable is now constructed so that it is orthogonal to the entropy density. To explicitly write the states we start with writing a_1, which is related to the entropy density, s_{op},

$$
a_1(\mathbf{q}) = \frac{s_{op}(\mathbf{q})}{[k_B m \rho C_p(\mathbf{q})]^{1/2}}.
\tag{30}
$$

Here s_{op} is defined as

$$
s_{op}(\mathbf{q}) = \frac{1}{T}\left[e_{op}(\mathbf{q}) - \frac{\langle e\rangle + \langle P\rangle}{\langle \rho\rangle}\rho_{op}(\mathbf{q})\right].
\tag{31}
$$

The next state, $|2, \mathbf{q}\rangle = a_2(\mathbf{q})|\rangle$, is such formed that it is orthogonal to $|1, \mathbf{q}\rangle$ and properly normalized:

$$
a_2(\mathbf{q}) = \frac{(m\rho\beta)^{1/2}}{\langle\rho\rangle}c_s(\mathbf{q})\rho_{op}(\mathbf{q}) + \left(\frac{1}{k_B m \rho}\left[\frac{1}{C_V(\mathbf{q})} - \frac{1}{C_p(\mathbf{q})}\right]\right)^{1/2} s_{op}(\mathbf{q}),
\tag{32}
$$

In a similar way the other three states are also formed:

$$a_3(\mathbf{q}) = g_{x\,\mathrm{op}}(\mathbf{q})(\beta/m\rho)^{1/2}, \tag{33}$$

$$a_4(\mathbf{q}) = g_{y\,\mathrm{op}}(\mathbf{q})(\beta/m\rho)^{1/2}, \tag{34}$$

$$a_5(\mathbf{q}) = g_{z\,\mathrm{op}}(\mathbf{q})(\beta/m\rho)^{1/2}. \tag{35}$$

The wavenumber-dependent thermodynamic quantities in the above equations reduce to their standard thermodynamic values at zero wavenumber. For all other wavenumbers they are defined in such a way that $a_1(\mathbf{q})$ and $a_2(\mathbf{q})$ are always orthonormal for each q. $C_p(\mathbf{q})$ is obtained from the normalization condition of $|1, \mathbf{q}\rangle = a_1(\mathbf{q})|\rangle$,

$$C_p(\mathbf{q}) = \langle|s_{\mathrm{op}}(-\mathbf{q})s_{\mathrm{op}}(\mathbf{q})|\rangle/k_B m\rho, \tag{36}$$

and reduces to specific heat at constant pressure, at $q = 0$. Similarly, $C_V(\mathbf{q})$ and $c_s(\mathbf{q})$ are obtained from the orthonormalization condition of $|1, \mathbf{q}\rangle$ and $|2, \mathbf{q}\rangle$:

$$C_V(\mathbf{q})[c_s(\mathbf{q})]^2 = \frac{T\langle\rho\rangle}{m^2\rho} \frac{\langle|s_{\mathrm{op}}(-\mathbf{q})s_{\mathrm{op}}(\mathbf{q})|\rangle}{\langle|\rho_{\mathrm{op}}(-\mathbf{q})\rho_{\mathrm{op}}(\mathbf{q})|\rangle} \tag{37}$$

and

$$\frac{\beta}{\langle\rho\rangle}c_s(\mathbf{q}) = \frac{\left(k_B\beta\left\{\frac{[C_p(\mathbf{q})]^2}{C_V(\mathbf{q})} - C_p(\mathbf{q})\right\}\right)^{1/2}}{\langle|\rho_{\mathrm{op}}(-\mathbf{q})s_{\mathrm{op}}(\mathbf{q})|\rangle}. \tag{38}$$

$C_V(\mathbf{q})$ reduces to the specific heat at constant volume and $c_s(\mathbf{q})$ reduces to the adiabatic sound velocity at $q = 0$.

C. Transport Processes

Once the local equilibrium states are constructed, we can now describe the transport processes. An eigenstate of L with eigenvalue z will relax in time as e^{-zt}. The transport modes of the systems are those states where these relaxation time goes to infinity (that is $z \to 0$) as $q \to 0$. Thus the transport eigenstates are mostly composed of the local equilibrium states.

In order to find the eigenvalues of L, let us define a projection operator P. This operator projects onto the local equilibrium states and is written as

$$P = \sum_{j=1}^{5} |j, \mathbf{q}\rangle\langle j, \mathbf{q}| \tag{39}$$

Now we define Q, which is the projection operator that rejects the local equilibrium states:

$$Q = 1 - P \qquad (40)$$

The equation for the νth right eigenstate of L corresponding to the eigenvalue z_ν is given by

$$z_\nu |\nu, \mathbf{q}\rangle_R = L|\nu, \mathbf{q}\rangle_R. \qquad (41)$$

In the above expression, $\nu = 1, 2, \ldots, \infty$ labels the eigenvalues of L. $\langle i, \mathbf{q}|$ (where $i = 1, 2, \ldots, 5$ labels the local equilibrium state) is operated from the left on Eq. (41), and standard projection operator technique [12] is used to obtain the following expression:

$$\sum_j [z_\nu \delta_{ij} - L_{ij}(\mathbf{q}) - U_{ij}(\mathbf{q}, z_\nu)]\langle j, \mathbf{q}|\nu, \mathbf{q}\rangle_R = 0, \qquad (42)$$

where

$$L_{ij}(\mathbf{q}) = \langle i, \mathbf{q}|L|j, \mathbf{q}\rangle, \qquad (43)$$

and

$$U_{ij}(\mathbf{q}, z) = \left\langle i, \mathbf{q}\left|LQ\frac{1}{z - QLQ}QL\right|j, \mathbf{q}\right\rangle. \qquad (44)$$

The eigenvalues of L are determined by imposing the condition that the matrix $z\delta_{ij} - L_{ij} - U_{ij}$ has zero determinant. If U_{ij} are set equal to zero, then only two nonzero eigenvalues (relaxation times) are obtained, which are purely imaginary. In the hydrodynamic limit (that is, $q \to 0$) they are given by

$$z_\pm = \pm i c_s q_x. \qquad (45)$$

These purely imaginary relaxation times reflect the oscillatory behavior of the undamped sound waves. Thus, L_{ij} can be identified with the set of thermodynamic derivatives which appear in the nondissipative part of the theory. This implies that the diffusive process must arise from U_{ij}, which now can be identified with the transport coefficients appearing in the dissipative part of the theory.

When Eq. (44) is rewritten in the following form,

$$U_{ij}(\mathbf{q}, z) = -\left\langle \left|\mathbf{q} \cdot \mathbf{j}_i(-\mathbf{q})Q\frac{1}{z - QLQ}Q\mathbf{q} \cdot \mathbf{j}_j(\mathbf{q})\right|\right\rangle, \qquad (46)$$

it becomes clear that U_{ij} is essentially the Kubo formula [38] for transport coefficient written in the frequency plane. $\mathbf{j}$ denotes the currents for the three conserved densities. With the U matrix, now the complete eigenvalues (that is both the nondissipative and the dissipative part) of L are obtained. The eigenvalue (relaxation time) of the heat flow mode (a_1), denoted as z_H, is given by

$$z_H = \frac{\lambda(\mathbf{q}, z)q^2}{m\rho C_P(\mathbf{q})}. \tag{47}$$

The viscous flow modes (a_4 and a_5) are degenerate, with two eigenvalues having the same value. The eigenvalue, z_η, is given by

$$z_\eta = \frac{\eta_s(\mathbf{q}, z)q^2}{m\rho}. \tag{48}$$

The other two local equilibrium states (a_2 and a_3) are coupled and gives sound wave propagation. When the sound wave damping is small compared to its rate of oscillation, the sound wave obeys a dispersion relation,

$$z_\pm = \pm q_x c_s(\mathbf{q}) + \tfrac{1}{2}q^2 D_s(\mathbf{q}, z), \tag{49}$$

where

$$D_s(\mathbf{q}, z) = \frac{\frac{4}{3}\eta_s(\mathbf{q}, z) + \eta_v(\mathbf{q}, z)}{m\rho} + \frac{\lambda(\mathbf{q}, z)}{m\rho}\left(\frac{1}{C_V(\mathbf{q}, z)} - \frac{1}{C_p(\mathbf{q}, z)}\right). \tag{50}$$

Note that for $q = 0$ all these eigenvalues are identical to that obtained from hydrodynamics.

In the expression of the eigenvalues, the wavevector- and frequency-dependent transport coefficients are present. As mentioned before, these are defined by the U matrix elements. Thus in the definition of all these transport coefficients there appears a structure of the form

$$X = \frac{1}{QLQ - z}, \tag{51}$$

which plays the role of the frequency denominators in the Kubo formula. To derive the expressions for the transport coefficients, we need to write X in a convenient form. To begin with, first L is represented in terms of its right eigenstates, $|v, \mathbf{q}\rangle_R$, its left eigenstates, $\langle v, \mathbf{q}|_L$, and its eigenvalues $z_v(\mathbf{q})$:

$$L = \sum_{v'} \int \frac{d^3 q'}{(2\pi)^3} |v', \mathbf{q}'\rangle_R z_{v'}(\mathbf{q}')\langle v', \mathbf{q}'|_L \tag{52}$$

The projection operator Q in X eliminates all the local equilibrium states, which leads to the elimination of the eigenstates of L with small eigenvalues (the transport states). Q leaves the remaining states almost untouched. Hence, the wavevector dependent X can be written as

$$X_{\mathbf{q}} = \sum_{v'=6}^{\infty} \frac{|v', \mathbf{q}\rangle_R \langle v', \mathbf{q}|_L}{z_{v'}(\mathbf{q}) - z}. \tag{53}$$

To study the divergence of the transport processes near the critical point, we need to choose a set of states to represent the eigenstates present in the above expression (that is, $|v', \mathbf{q}\rangle_R$ and $\langle v', \mathbf{q}|_L$ for $v = 6, 7, \ldots, \infty$). The divergence can be expected to arise from states that give small eigenvalues. Although linear combination of these low eigenvalue states are ruled out due to the presence of the projection operator Q, *bilinear products of these states form a set of suitable candidates.* Thus, we need to consider states involving multiple independent transport processes with long wavelength. In the lowest order it will just be bilinear products of the transport states. The product of two noninteracting disturbances have an inverse relaxation time which is the sum of the inverse relaxation times for the individual disturbances. Thus, the eigenvalues of such product states will be just the sum of the eigenvalues of the individual states. With this logic the wavenumber-dependent frequency denominator, $X_{\mathbf{q}}$, can be written as

$$X_{\mathbf{q}} = \frac{1}{2!} \sum_{vv'} \int \frac{d^3 q'}{(2\pi)^3} \frac{a_v(\mathbf{q}')a_{v'}(\mathbf{q} - \mathbf{q}')|\rangle\langle|a_v(-\mathbf{q}')a_{v'}(\mathbf{q}' - \mathbf{q})}{z_v(\mathbf{q}') + z_{v'}(\mathbf{q} - \mathbf{q}') - z} + \cdots, \tag{54}$$

where a_v's are the linear combinations of the a_j's which generate specific transport processes. With the expression of U and X it is now easy to calculate the contributions from different modes to the transport coefficients, the thermal conductivity, the shear viscosity, and the longitudinal viscosity. In this chapter we will not present the detailed calculation of these contributions but state some important results. The essence of the calculation and also some physical understanding can be gained from the following steps.

Let us first write down the formula for the shear viscosity obtained from the expression of U:

$$-q^2 \eta_s(\mathbf{q}, z) = \langle|g_{y\,\mathrm{op}}(-\mathbf{q})LQXQLg_{y\,\mathrm{op}}(\mathbf{q})|\rangle \beta \tag{55}$$

The contribution to the viscosity from two heat modes is obtained by replacing X by the first term on the right-hand side of Eq. (54), where both a_v and $a_{v'}$ are now the heat modes. The projection operator Q in the expression of the shear viscosity can now be replaced by unity because $Q - 1$ makes no contribution. Therefore the heat mode contribution to the shear viscosity, η_{sHH}, is given by

$$-q^2\eta_{sHH}(\mathbf{q},z) = \frac{\beta}{2k_B^2}\int \frac{d^3q'}{(2\pi)^3} \quad \frac{\langle|g_{y\,\mathrm{op}}(-\mathbf{q})La_1(\mathbf{q}')a_1(\mathbf{q}-\mathbf{q}')|\rangle}{[z_H(\mathbf{q}') + z_H(\mathbf{q}-\mathbf{q}') - z]}$$
$$\times \langle|a_1(\mathbf{q}'-\mathbf{q})a_1(-\mathbf{q}')Lg_{y\,\mathrm{op}}(\mathbf{q})|\rangle, \quad (56)$$

Since

$$\langle|a_1(\mathbf{q}'-\mathbf{q})a_1(-\mathbf{q}')Lg_{y\,\mathrm{op}}(\mathbf{q})|\rangle = -\langle|g_{y\,\mathrm{op}}(-\mathbf{q})La_1(\mathbf{q}')a_1(\mathbf{q}-\mathbf{q}')|\rangle^*, \quad (57)$$

Eq. (56) can be rewritten as

$$q^2\eta_{sHH}(\mathbf{q},z)$$
$$= \frac{\beta}{2k_B^2}\int \frac{d^3q'}{(2\pi)^3} \frac{|M_{\mathbf{q},\mathbf{q}'}|^2}{[m\rho C_p(\mathbf{q}')m\rho C_p(\mathbf{q}-\mathbf{q}')][z_H(\mathbf{q}') + z_H(\mathbf{q}-\mathbf{q}') - z]},$$
$$(58)$$

with

$$M_{\mathbf{q},\mathbf{q}'} = \langle|g_{y\,\mathrm{op}}(-\mathbf{q})Ls_{\mathrm{op}}(\mathbf{q}')s_{\mathrm{op}}(\mathbf{q}-\mathbf{q}')|\rangle. \quad (59)$$

$M_{\mathbf{q},\mathbf{q}'}$ can be considered to be the matrix element, which represents a process in which a viscous mode of wavevector $\mathbf{q}$ is annihilated and two thermal (heat) modes of wavevectors $\mathbf{q}'$ and $\mathbf{q} - \mathbf{q}'$ are created. Thus in this treatment the transport modes are assumed to be coupled to each other nonlinearly, and the disturbances in the fluid can be transmitted back and forth between the various modes.

The expression for the thermal conductivity is given by

$$q^2\lambda(\mathbf{q},z) = \langle|s_{\mathrm{op}}(-\mathbf{q})LQXQLs_{\mathrm{op}}(\mathbf{q})|\rangle/k_B. \quad (60)$$

The contribution to $\lambda(\mathbf{q},z)$ from one viscous mode and one heat mode (that is one of the a_v in Eq. 54 is replaced by s_{op} and the other by $\mathbf{g}_{\mathrm{op}}$) can be obtained following similar procedure as mentioned above, and is given by

$$q^2\lambda(\mathbf{q},z) = \frac{\beta}{k_B^2}\int \frac{d^3q'}{(2\pi)^3} \frac{|N_{\mathbf{q},\mathbf{q}'}|^2}{m^2\rho^2 C_p(\mathbf{q}')[z_H(\mathbf{q}') + z_\eta(\mathbf{q}-\mathbf{q}') - z]}, \quad (61)$$

where

$$N_{\mathbf{q},\mathbf{q}'} = \langle |s_{\mathrm{op}}(-\mathbf{q})L\hat{n}\mathbf{g}_{\mathrm{op}}(\mathbf{q}-\mathbf{q}')s_{\mathrm{op}}(\mathbf{q}')| \rangle. \tag{62}$$

Here the momentum in the intermediate state is described by the unit vector $\hat{n}$. $N_{\mathbf{q},\mathbf{q}'}$ can be considered to represents a process in which a heat mode of wavevector $\mathbf{q}$ is annihilated and a thermal (heat) mode of wavevector $\mathbf{q}'$ and a viscous mode of wavevector $\mathbf{q}-\mathbf{q}'$ are created.

In a similar way the contribution for all the different modes to the three transport coefficients can be calculated. Equations (58) and (61) are the classic mode coupling theory expressions that provide general expressions for the shear viscosity and thermal conductivity, respectively. Using these general expressions and the ideas of static scaling laws, Kadanoff and Swift have calculated the transport coefficients near the critical point.

Finally, note that the method used by Kadanoff and Swift is a very general scheme. For example, the expression of η_{sHH} is similar to the expression of viscosity derived later by Geszti [39]. In addition, the projection operator technique used in their study is the same used to derive the relaxation equation [20], and the expression of L_{ij} and U_{ij} are equivalent to the elements of the frequency and memory kernel matrices, respectively.

VI. RENORMALIZED KINETIC THEORY: MICROSCOPIC APPROACH TO THE MODE COUPLING THEORY

The mode coupling theory discussed in the previous section was developed from a hydrodynamic approach. The same has also been developed from a microscopic point of view. This later approach is commonly known as the renormalized kinetic theory approach. This is actually an extension of the kinetic theory approach where along with the Enskog collision term, correlated collisions and also the longer range interactions are included. These longer-range interactions take into account the effect of the slow hydrodynamical variables. The renormalized kinetic theory actually interpolates between the kinetic theory (which takes into account only the short-range interactions) and the hydrodynamic theory (which can describe the collective dynamics). If properly implemented, such an approach can, in principle, describe the dynamics of the liquid over the whole time and length scale.

The renormalized kinetic theory has been developed almost simultaneously by many authors [5–9, 40–43]. The main idea of this theory was to include the effect of the collective dynamics along with the binary collision term that, in the case of hard-sphere systems, is given by Enskog collision. These theories are often referred to as ring collision theory because, along with the Enskog term, it takes into account the ring events. The ring events can be described in the following way. After colliding, a solute and a solvent travel independently in the field of other particles and then the same solute collides either with the same solvent molecules or with any other solvent molecule which is dynamically correlated to the first one. Thus, the ring event describes two binary collisions that are dynamically correlated. This theory has been successful in obtaining the long-time tail in the VACF and can also describe the self-diffusion at low density which agrees reasonably with the computer simulation results [5–7]. But the theory fails at high density where much larger class of dynamical events other than the ring events contribute to the collective effect. These dynamical events can be taken care of by considering a larger number of binary collisions to be correlated, thus considering multiple ring events. This theory is referred to as repeated ring theory.

Repeated ring theory was first presented by Ernst and Dorfman [44] to study the dispersion relation in gases and by Dorfman, van Beijeren and McClure [45] to study the Stokes law for infinitely massive particle. In these above-mentioned works the equilibrium correlations were neglected. These equilibrium correlations were included in the repeated ring theory for binary hard-sphere fluid by Mehaffey and Cukier [8]. In their theory all collisions are expressed in terms of the Enskog binary collision operator in which the static structure of the fluid is incorporated through the radial distribution function. In their formulation the intermediate propagation between two binary collisions are disconnected. This means that the particles that collide move independently after collision, in the field of other particles. They have primarily concentrated in studying the diffusion of a particle which has a much larger radius compared to the solvent particles. In their theory the intermediate propagation between two binary interactions are given by the Enskog term. Thus further renormalization of this theory is required so that the intermediate propagation is described by the full many-body intermediate propagators.

In the real world, however, *the interaction potential between molecules cannot be described by the hard-sphere potential.* It is continuous in nature. This makes the calculations difficult, and even an exact calculation of the binary collision term for a continuous potential is numerically formidable [46]. Sjogren and Sjolander have developed a repeated ring kinetic theory for a one-component system where the interaction is described by a continuous potential [9]. They have also included the effect of the full many-body propagators in describing the intermediate propagation.

The numerical calculations show that this theory can describe the velocity autocorrelation function (VACF) for liquid argon and rubidium [47] fairly well, and the agreements with the computer simulation studies [48, 49] were satisfactory. However, the numerical calculations needed the VACF as an input which was obtained from the respective computer simulation studies. This limits the use of the formulation to study such systems where computer simulated VACFs are already available.

VII. THE GENERAL RELAXATION EQUATION

In the discussion of the hydrodynamic theory we find that along with the conservation equations we need several phenomenological relations (usually known as the constitutive relation) to derive the final expressions of the time correlation functions. In this section we present a microscopic derivation of a general expression, commonly known as the generalized relaxation equation (GRE). This relaxation equation is exact and provides a powerful formalism that can be used to compute the time-correlation functions under circumstances where the phenomenological equations do not apply. The relaxation equation plays a role that is analogous to the Langevin equation (LE). While LE is phenomenological, all the terms in GRE are expressed in terms of the Liouville operator, namely, the relaxation frequency, the memory kernel, and the random force term.

Zwanzig showed how a powerful but simple technique, known as the projection operator technique, can be used to derive the relaxation equations [12]. Let us consider a vector $A(t)$, which represents an arbitrary property of the system. Since the time evolution of the system is given by the Liouville operator, the time evolution of the vector can be written as $A(\mathbf{q}, t) = e^{iLt}A(\mathbf{q})$, where A is the initial value. A projection operator P is defined such that it projects an arbitrary vector on $A(\mathbf{q})$. P can be written as

$$P \equiv (\cdots |A^*(\mathbf{q}))(A(\mathbf{q})|A^*(\mathbf{q}))^{-1}A(\mathbf{q}), \qquad (63)$$

where the brackets denote the classical scalar product, as has been defined in the derivation of viscosity. Another projection operator Q is defined in such a way that it projects an arbitrary vector on the subspace which is orthogonal to A. Thus Q can be written as

$$Q \equiv 1 - P. \qquad (64)$$

The time derivative of $A(\mathbf{q}, t)$ can be written as

$$\frac{\partial A(\mathbf{q}, t)}{\partial t} = e^{iLt}iLA(\mathbf{q}) \qquad (65)$$

The identity, $P + Q = 1$ when inserted in the above equation, it can be rewritten as

$$\frac{\partial A(\mathbf{q}, t)}{\partial t} = e^{iLt}(P + Q)iLA(\mathbf{q}). \qquad (66)$$

An important step in the derivation of relaxation equation is to apply the identity satisfied by the propagator e^{iLt}:

$$e^{iLt} = e^{iQLt} + \int_0^t d\tau e^{iL(t-\tau)} iPL e^{iQL\tau} \tag{67}$$

Using this identity, Eq. (66) can be rewritten in the following form:

$$\frac{\partial A(\mathbf{q}t)}{\partial t} = e^{iLt} PiLA(\mathbf{q}) + e^{iQLt} QiLA(\mathbf{q}) + \int_0^t d\tau e^{iL(t-\tau)} iPL e^{iQL\tau} QiLA(\mathbf{q}). \tag{68}$$

Let us now define the following quantities. The frequency $\Omega(\mathbf{q})$ is

$$\Omega(\mathbf{q}) \equiv (LA(\mathbf{q})|A^*(\mathbf{q}))(A(\mathbf{q})|A^*(\mathbf{q}))^{-1}. \tag{69}$$

The random force $F(\mathbf{q}, \tau)$ is defined as

$$F(\mathbf{q}, \tau) \equiv e^{iQL\tau} QiLA(\mathbf{q}) = Qe^{iQL\tau} QiLA(\mathbf{q}) = QF(\mathbf{q}, \tau). \tag{70}$$

This implies that random force remains orthogonal to $A(\mathbf{q})$ at all times.

The last integral in Eq. (68) involves the term $iPLF(\mathbf{q}, \tau)$, which can be rewritten as

$$iPLF(\mathbf{q}, \tau) = iPLQF(\mathbf{q}, \tau) = (iLQF(\mathbf{q}\tau)|A^*(\mathbf{q}))(A(\mathbf{q})|A^*(\mathbf{q}))^{-1} A(\mathbf{q}). \tag{71}$$

Since Q and L are both Hermitian, $iPLF(\mathbf{q}, \tau)$ can be written as

$$
\begin{aligned}
iPLF(\mathbf{q}, \tau) &= (iLQF(\mathbf{q}, \tau)|A^*(\mathbf{q}))(A(\mathbf{q})|A^*(\mathbf{q}))^{-1} A(\mathbf{q}) \\
&= -(F(\mathbf{q}, \tau)|(QiLA(\mathbf{q}))^*)(A(\mathbf{q})|A^*(\mathbf{q}))^{-1} A(\mathbf{q}) \\
&= -(F(\mathbf{q}, \tau)|F^*(\mathbf{q}, 0))(A(\mathbf{q})|A^*(\mathbf{q}))^{-1} A(\mathbf{q}). \tag{72}
\end{aligned}
$$

Now we define the memory function $\Gamma(\tau)$ as

$$\Gamma(\mathbf{q}, \tau) \equiv +(F(\mathbf{q}, \tau)|F^*(\mathbf{q}, 0))(A(\mathbf{q})|A^*(\mathbf{q}))^{-1}. \tag{73}$$

Note that in the above expression the memory function is proportional to the auto-correlation function of the random force. This is the well-known second fluctuation–dissipation theorem.

With the definition of the frequency and memory function, Eq. (68) can be rewritten as

$$\frac{\partial A(\mathbf{q}, t)}{\partial t} = i\Omega(\mathbf{q})A(\mathbf{q}, t) - \int_0^t d\tau \Gamma(\mathbf{q}, \tau)A(\mathbf{q}, t - \tau) + F(\mathbf{q}, t). \qquad (74)$$

Thus the equation for the time correlation function can be written as

$$\frac{\partial C(\mathbf{q}, t)}{\partial t} = i\Omega(\mathbf{q})C(\mathbf{q}, t) - \int_0^t d\tau \Gamma(\mathbf{q}, \tau)C(\mathbf{q}, t - \tau), \qquad (75)$$

where $C(\mathbf{q}, t) = (A(\mathbf{q}, t)|A^*(\mathbf{q}))$. Equation (75) is the generalized relaxation equation and is also known as the memory function equation.

In the above discussion, only a single variable was considered. This can be extended to evaluate the time evolution of many coupled variables—for example, the five hydrodynamical variables. In that case, A is not a single variable but a column matrix and $C(\mathbf{q}, t) = (A(\mathbf{q}, t)|A^+(\mathbf{q}))$ is now the correlation matrix. $\Omega(\mathbf{q})$ and $\Gamma(\mathbf{q}, \tau)$ are the frequency and the memory function matrices, respectively [20].

The generalized relaxation equation in terms of its components and in the frequency plane can be written as

$$zC_{\mu\nu}(\mathbf{q}, z) - i\Omega_{\mu\lambda}(\mathbf{q})C_{\lambda\nu}(\mathbf{q}, z) + \Gamma_{\mu\lambda}(\mathbf{q}, z)C_{\lambda\nu}(\mathbf{q}, z) = \tilde{C}_{\mu\nu}(\mathbf{q}). \qquad (76)$$

In the above expression, summation over repeated indices is implied. $\tilde{C}_{\mu\nu}(\mathbf{q}) = \tilde{C}_{\mu\nu}(\mathbf{q}, z = 0)$. The matrix elements of the frequency and memory function are given by

$$\Omega_{\nu\lambda}(\mathbf{q}) = \sum_k (LA_\nu(\mathbf{q})|A_k^*(\mathbf{q}))(A_k(\mathbf{q})|A_\lambda^*(\mathbf{q}))^{-1} \qquad (77)$$

and

$$\Gamma_{\nu\lambda}(\mathbf{q}, z) = \sum_k (F_\nu(\mathbf{q}, z)|F_k^*(\mathbf{q}))(A_k(\mathbf{q})|A_\lambda^*(\mathbf{q}))^{-1}. \qquad (78)$$

Equation (76) is commonly written in the following form:

$$C(\mathbf{q}, z) = \frac{\tilde{C}(\mathbf{q})}{z\mathbf{1} - i\Omega(\mathbf{q}) + \Gamma(\mathbf{q}, z)}. \qquad (79)$$

A.   Comments on the Relaxation Equation

1. The frequency matrix Ω_{ij} and the memory function matrix Γ_{ij}, in the relaxation equation are equivalent to the Liouville operator matrix L_{ij} and the U_{ij} matrix, respectively. The later two matrices were introduced by Kadanoff and Swift [37] (see Section V). Thus the frequency matrix can be identified with the static variables (the wavenumber-dependent thermodynamic quantities) associated with the nondissipative part, and the memory kernel matrix can be identified with the transport coefficients associated with the dissipative part.

2. The equations of motion in the extended hydrodynamic theory (Section IV) are obtained from the relaxation equation, where the correlation function is normalized. As mentioned before, in the extended hydrodynamic theory, the memory kernel matrix is considered to be independent of frequency; thus the transport coefficients are replaced by their corresponding Enskog values.

3. In the critical phenomena, the contribution from the different hydrodynamic modes to the transport coefficients are calculated. On the other hand, in the extended hydrodynamic theory, only the Enskog values of the transport coefficients are used. Thus, while the critical phenomena considers only the *long-time part* of the memory function, the extended hydrodynamic theory uses only the *short-time part* of the memory function. None of the theories involve any self-consistent calculation.

4. Finally, note that the relaxation equation [Eq. (76)] is usually written in terms of the hydrodynamic modes. In many problems of chemical interest, nonhydrodynamic modes such as intramolecular vibration, play an important role [50]. Presence of such coupling creates an extra channel for dissipation. Thus, the memory kernel, Γ, gets renormalized and acquires an additional frequency-dependent term [16, 43].

VIII.   STRUCTURE OF MODE COUPLING THEORY AS APPLIED TO LIQUID-STATE DYNAMICS

In Sections V and VI, a brief history of the developments of the MCT from the hydrodynamic approach (Critical Phenomena) and the renormalized kinetic theory approach has been presented. *The basic concept of MCT is to use the product of the slow (hydrodynamic) variables to span the orthogonal subspace of the fast variables.*

In the previous section we find that the time correlation function can be obtained from the relaxation equation. When A is not a single variable but represents a column matrix containing a set of coupled variables, then the relaxation equation leads to a set of coupled equations. As mentioned before, the frequency matrix and the memory kernel matrix in the relaxation equation can be identified with the static correlations and the transport co-efficients, respectively. Thus, the time correlation functions are expressed in terms of static quantities and the transport coefficients. On the other hand, the calculation of the transport coefficients require the knowledge of the different time correlation functions. This calls for a self-consistent calculation.

In the formulations developed from the renormalized kinetic theory approach, these self-consistencies were avoided either by using values obtained from computer simulation and experiments or by using some exactly known limiting values for the transport coefficient. For example, in the treatment of Mazenko [5–7], and of Mehaffey and Cukier's [8] the transport coefficients are replaced by their Enskog values. In the theory developed by Sjogren and Sjolander [9], the velocity autocorrelation function is required as an input that was obtained from the computer simulated values. This limits the validity of the theories only to certain regimes and for certain systems where the experimental or computer-simulated results are available.

In the critical phenomena, the static correlations play a major role. Thus, in the calculation of the critical exponents of the transport coefficients, the self-consistency was not required [37]. The liquid-glass transition, on the other hand, is a purely dynamic transition, and this requires self-consistent calculation of the correlation functions and transport coefficients. The first such self-consistent calculation was done by Geszti [39] to describe the growth of viscosity in a previtrification region. The self-consistent calculation allows a feedback mechanism (otherwise absent) when the left-hand side of an equation depends on the right-hand side (that is the quantity to be determined). Geszti used the viscosity feedback mechanism to calculate the growth of viscosity and derived the Batchinski–Hildebrand equation from MCT. Later Leutheusser [34], Kirkpatrick [30] and Gotze et al. [3,4] have performed self-consistent calculations of the dynamic structure factor near glass transition to explain the divergence of the dynamic structure factor and the viscosity.

Here we will present a pictorial description of the self-consistencies, involved in the calculations of the viscosity and friction/diffusion. No explicit expressions will be presented here, since they will be given in details in separate sections.

The friction on a tagged particle is expressed in terms of the time-dependent force–force autocorrelation function. Although the bare, short-time part of the friction that arises from binary collisions can be calculated from kinetic theory, the long-time part needs the knowledge of the solvent and the solute dynamics and the coupling between them. The solvent dynamic quantities that determine the friction on a tagged particle are the dynamic structure factor and the transverse and longitudinal current autocorrelation functions of the liquid. The dynamics of the solute is given by the self-dynamic structure factor. The coupling between the solute and the solvent dynamics is given in terms of static quantities. Similarly, in the calculation of the viscosity, the knowledge of the dynamic structure factor, the transverse current autocorrelation function, and other static quantities are required.

As mentioned before, due to the coupling between different modes of the solvent, all solvent dynamic quantities are interdependent and need to be calculated self-consistently. An additional self-consistency is present in the calculation of friction, since the self-dynamic structure factor itself depends on the friction.

The full self-consistent scheme, required to calculate the friction and the viscosity, is pictorially depicted in Fig. 2.

From the above discussion, it is obvious that the mode coupling theory calculations are quite involved and numerically formidable. Balucani et al. [16] have made some simple approximations to incorporate the self-consistency between the self-dynamic structure factor and the friction. This required the knowledge of only the zero frequency friction. The full self-consistent calculation is more elaborate and will be discussed later in this chapter.

A. Viscoelastic Models

To make MCT calculations simpler, the solvent dynamic quantities are sometimes modeled in such a way that the self-consistency is avoided. This is where the viscoelastic models (VEMs) play an important role. The VEM is usually expressed as a Mori continued fraction where the frequencies are de-

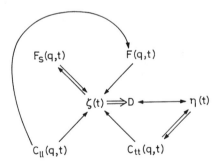

Figure 2. A pictorial representation of the mode coupling theory scheme for the calculation of the time-dependent friction $\zeta(t)$ on a tagged molecule at time t. The rest of the notation is as follows: $F_s(q,t)$, self-scattering function; $F(q,t)$, intermediate scattering function; D, self-diffusion coefficient; $\eta_s(t)$, time-dependnet shear viscosity; $C_{ll}(q,t)$, longitudinal current correlation function; $C_{tt}(q,t)$, longitudinal current correlation function.

termined from the moments of the relevant time correlation function. The VEM can also be obtained from the general relaxation equation (GRE), where the last term in the continued fraction is the memory function. Note that the self-consistency between the correlation function and the transport coefficient enters through the memory kernel in the relaxation equation. The idea behind VEM is that instead of writing the memory function in terms of the transport coefficients, if it can be modeled in some other way, then the self-consistency can be avoided. The exact value of the memory function at $t = 0$ is known from the calculation of the moments for the corresponding correlation function. The simplest approximation for the time dependence of the memory function is an exponential decay. The time scale of decay is constructed in such a way that the correlation functions agree with the computer simulation and experimental results. Such expression was first derived by Akeasu and Daniels [51] to express the time autocorrelation function of the transverse current. The same for the dynamic structure factor was derived by Lovesey [52]. Although the viscoelastic model can properly describe the correlation function in the normal liquid regime, it fails to describe the correlation functions in the supercooled liquid regime where the long-time part of the memory function becomes important. That is, the VEMs fail to describe the slow long-time decay of the time correlation functions in the supercooled liquid.

IX. THEORY OF DIFFUSION AND FRICTION

The diffusion of a tagged particle or solute is given in terms of the time-dependent velocity autocorrelation function (VACF) $C_v(t)$ and is given by

$$D = \frac{1}{3} \int_0^\infty d\tau C_v(\tau).$$ (80)

The velocity autocorrelation function can be obtained from the relaxation equation [Eq. (76)], where $C_v(z) = C_{11}^s(\mathbf{q} = 0z)$. Here the suffix "$s$" stands for single-particle property. For zero wavenumber, there is no contribution from the frequency matrix [that is, $\Omega_{\mu\nu}^s(\mathbf{q} = 0) = 0$] and the memory function matrix becomes diagonal. If we write $\zeta(z) = \Gamma_{11}^s(\mathbf{q} = 0z)$, then the VACF in the frequency plane can be written as

$$C_v(z) = \frac{k_B T}{m(z + \zeta(z))}.$$ (81)

Note that the above expression is known as the generalized Einstein equation and that the memory function, $\zeta(z)$, is the frequency-dependent friction.

The friction is given in terms of the force–force time correlation functions and in the frequency plane can be written as

$$\zeta(z) = \left\langle \mathbf{F}(0) \frac{1}{z - iQLQ} \mathbf{F}(0) \right\rangle, \tag{82}$$

where L is the Liouville operator that describes the time evolution of the system and Q is the projection operator that rejects all the slow variables. The force on a tagged particle can be expressed in terms of the gradient of the pair potential. As was done in the study of critical phenomena, the frequency denominator $1/z - iQLQ$ now needs to be constructed.

Kirkwood's formula for the frequency-dependent friction,

$$\zeta(z) = \left\langle \mathbf{F}(0) \frac{1}{z - iL} \mathbf{F}(0) \right\rangle,$$

involves the unprojected force. Such an approach could be valid when the mass of the solute is infinitely large. This is because in this limit, the dynamics of the solute is totally quenched and all the dynamics of the solvent can be considered to take place in the subspace orthogonal to solute dynamics.

There is an alternative way of determining the friction. This was done by Sjogren [43], who expressed the propagator $\exp(+iQLQt)$ in terms of the Greens function. The expression for the friction in terms of the Greens function is given by [5, 7, 9, 41, 43],

$$\zeta(z) = \frac{1}{k_B TmV} \int d1 \ldots d2' [\hat{\mathbf{q}} \cdot \nabla_{\mathbf{r}_1} v(\mathbf{r}_1 - \mathbf{r}_2)] G^s(12; 1'2', z)[\hat{\mathbf{q}} \cdot \nabla_{\mathbf{r}_1'} v(\mathbf{r}_1' - \mathbf{r}_2')], \tag{83}$$

where the four-point function $G^s(12; 1'2', t')$ describes the correlated motion of the solute and the solvent particles. It describes the time-dependent probability that the solute moves from the position (r_1', p_1') at t' to position (r_1, p_1) at t and a solvent particle is located at (r_2', p_2') at t' and the same or some other solvent particle is found at (r_2, p_2) at t. $G^s(12; 1'2', t')$ also contains information on the static correlation between the tagged particle and the solvent particles through its initial value $\tilde{G}^s(12; 1'2')$. z is the Laplace frequency and $G^s(12; 1'2', z)$ is obtained from the usual Laplace transformation of $G^s(12; 1'2', t')$.

An exact equation of $G^s(12; 1'2', z)$ is given by [41, 5, 7, 43]

$$zG^s(12; 1'2', z) - \int d1''d2''\Omega^s(12; 1''2'')G^s(1''2''; 1'2', z)$$

$$+ \int d1''d2''\zeta^s(12; 1''2'', z)G^s(1''2''; 1'2', z)$$

$$= \tilde{G}^s(12; 1'2'),\tag{84}$$

where $\Omega^s(12; 1'2')$ in the lowest order in density is written as

$$\Omega^s(12; 1'2') = -[(1/m)\mathbf{p_1}.\nabla_{\mathbf{r_1}} + (1/m)\mathbf{p_2}.\nabla_{\mathbf{r_2}}$$
$$- \nabla_{\mathbf{r_1}}v(\mathbf{r_1} - \mathbf{r_2}) \cdot (\nabla_{\mathbf{p_1}} - \nabla_{\mathbf{p_1}})]\delta(11')\delta(22').\tag{85}$$

Now the propagator for G^s can symbolically be written as

$$R^s = (z - \Omega^s + \zeta^s)^{-1}.\tag{86}$$

Thus one can write

$$G^s = R^s\tilde{G}^s.\tag{87}$$

For a very short time, ζ^s (which accounts for the memory function effect) may approximately be neglected and the evolution of G^s is largely determined by Ω^s. This instantaneous response corresponds to a binary collision event where the density effects are taken into account only as an average mean-field background given by the $\nabla v(r)$ term in Ω^s. However, the occurrence of the binary collisions causes rapid rearrangements amongst the neighboring particles; this is accounted for by ζ^{sB}, the rapidly varying part in the memory function, ζ^s.

Thus the full propagator can now be split into parts. Let R^{sB} represent a single binary collision event, and $\zeta_1 = \zeta^s - \zeta^{sB}$ is responsible for the correlation effects between these collisions. Thus propagator R^{sB} can be defined as

$$R^{sB} = (z - \Omega^s + \zeta^{sB})^{-1}.\tag{88}$$

The earlier authors [9] have considered that ζ^{sB} contributes to the rapid renormalization of the medium and also includes the asymptotic part of the liquid; this implies that R^{sBD}, the disconnected part (considering that the solute and the solvent motions are disconnected) of R^{sB} contains the free motion of the solute and the full motion of the medium. The present formulation differs from the earlier one [9] in the definition of ζ^{sB}. It is considered that ζ^{sB}

includes that part of ζ^s which *only* contributes to the rapid renormalization of the medium due to a binary collision. This implies that R^{sBD} represents the free motion of the solute and the collective short-time dynamics of the medium. Since R^{sB} is supposed to describe the propagation during the binary collision, it is more appropriate to consider the short-time part of the solvent dynamics instead of the full solvent dynamics.

This point becomes clear with the arguments based on time scale, presented below. First let us expand R^s in powers of ζ_1:

$$R^s = R^{sB} - R^{sB}\zeta_1 R^{sB} + R^{sB}\zeta_1 R^{sB}\zeta_1 R^{sB} + \cdots = R^{sB} - R^{sB}\zeta_1 R^s. \quad (89)$$

The above equation represents a series of binary collisions of the solute which are correlated by ζ_1. From Eq. (86) and Eq. (88), it can be written

$$\zeta_1 = (R^s)^{-1} - (R^{sB})^{-1} = -(R^{sB})^{-1}(R^s - R^{sB})(R^s)^{-1}. \quad (90)$$

The logic behind splitting ζ^s into ζ^{sB} and ζ_1 is that $(R^s - R^{sB})$ should essentially be zero within the time taken for a binary collision. The time scale of the binary collision is determined by the time scale of the free-particle motion of the solute and the solvent. One of the factors that determines the time scale of a free-particle motion is its mass. Thus in the case where the solute and the solvent has the same mass the time scale can be assumed to be determined by the solute mass alone. On the other hand if the solute and the solvent masses are different, then the binary time scale will be determined by the reduced mass and primarily by the mass of the lighter particle. Thus for massive solute the binary collision time is determined by the mass of the solvent. Now let us turn our attention to the time scale of R^{sBD}. According to Sjogren and Sjolander [9] R^{sBD} contains the free motion of the tagged particle and the full motion of the medium. In that case, $(R^{sD} - R^{sBD})$ remains zero as long as the free motion of the solute continues, since the full motion of the medium usually has a much longer time scale. Now for massive particle the free inertial motion of the solute will continue for a much longer time compared to the binary collisional time scale (which is essentially determined by the solvent mass). This means that $(R^{sD} - R^{sBD})$ remains zero even after the binary collision is over, thus leading to unphysical results. According to the present definition, R^{sBD} contains the free motion of the solute and the short-time collective motion of the solvent. Thus the time scale of R^{sBD} is determined either by the time scale of the free motion of the solute or by the time scale of the short-time collective motion of the solvent, whichever is smaller. This implies that for a massive solute, $(R^{sD} - R^{sBD})$ remains zero only as long as the short-time collective motion of the solvent continues, which will also be the time scale of the binary collision. This is precisely the criteria for the decomposition of ζ^s into ζ^{sB} and ζ_1.

It should be noted that although in Eq. (90) only the connected motion of the solute and the solvent is retained, in the argument presented on the time scale it is the disconnected parts which have been considered. This is because in the latter part, for the derivation of the expression of ζ_1, the solute and the solvent motions are assumed to be disconnected. This assumption is the same as those made in the density functional theory and also in mode coupling theories where a four-point correlation function is approximated as the product of two two-point correlation functions. This approximation when incorporated in ζ_1, means that after the binary collision takes place, the disturbances in the medium will propagate independently. A more exact calculation would be to consider the whole four-point correlation function, thus considering the dynamics of the solute and the solvent to be correlated even after the binary collision is over. Such a calculation is quite cumbersome and has not been performed yet.

Next, in the expression of ζ_1 [Eq. (90)], R^s is replaced by R^{sD} while R^{sB} is replaced by R^{sBD}. We substitute the resulting expression in Eq. (89) to find

$$R^s = R^{sB} + R^{sB}(R^{sBD})^{-1}(R^{sD} - R^{sBD})(R^{sD})^{-1}R^s. \tag{91}$$

From the above expression it is clear that it is only the propagation which is considered to be disconnected whereas R^{sB} and R^s at the ends of the second term contains the full interaction between the solute and the solvent. If R^s is replaced by R^{sB} at the right-hand side, then the second term on the right-hand side represents two binary collisions. Presence of the full R^s represents a series of binary collisions.

The rest of the derivation is performed following Sjogren and Sjolander [9]. One can rewrite $R^{sB}(R^{sBD})^{-1}$ in the following form:

$$R^{sB}(R^{sBD})^{-1} = \tilde{G}^s(R^{sBD})^\dagger[(R^{sBD})^{-1}]^\dagger(\tilde{G}^{sD})^{-1}. \tag{92}$$

Using Eq. (87), $(R^{sD} - R^{sBD})$ can be written as

$$R^{sD} - R^{sBD} = (G^{sD} - G^{sBD})(\tilde{G}^{sD})^{-1}. \tag{93}$$

The following integration, when simplified, yields [43]

$$\int d1'd2'\tilde{G}^s(12; 1'2')\nabla_{\mathbf{r}_1'}v(\mathbf{r}_1' - \mathbf{r}_2') = \left(-\frac{\rho}{\beta}\right)\phi_M(\mathbf{p}_1)\phi_M(\mathbf{p}_2)\nabla_{\mathbf{r}_1}g(\mathbf{r}_1 - \mathbf{r}_2),$$

$$\tag{94}$$

where $\beta = (k_B T)^{-1}$ and $\phi_M(\mathbf{p})$ is the Maxwellian distribution given by

$$\phi_M(\mathbf{p}) = (\beta/2\pi m)^{3/2} \exp(-\beta p^2/2m). \tag{95}$$

Using the above-mentioned expressions, we can write the friction given by Eq. (83) as

$$\zeta(z) = \zeta^B(z) + \zeta^R(z). \tag{96}$$

In the above expression the superscript R indicates the contribution to the friction from the recollision events. The expressions for $\zeta^B(z)$ and $\zeta^R(z)$ are given by

$$\zeta^B(z) = -\left(\frac{\rho}{mV}\right) \int d1 \cdots d2' (\hat{\mathbf{q}} \cdot \nabla_{\mathbf{r_1}} v(\mathbf{r_1} - \mathbf{r_2})) R^{sB}(12; 1'2'|z) \phi_M(\mathbf{p'_1}) \phi_M(\mathbf{p'_2})$$
$$\times \left[\hat{\mathbf{q}} \cdot \nabla_{\mathbf{r'_1}} g(\mathbf{r'_1} - \mathbf{r'_2}) \right] \tag{97}$$

and

$$\zeta^R(z) = \int \frac{d\mathbf{q'}}{(2\pi)^3} T^{B\dagger}_{\mu\nu}(\mathbf{q'}z) \tilde{C}^{-1}_{\nu\nu}(\mathbf{q'}) \left[\int_0^\infty dt e^{-zt} \left[C^s_{\mu\lambda}(-\mathbf{q'}, t) C_{\nu\rho}(\mathbf{q'}, t) \right. \right.$$
$$\left. \left. - C^s_{0\mu\lambda}(-\mathbf{q'}, t) C_{0\nu\rho}(\mathbf{q'}, t) \right] \right] \times \tilde{C}^{-1}_{\rho\rho}(\mathbf{q'}) T_{\lambda\rho}(\mathbf{q'}z). \tag{98}$$

While writing the above expression, the facts that have been used are $G^{sD} = C^s C$, $G^{sBD} = C^s_0 C_0$, and $(\tilde{G}^{sD})^{-1} = (\tilde{C}^s)^{-1} \tilde{C}^{-1}$. Note that the definition of G^{sBD} is different from the previous formulation [9]; this follows from the difference in definition of ζ^{sB}, which has been discussed before. Here C^s and C are the phase space correlation functions defined as

$$\tilde{C}^s(11') = V\langle \delta\rho^s_1(1, t) \delta\rho^s_1(1', t) \rangle$$
$$\tilde{C}^s(12; 1') = V\langle \delta\rho^s_2(12, t) \delta\rho^s_1(1', t) \rangle$$
$$\tilde{C}^s(12; 1'2') = V\langle \delta\rho^s_2(12, t) \delta\rho^s_2(1'2', t) \rangle \tag{99}$$

and, similarly,

$$\tilde{C}(11') = V\langle \delta\rho_1(1, t) \delta\rho_1(1', t) \rangle, \tag{100}$$

where $\delta\rho = \rho - \langle\rho\rangle$ and ρ^s and ρ are the microscopic phase space density of the tagged particle and the fluid (without the tagged particle) respectively.

$$\rho_1^s(1t) = \delta(1 - q_1(t)),$$

$$\rho_2^s(12t) = \delta(1 - q_1(t)) \sum_{i=2}^{N} \delta(2 - q_i(t)),$$

$$\rho_1(1t) = \sum_{i=1}^{N} \delta(1 - q_i(t)), \tag{101}$$

where "1" stands for $(\mathbf{r}_1, \mathbf{p}_1)$, and $q(t) = (\mathbf{r}(t), \mathbf{p}(t))$ denotes the actual coordinates of the particle.

$(\tilde{C})^{-1}$ in Eq. (98) is defined by

$$\int d1'' \tilde{C}(11'')(\tilde{C})^{-1}(1''1') = \delta(11') \tag{102}$$

and similarly $(\tilde{C}^s)^{-1}$, where the integration runs over the entire six-dimensional phasespace.

The phase space correlation function can be expressed in terms of Maxwellian distribution and static pair correlation function and can be written as

$$\tilde{C}^s(11') = \phi_M(\mathbf{p_1})\delta(11'),$$

$$\tilde{C}^s(12; 1') = \rho\phi_M(\mathbf{p_1})\phi_M(\mathbf{p_2})g(\mathbf{r_1} - \mathbf{r_2})\delta(11'),$$

$$\tilde{C}(11') = \rho\phi_M(\mathbf{p_1})\delta(11') + \rho^2\phi_M(\mathbf{p_1})\phi_M(\mathbf{p_1'})[g(\mathbf{r_1} - \mathbf{r_1'}) - 1]. \tag{103}$$

Thus from Eqs. (102) and (103) one can write

$$(\tilde{C}^s)^{-1}(11') = \delta(11')/\phi_M(\mathbf{p_1}),$$

$$\tilde{C}^{-1}(11') = \delta(11')/\rho\phi_M(\mathbf{p_1}) - c(\mathbf{r_1} - \mathbf{r_1'}), \tag{104}$$

where $c(r)$ is the direct correlation function.

In Eq. (98), C_0^s represents the free inertial motion of the tagged particle, C^s contains its full motion, C describes the complete disconnected motion of the surrounding fluid, and C_0 describes the short-time dynamics of this disconnected motion of the fluid. The T-matrix in Eq. (98) is given by [9]

$$T_{\mu\nu}(\mathbf{q}', z) = -\left(\frac{1}{m\beta}\right)^{1/2}\left(\frac{\rho}{V}\right)\int d1\ldots d2' H_\mu(\mathbf{p_1})H_\nu(\mathbf{p_2})\exp[i\mathbf{q}'\cdot(\mathbf{r_1} - \mathbf{r_2})]$$
$$\times [R^{sD}(12;1''2''|z)]^{-1}R^s(1''2''; 1'2'|z)\phi_M(\mathbf{p_1'})\phi_M(\mathbf{p_2'})\hat{\mathbf{q}}\cdot\nabla_{\mathbf{r_1'}}g(\mathbf{r_1'} - \mathbf{r_2'}). \tag{105}$$

In the above expression, $H_\mu(\mathbf{p})$ denotes the Hermite polynomials. The expression for T^B is obtained by replacing $(R^{sD})^{-1}$ by $(R^{sBD})^{-1}$ and R^s by R^{sB} in Eq. (105). Also one can write

$$T^{B\dagger}_{\mu\nu}(\mathbf{q}',z) = T^B_{\mu\nu}(-\mathbf{q}',-z). \tag{106}$$

The recollision term is found to start as t^4. At high density the dominant contribution comes from the first-order recollision term—that is, the ring term—and is obtained by replacing $T_{\lambda\rho}$ in Eq. (98) by $T^B_{\lambda\rho}$. These matrices contain the information about which modes are excited in the surrounding fluid and also the strength of the coupling.

In calculation of the recollision term, only the coupling to the density and the longitudinal and transverse current modes of the fluid are considered. The temperature mode is also hydrodynamic; but as discussed by Mehaffey et al. [53] and Sjogren and Sjolander [9], the coupling to it occurs through higher power of t, and it is also unimportant for asymptotic times. Thus this coupling is neglected.

For the tagged particle, only density will be the conserved variable. All other modes will decay rapidly. The nonhydrodynamic modes should be included mainly in R^{sB}.

Thus in Eq. (98), to consider the modes mentioned above, $\mu = \lambda = 0$ and $\nu, \rho = 0, 1, 2, 3$ need to considered, where 0 represents the density, 1 the longitudinal current, 2 the transverse current, and 3 the temperature. Please note that the subscript ρ is a dummy index and is not to be confused with the density. $\tilde{C}^{-1}_{\nu\nu}$ for different values of ν are

$$\tilde{C}^{-1}_{00}(\mathbf{q}') = 1/\rho S(q')$$

$$\tilde{C}^{-1}_{\nu\nu}(\mathbf{q}') = 1/\rho, \qquad \nu \neq 0 \tag{107}$$

Here $S(q)$ is the static structure factor.

Note that the MCT treatment presented above is quite general and can be extended to describe relaxation in many different systems, such as orientational relaxation in dipolar liquids [54]. This approach can also be extended to multicomponent systems, in particular to describe transport properties of electrolyte solutions [55]. The usefulness and the simplicity of the expressions lies in the separation between the single particle and collective dynamics (as in Eq. 98). Actually this sepration allows one to make connections with hydrodynamic (or continuum framewrk) models where only the collective dynamics is included but the single particle motion is ignored. However, the same separation is also the reason for the failure of the Sjogren-Sjolander scheme because in many situations such separation is not possible. This will be addressed to later in this section.

A. Discussion on the Gaussian Approximation

In writing Eq. (98) the propagator has been written considering a disconnected approximation. This implies that the solute and the solvent motion in the propagator has been considered to be disconnected. An analysis of the approximation is presented here.

The expression for Green's function is given by

$$G^s(12t; 1'2't') = \langle \delta\rho_2^s(12,t)\delta\rho_2^s(1'2',t')\rangle. \tag{108}$$

When the expression for $\delta\rho_2^s(12,t)$ is substituted in the above expression and cumulant expansion is performed, Green's function is written as

$$
\begin{aligned}
G^s(12t; 1'2't') &= \langle \delta\rho_1^s(1,t)\delta\rho_1(2,t)\delta\rho_1^s(1',t')\delta\rho_1(2',t')\rangle \\
&= \langle \delta\rho_1^s(1,t)\delta\rho_1^s(1',t')\rangle\langle \delta\rho_1(2,t)\delta\rho_1(2',t')\rangle \\
&\quad + \langle \delta\rho_1^s(1,t)\delta\rho_1(2,t)\rangle\langle \delta\rho_1^s(1',t')\delta\rho_1(2',t')\rangle \\
&\quad + \langle \delta\rho_1^s(1,t)\delta\rho_1(2',t')\rangle\langle \delta\rho_1(2,t)\delta\rho_1^s(1',t')\rangle \\
&\quad + \langle \delta\rho_1^s(1,t)\delta\rho_1(2,t)\delta\rho_1^s(1',t')\delta\rho_1(2',t')\rangle_c
\end{aligned}
\tag{109}
$$

The joint probability distribution for ρ_1^s and ρ_1 is given by

$$P\{\rho_1^s, \rho_1\} = exp(-\beta F\{\rho_1^s, \rho_1\}). \tag{110}$$

The equation of the probability distribution functional can be written as

$$
\begin{aligned}
\frac{\partial P(\{\rho_1^s, \rho_1\}, t)}{\partial t} &= -D_0 \int d1 \left\{ \frac{\delta}{\delta_{\rho_1}(\mathbf{r})} \nabla \cdot \rho_1(1)\nabla \left[\frac{\delta}{\delta\rho_1(1)} + \beta\frac{\delta F\{\rho_1^s, \rho_1\}}{\delta\rho_1(1)} \right] \right. \\
&\quad \left. + \frac{\delta}{\delta\rho_1^s(1)} \nabla \cdot \rho_1^s(1)\nabla \left[\frac{\delta}{\delta\rho_1^s(1)} + \beta\frac{\delta F\{\rho_1^s, \rho_1\}}{\delta\rho_1^s(1)} \right] \right\} \\
&\quad \times P(\{\rho_1^s, \rho_1\}(2, t)
\end{aligned}
\tag{111}
$$

where $F\{\rho_1^s\rho_1\}$ is the free energy functional and D_0 is the microscopic diffusion. According to the density functional theory the free energy functional is given by [54, 55],

$$
\beta F\{\rho_1^s, \rho_1\} = \int d1\rho_1(1, t)\left[\ln\frac{\rho_1(1, t)}{\rho} - 1\right]
$$

$$
+ \int d1\rho_1^s(1, t)\left[\ln\frac{\rho_1^s(1, t)}{\rho^s} - 1\right]
$$

$$
- \int d1d2c_{12}(\mathbf{r}_1 - \mathbf{r}_2)\delta\rho_1^s(1, t)\delta\rho_1(2, t)
$$

$$
- \frac{1}{2}\int d1d2c(\mathbf{r}_1 - \mathbf{r}_2)\delta\rho_1(1, t)\delta\rho_1(2, t)
$$

$$
+ \text{higher-order terms}, \tag{112}
$$

where ρ^s is the density of the tagged particle and ρ is the average density of the solvent. $c_{12}(r)$ is the solute–solvent and $c(r)$ is the solvent–solvent two–particle direct correlation function. Since there is only one solute, the solute–solute interaction is neglected. As the solute and the solvent motions are considered to be disconnected, the two-particle correlation between the solute–solvent and higher-order correlations between the solute–solvent are neglected. With the above-mentioned approximations and neglecting the connected term in Eq. (109), the only term that survives in the cumulant expansion of Green's function, G^s, is G^{sD} and can be written as

$$
G^s(12t; 1'2't') \simeq G^{sD}(12t; 1'2't') = C^s(1t; 1't')C(2t; 2't'). \tag{113}
$$

It should be mentioned that the probability is still not Gaussian due to the presence of the logarithmic term and also the three-particle and higher-order correlation terms. If the logarithmic term is expanded and terms that are only quadratic in density are retained and also the three-particle and higher-order correlations are neglected, then an Gaussian approximation is achieved. Thus the Gaussian approximation considers a decoupling between the solute and the solvent motion by ignoring the collective dynamics of the four particles and also ignores higher-order correlations. This disconnected approximation might lead to serious errors where the solute is strongly coupled to the solvent. In this theoretical formulation, since the approximation is introduced only in the derivation of the intermediate propagator, it might not lead to a large error in the final calculation of the friction or diffusion.

B. Derivation of the Binary Term

The exact initial value of the friction is known and is given by the well-known Einstein frequency ω_0:

$$
\zeta^B(t = 0) = \omega_0^2 = \frac{\rho}{3m}\int d\mathbf{r}g(r)\nabla^2 v(r). \tag{114}
$$

As mentioned before, the recollision term begins as t^4, and thus the binary collision term contains all the contributions to order t^2. Also only even powers in t appear. Thus ζ^B can be assumed to be given by Gaussian approximation and can be written as

$$\zeta^B(t) = \omega_0^2 \exp(-t^2/\tau_\zeta^2). \tag{115}$$

In the above expression the relaxation time τ_ζ is determined from the second derivative of $\zeta^B(t)$ at $t = 0$.

The derivation of τ_ζ is given in the subsequent steps. For the derivation of τ_ζ a more general expression for the binary part of the friction is considered, which can be written as

$$\zeta_{11}^{sB}(\mathbf{q}z) = \left(\frac{\beta}{mV}\right)\int d1 \cdots d2'' \exp(-i\mathbf{q}\cdot\mathbf{r}_1)(\hat{\mathbf{q}}\cdot\nabla_{\mathbf{r}_1}v(\mathbf{r}_1 - \mathbf{r}_2))R^{sB}(12:1'2'|z)$$
$$\times \tilde{G}^s(1'2';1''2'') \times [\hat{\mathbf{q}}\cdot\nabla_{\mathbf{r}_1''}v(\mathbf{r}_1'' - \mathbf{r}_2'')]\exp(i\mathbf{q}\cdot\mathbf{r}_1''), \tag{116}$$

where $\zeta_{11}^{sB}(\mathbf{q} = 0z) = \zeta^B(z)$ [9]. R^{sB} in the above expression can be expanded in powers of $1/z$:

$$R^{sB}(z) = [z - \Omega^s + \zeta^{sB}(z)]^{-1}$$
$$= (1/z) + (1/z^2)\Omega^s + (1/z^3)[\Omega^s\Omega^s - \zeta^{sB}(t = 0)] + \cdots \tag{117}$$

It is found that when the above expansion is substituted in Eq. (116), the contribution from $\zeta^{sB}(t = 0)$ vanishes and also contribution from the $(1/z^2)$ term vanishes. ζ^{sB} contributes only to order $(1/z^5)$; thus changing the definition of ζ^{sB} does not affect the expression of the binary time constant. The expression for $\tilde{G}^s(12;1'2')$ is given by [9]

$$\tilde{G}^s(12;1'2') = \phi_M(\mathbf{p}_1)\delta(11')[\phi_M(\mathbf{p}_2)\rho g(\mathbf{r}_1 - \mathbf{r}_2)\delta(22') + \phi_M(\mathbf{p}_2)\phi_M(\mathbf{p}_2')$$
$$\times \rho^2[g_3(\mathbf{r}_1,\mathbf{r}_2,\mathbf{r}_2') - g(\mathbf{r}_1 - \mathbf{r}_2)g(\mathbf{r}_1 - \mathbf{r}_2')]]. \tag{118}$$

The expressions for Ω^s [given by Eq. (85)] and $\tilde{G}^s$ are substituted in Eq. (116), and after some simplifications the following expression is obtained:

$$\zeta_{11}^{sB}(\mathbf{q}z) = \frac{1}{z}\omega_0^2 - \frac{1}{z^3}\left(\frac{q^2}{m\beta}\omega_0^2 + \frac{2\rho}{m^2}\int d\mathbf{r}[\hat{q}^\beta\nabla_\mathbf{r}^\beta\nabla_\mathbf{r}^\alpha v(\mathbf{r})]g(\mathbf{r})[\hat{q}^\nu\nabla_\mathbf{r}^\nu\nabla_\mathbf{r}^\alpha v(\mathbf{r})]\right.$$
$$+ \frac{\rho^2}{m^2}\frac{1}{V}\int d\mathbf{r}_1 d\mathbf{r}_2 d\mathbf{r}_2'[\hat{q}^\beta\nabla_{\mathbf{r}_1}^\beta\nabla_{\mathbf{r}_1}^\alpha v(\mathbf{r}_1 - \mathbf{r}_2)]$$
$$\left. \times \left[g_3(\mathbf{r}_1,\mathbf{r}_2,\mathbf{r}_2') - g(\mathbf{r}_1 - \mathbf{r}_2)g(\mathbf{r}_1 - \mathbf{r}'^2)\right][\hat{q}^\nu\nabla_{\mathbf{r}_1}^\nu\nabla_{\mathbf{r}_1}^\alpha v(\mathbf{r}_1 - \mathbf{r}_2')]\right), \tag{119}$$

where summation over repeated indices is implied.

For g_3, Kirkwood's superposition approximation is assumed. Next a double derivative of Eq. (115) is taken and Laplace inversion is performed. Now equating the terms in this expression and in Eq. (119) (after setting $q = 0$) of the order of $(1/z^3)$ the following expression for the binary time constant is obtained:

$$\omega_0^2/\tau_\zeta^2 = (\rho/3m^2) \int d\mathbf{r}(\nabla^\alpha \nabla^\beta v(\mathbf{r}))g(\mathbf{r})(\nabla^\alpha \nabla^\beta v(\mathbf{r}))$$

$$+ (1/6\rho) \int [d\mathbf{q}/(2\pi)^3] \gamma_d^{\alpha\beta}(\mathbf{q})(S(q) - 1)\gamma_d^{\alpha\beta}(\mathbf{q}), \qquad (120)$$

where summation over repeated indices is implied. Here $S(q)$ is the static structure factor. The expression for $\gamma_d^{\alpha\beta}(\mathbf{q})$ is written as a combination of the distinct parts of the second moments of the longitudinal and transverse current correlation functions $\gamma_d^l(\mathbf{q})$ and $\gamma_d^t(\mathbf{q})$, respectively:

$$\gamma_d^{\alpha\beta}(\mathbf{q}) = -(\rho/m) \int d\mathbf{r} \, \exp(-i\mathbf{q} \cdot \mathbf{r})g(\mathbf{r})\nabla^\alpha \nabla^\beta v(\mathbf{r})$$

$$= \hat{q}^\alpha \hat{q}^\beta \gamma_d^l(\mathbf{q}) + (\delta_{\alpha\beta} - \hat{q}^\alpha \hat{q}^\beta)\gamma_d^t(\mathbf{q}), \qquad (121)$$

where $\gamma_d^l(\mathbf{q}) = \gamma_d^{zz}(\mathbf{q})$ and $\gamma_d^t(\mathbf{q}) = \gamma_d^{xx}(\mathbf{q})$.

C. Derivation of the Ring Collision Terms

For the derivation of the ring collision term the expression for the T matrix is required.

As shown by Sjogren and Sjolander [9], the exact calculation of the T matrix is quite difficult and can be performed only at certain limits.

Over here the final expressions for the components of the T matrix are only presented:

$$T_{00}(\mathbf{q}'z) = i(1/m\beta)^{1/2}[S(q') - 1](\hat{\mathbf{q}} \cdot \hat{\mathbf{q}}'). \qquad (122)$$

The above expression suggests that the coupling to the density of the medium surrounding the tagged particle is independent of z; that is, the coupling is instantaneous. The calculation of the other components show certain time dependence, but it is not possible to calculate them exactly. The initial values can be calculated exactly and are given by [9]

$$T_{0\alpha}(\mathbf{q}', t = 0) = -(\hat{\mathbf{q}} \cdot \hat{\mathbf{q}}')\hat{q}'^\alpha[\gamma_d^l(q') + (\rho q'^2/m\beta)c(q')]$$

$$- [\hat{q}^\alpha - (\hat{\mathbf{q}} \cdot \hat{\mathbf{q}}')\hat{q}'^\alpha]\gamma_d^t(q'), \qquad (123)$$

where $\alpha = 1, 2, 3$. For the binary parts it is found that $T_{00}^B = T_{00}$, and the initial values of $T_{0\alpha}^B$ are the same as for $T_{0\alpha}$.

Also the time/frequency dependence of the T-matrix elements can be calculated at $q' = 0$:

$$T_{0\alpha}(\mathbf{q}' = 0, z) = \zeta(z)\hat{q}^\alpha \qquad (124)$$

and, similarly,

$$T_{0\alpha}^B(\mathbf{q}' = 0, z) = \zeta^B(z)\hat{q}^\alpha. \qquad (125)$$

Thus the T-matrix components have the same time dependence as the friction.

Considering that the two particles in R^s and $(R^{sD})^{-1}$ move independently, from Eq. (105) for large q' values the following result is obtained:

$$T_{0\alpha}^B(\mathbf{q}', t) = T_{0\alpha}^B(\mathbf{q}', t = 0)\exp(-q'^2 t^2/m\beta) \qquad (126)$$

Combining this with what has been obtained for $q' = 0$, one can make a more realistic *ansatz* and write

$$T_{0\alpha}^B(\mathbf{q}', t) = T_{0\alpha}^B(\mathbf{q}', t = 0)\exp(-q'^2 t^2/m\beta)\exp(-t^2/\tau_\zeta^2), \qquad (127)$$

thus satisfying both the limits of small and large q' and interpolating between these two limits. It has been shown by Sjogren and Sjolander [9] that the dependence on q' is weak; thus one can approximate as follows:

$$T_{0\alpha}^B(\mathbf{q}', t) = \frac{T_{0\alpha}^B(\mathbf{q}', t = 0)}{\omega_0^2}\zeta^B(t). \qquad (128)$$

Similarly for $T_{0\alpha}(\mathbf{q}', t)$, one can write

$$T_{0\alpha}(\mathbf{q}', t) = \frac{T_{0\alpha}(\mathbf{q}', t = 0)}{\omega_0^2}\zeta(t). \qquad (129)$$

From Eq. (106) it is found that

$$T_{0\alpha}^{B\dagger}(\mathbf{q}'z) = -T_{0\alpha}^B(\mathbf{q}'z). \qquad (130)$$

When the components of the T matrix are substituted in Eq. (98), the ring collision term can be written as

$$\zeta^R(z) = R_{\rho\rho}(z) - \zeta^B(z)R_{\rho L}(z) - R_{\rho L}(z)\zeta(z) - \zeta^B(z)[R_{LL}(z) + R_{TT}(z)]\zeta(z), \qquad (131)$$

where $R_{\rho\rho}(z)$ contains the coupling to the density and is given by

$$
R_{\rho\rho}(t) = \frac{\rho}{m\beta} \int [d\mathbf{q}'/(2\pi)^3](\hat{\mathbf{q}} \cdot \hat{\mathbf{q}}')^2 q'^2 [c(q')]^2 [F^s(q',t)F(q',t)
$$
$$
- F_0^s(q',t)F_0(q',t)], \tag{132}
$$

where $F^s = C_{00}^s$, $F_0^s = C_{000}^s$, $F = C_{00}/\rho$, and $F_0 = C_{000}/\rho$.

$R_{\rho L}(z)$ contains the combined coupling to the density and the longitudinal current term and can be written as

$$
R_{\rho L}(t) = \int [d\mathbf{q}'/(2\pi)^3](\hat{\mathbf{q}} \cdot \hat{\mathbf{q}}')^2 c(q')[\gamma_d^l(q') + \frac{\rho q'^2}{m\beta}c(q')]\omega_0^{-2}
$$
$$
\times [F^s(q',t)\frac{\partial}{\partial t}F(q',t) - F_0^s(q',t)\frac{\partial}{\partial t}F_0(q',t)]. \tag{133}
$$

R_{LL} and R_{TT} gives the coupling to the longitudinal and transverse current modes, respectively, and can be written as

$$
R_{LL}(t) = \frac{1}{\rho} \int [d\mathbf{q}'/(2\pi)^3](\hat{\mathbf{q}} \cdot \hat{\mathbf{q}}')^2 [\gamma_d^l(q') + \frac{\rho q'^2}{m\beta}c(q')]^2 \omega_0^{-4}
$$
$$
\times [F^s(q',t)C_{ll}(q't) - F_0^s(q',t)C_{ll0}(q't)]. \tag{134}
$$

and

$$
R_{TT}(t) = \frac{1}{\rho} \int [d\mathbf{q}'/(2\pi)^3][1 - (\hat{\mathbf{q}} \cdot \hat{\mathbf{q}}')^2][\gamma_{d12}^t(q')]^2 \omega_{012}^{-4}
$$
$$
\times [F^s(q',t)C_{tt}(q',t) - F_0(q',t)C_{tt0}(q',t)]. \tag{135}
$$

where $C_{ll}(q,t)$ and $C_{tt}(q,t)$ denote the longitudinal and the transverse current–current correlation functions, respectively, and $C_{ll0}(q,t)$ and $C_{tt0}(q,t)$ denote the short time parts of the same.

D. Derivation of the Final Expression for the Frequency-Dependent Friction

Note that in Eq. (131), the full friction $\zeta(z)$ is on the right-hand side. By replacing the expression of $\zeta^R(z)$ in Eq. (96), the following expression for the friction is obtained:

$$
\zeta(z) = \frac{\zeta^B(z) + R_{\rho\rho}(z) - \zeta^B(z)R_{\rho L}(z)}{1 + R_{\rho L}(z) + \zeta^B(z)[R_{LL}(z) + R_{TT}(z)]}. \tag{136}
$$

The $\zeta^B(z)$ term goes as t^2, the $R_{\rho\rho}$ term starts as t^4, and all the other terms start as t^6.

It is well known that the velocity autocorrelation function decays as $t^{-3/2}$ in the asymptotic limit due to the coupling between the tagged particle motion and the transverse current mode of the solvent [23, 56, 57]. The asymptotic limit of the R_{TT} term can be calculated by assuming that $F^s(q,t)$ and $C_{tt}(q,t)$ have simple diffusive behavior. Thus the expression for R_{TT} in this limit takes the following form:

$$R_{TT}(t) \to \frac{1}{12\rho}[\pi(D + \frac{\eta_s}{\rho m})t]^{-3/2}, \tag{137}$$

where D is the diffusion coefficient and η_s is the shear viscosity. It is found that the prefactor to the $t^{-3/2}$ term is not correct. As discussed by Sjogren and Sjolander [9], this happens because for asymptotic times both the density fluctuations and the longitudinal current are rapidly decaying modes compared to the transverse current. Thus they should contribute in the renormalization of the coupling constant to the transverse current, whereas it has been assumed that only the binary collision enters into the renormalization process. Thus the problem is reformulated by assuming that in Eq. (96), ζ^B is replaced by ζ^C, which includes all the modes except the transverse current mode. Now the expression of friction can be written as

$$\zeta(z) = \frac{\zeta^C(z)}{(1 + \zeta^C(z)R_{TT}(z))}. \tag{138}$$

ζ^C can be derived in a similar way by separating it into binary and recollision parts (neglecting the transverse current mode). Thus ζ^C will have an expression which is analogous to $\zeta(z)$ [given by Eq. (136)]. The expression for ζ^C can be written as

$$\zeta^C(z) = \frac{\zeta^B(z) + R_{\rho\rho}(z) - \zeta^B(z)R_{\rho L}(z)}{1 + R_{\rho L}(z) + \zeta^B(z)R_{LL}(z)}. \tag{139}$$

When the expression for ζ^C is substituted in Eq. (138), the expression for the friction can be written as

$$\zeta(z) = \frac{\zeta^B(z) + R_{\rho\rho}(z) - \zeta^B(z)R_{\rho L}(z)}{1 + R_{\rho L}(z) + \zeta^B(z)R_{LL}(z) + [\zeta^B(z) + R_{\rho\rho}(z) - \zeta^B(z)R_{\rho L}(z)]R_{TT}(z)}. \tag{140}$$

It has been found that the contributions from the $R_{\rho L}$ and R_{LL} terms are very small and can be neglected. Thus the final expression for the frequency-dependent friction reduces to

$$\frac{1}{\zeta(z)} = \frac{1}{\zeta^B(z) + R_{\rho\rho}(z)} + R_{TT}(z). \tag{141}$$

Although the above expression is derived based on the logic of separation of time scale, it can be argued that there also exists a separation of length scale. The binary contribution to the friction is known to be determined by the nearest neighbor. The difference in length scale of the events that determine the density and the current mode contribution can be understood from the following argument. As discussed earlier, the dynamic structure factor exhibits a slowing down in the intermediate wavenumbers (de Gennes's narrowing). Also note that the integrand in Eq. (132) has a q^4 term that makes the small q less important. These two effects combine to provide the major contribution to $R_{\rho\rho}$ from the intermediate wavenumbers. The major contribution to R_{TT}, on the other hand, comes from smaller wavenumbers. This is because there is no such slowing down of the current autocorrelation function and also the integrand in Eq. (135) contains a q^2 term.

It is curious to note that in certain limits (neglecting the contribution from the density mode) the present formulation of the friction has the same expression as obtained separately by Hynes, Kapral, and Weinberg [58] and by Mehaffey and Cukier [8]. Thus this formulation provides a microscopic justification of the semiempirical expression given by Hynes et al. [58]. The density term is known [59] to make a large contribution and cannot be neglected for neat liquids. It will be shown later that for large solutes (keeping the solute–solvent interaction potential the same as the solvent–solvent one), the contribution from the density term vanishes. Thus Eq. (141) reduces to the form given by Hynes et al. and by Mehaffey and Cukier.

E. Calculational Method for Friction

The system studied consists of one solute molecule that is tagged and N solvent molecules, each of mass m. Since we are interested in a spherically symmetrical potential, the pair potential of the solvent–solvent pair and the solute–solvent pair is assumed to be given by the simple Lennard-Jones 12-6 potential

$$v(r) = 4\epsilon\left[\left(\frac{\sigma}{r}\right)^{12} - \left(\frac{\sigma}{r}\right)^6\right], \tag{142}$$

where ϵ is the energy scale of the pairwise interaction, σ is the diameter of both the solute and the solvent, and r is the distance between two molecules.

The system is characterized by two dimensionless parameters, reduced density $\rho^* = \rho\sigma^3$, and reduced temperature $T^* = k_B T/\epsilon$.

For the calculation of the binary collision term the radial distribution function $g(r)$ and the static structure factor of the solute $S(q)$ is required.

$g(r)$ is calculated from the HMSA scheme [60], and the well-known Ornstein–Zernike relation is used to calculate $S(q)$ from $c(q)$, which is also obtained from the HMSA scheme.

In calculating the contribution from the density mode given by Eq. (132), the required solvent dynamical variables are $F(q,t)$, the intermediate scattering function, and $F_0(q,t)$, the short-time part of the intermediate scattering function.

$F(q,t)$ is obtained from its Laplace transform form $F(q,z)$. By using the following well-known Mori continued-fraction expansion and truncating at second order, the viscoelastic expression for $F(q,z)$ can be written as [16, 21, 22]

$$F(q,z) = \cfrac{S(q)}{z + \cfrac{\langle \omega_q^2 \rangle}{z + \cfrac{\Delta_q}{z + \tau_q^{-1}}}}, \tag{143}$$

where $\langle \omega_q^2 \rangle = k_B T q^2/mS(q)$ and $\tau_q^{-1} = 2\sqrt{\Delta_q/\pi}$. $\Delta_q = \omega_l^2(q) - \langle \omega_q^2 \rangle$, where $\omega_l^2(q)$ is the second moment of the longitudinal current correlation function given by [16, 21, 22]

$$\omega_l^2(q) = 3q^2 \frac{k_B T}{m} + \omega_0^2 + \gamma_d^l(q), \tag{144}$$

where $\gamma_d^l(q)$ is the longitudinal component of the vertex function given by Eq. (121).

The inertial part of the intermediate scattering function, $F_0(q,t)$ is given by

$$F_0(q,t) = S(q)\exp\left(-\frac{q^2 t^2}{2mS(q)}\right). \tag{145}$$

In calculating the contribution from the current term given by Eq. (135), the required solvent dynamical variables are $C_{tt}(q,t)$, the current autocorrelation function of the solvent, and $C_{tt0}(q,t)$, the short-time part of the same.

The transverse current correlation function is assumed to be given by the viscoelastic expression

$$C_{tt}(q,z) = \cfrac{1}{z + \cfrac{\omega_t^2(q)}{z + \tau_t^{-1}(q)}}, \tag{146}$$

where $\omega_t^2(q)$ is the second moment of the transverse current correlation function which is given by [16, 22]

$$\omega_t^2(q) = q^2 \frac{k_B T}{m} + \omega_0^2 + \gamma_d^t(q); \qquad (147)$$

here $\gamma_d^t(q)$ is the transverse component of the vertex function, given by Eq. (121). For $\tau_t(q)$ the expression proposed by Akeazu and Daniels [51] is used:

$$\tau_t^{-2}(q) = 2\omega_t^2(q) + \frac{\tau_t^{-2}(0) - 2\omega_t^2(q) + 2q^2 \frac{k_B T}{m}}{1 + (q/q_0)^2}, \qquad (148)$$

where q_0 is an adjustable parameter that actually determines the transition of the behavior of $C_{tt}(q,z)$ from "small q" to "large q." For argon, $q_0 = 1.5 \, \text{Å}^{-1}$. $\tau_t^{-1}(0) = \lim_{q \to 0} [m\rho\omega_t^2(q)]/q^2\eta_s$. Here η_s is the zero frequency shear viscosity which is calculated using the mode coupling theory. The derivation and the calculational details of η_s have been presented in the next section.

The inertial part of the the current autocorrelation function $C_{tt0}(q,t)$ is given by

$$C_{tt0}(q,t) = \frac{k_B T}{m} \exp\left(-\frac{\omega_t^2(q)t^2}{2}\right). \qquad (149)$$

The solute dynamic variables required to calculate the density and current contribution are the self-dynamic structure factor, $F^s(q,t)$ and inertial part of the self-intermediate structure factor, $F_0^s(q,t)$. $F_0^s(q,t)$ is given by

$$F_0^s(q,t) = \exp\left(-\frac{k_B T}{m} \frac{q^2 t^2}{2}\right) \qquad (150)$$

F.   Self-Consistent Scheme to Calculate $\zeta(z)$

Assuming Gaussian approximation the expression for $F^s(q,t)$ can be written as

$$F^s(q,t) = \exp\left(-\frac{q^2 \langle \Delta r^2(t) \rangle}{6}\right), \qquad (151)$$

where $\langle \Delta r^2(t) \rangle$ is the mean square displacement (MSD) which can be obtained from the time-dependent velocity autocorrelation function (VACF),

$C_v(t)$, through the following expression:

$$\langle \Delta r^2(t) \rangle = 2 \int_0^t d\tau C_v(\tau)(t - \tau). \tag{152}$$

The time-dependent VACF is obtained by numerically Laplace inverting the frequency-dependent VACF, which is related to the frequency-dependent friction through the following generalized Einstein relation given by Eq. (81):

$$C_v(z) = \frac{k_B T}{m(z + \zeta(z))} \tag{153}$$

Thus in this scheme the frequency-dependent friction has been calculated self-consistently with the MSD.

The self-consistency is implemented through the following iterative scheme. First, the VACF is obtained from Eq. (153) by replacing the total frequency-dependent friction, $\zeta(z)$ by its binary part, $\zeta^B(z)$. The VACF thus obtained is used to calculate the MSD through Eq. (152). Now this MSD is used to calculate $R_{\rho\rho}(t)$ and $R_{TT}(t)$ and thus $\zeta(z)$. This total friction is used to calculate the new VACF, which again is used to determine MSD and thus $\zeta(z)$. This iterative process is continued until the VACF obtained from two consecutive steps overlap.

Once the VACF is obtained self-consistently, the diffusion coefficient, D, is calculated using the following relation between the diffusion coefficient and the time-dependent VACF given by Eq. (80)

$$D = \frac{1}{3} \int_0^\infty d\tau C_v(\tau). \tag{154}$$

X. THEORY OF VISCOSITY

The expression for shear viscosity, η_s, can be written as [21]

$$\eta_s = \lim_{\epsilon \to 0} \lim_{q \to 0} \frac{m^2}{q^2 V} \int_0^\infty dt \langle \dot{j}^x(\mathbf{q}, t) \dot{j}^x(-\mathbf{q}, 0) \rangle \exp(i\epsilon t). \tag{155}$$

when the expression

$$\dot{j}^x(\mathbf{q}, t) + \frac{i}{m} q\sigma^{xz}(\mathbf{q}, t) = 0, \tag{156}$$

which has been derived from the conservation equation, is used in Eq. (155), the time-dependent shear viscosity can be expressed in terms of the stress autocorrelation function and is given by

$$\eta_s(t) = (Vk_BT)^{-1} \langle \sigma^{xz}(0)\sigma^{xz}(t) \rangle, \tag{157}$$

where in Eq. (156) it has been assumed that $\mathbf{q}$ has only one component along the z-axis. $j_{\mathbf{q}}^x$ is the x-component of the particle current density. σ^{xz} is the off-diagonal element of the stress tensor and is given by

$$\sigma^{xz} = \sum_{j=1}^{N} [(p_j^x p_j^z / m) + F_j^z x_j]. \tag{158}$$

Here F_j^z is the z-component of the force acting on the jth molecule, p_j^x is the x-component of the momentum of the jth molecule, and the corresponding position is x_j.

The high-frequency shear modulus is given by

$$G_\infty = (Vk_BT)^{-1} \langle (\sigma^{xz}(0))^2 \rangle. \tag{159}$$

After a few steps of algebra the above equation reduces to the following exact expression [61]:

$$G_\infty = \rho k_BT + \frac{2\pi}{15}\rho^2 \int_0^\infty dr\, g(r) \frac{d}{dr} \left[r^4 \frac{dv(r)}{dr} \right], \tag{160}$$

where $g(r)$ is the radial distribution function of the liquid.

By invoking the separation of time scales between the initial fast and the later slow decay (as has been done in case of friction), the time-dependent viscosity of a liquid can be written as the sum of two different terms. The initial fast part arises due to the dynamics within the cage formed by the surrounding molecules and is expressed in terms of the static correlations. The fast part is followed by a slow long-time part that arises from the dynamic correlations and basically describes the relaxation of the cage due to the presence of the hydrodynamic modes like the density and the current. As discussed at length by Geszti, in dense liquids it is the density mode which primarily contributes to the long-time viscosity [39]. The time-dependent viscosity can thus be written as

$$\eta_s(t) = \eta_s^B(t) + \eta_{s\rho\rho}(t). \tag{161}$$

In the above expression, $\eta_s^B(t)$ is the short-time part that arises from the static correlations and $\eta_{s\rho\rho}(t)$ is the long-time part that arises from the density mode contribution.

A. Derivation of the Binary Viscosity

The calculation of $\eta_s^B(t)$ is described first. The exact initial value of the viscosity is known and is given by G_∞. The collective term contribution to the viscosity starts as t^4. Thus all the contribution from t^2 is included in the binary term. As only even powers of t appear in $\eta_s(t)$, $\eta_s^B(t)$ is approximated to be expressed in terms of a Gaussian function and written as

$$\eta_s^B(t) = G_\infty \exp(-t^2/\tau_{\eta_s}^2). \tag{162}$$

In the above equation, τ_{η_s} is determined from the second derivative of $\eta_s(t)$.

$$\tau_{\eta_s} = \sqrt{\frac{-2G_\infty}{\ddot{\eta}_s(t=0)}}. \tag{163}$$

In the liquid range $\eta_s(t)$ is dominated purely by its potential part and thus the expression for viscosity reduces to

$$\eta_s(t) = (Vk_BT)^{-1}\left\langle \sum_i F_i^z x_i \sum_j F_j^z(t)x_j(t) \right\rangle. \tag{164}$$

The force is described in terms of derivative of the pair potential $\mathbf{F}_i = -\sum_{i(\neq j)}\nabla_i v(\mathbf{r}_{ij})$. Using the identity $\sum_i x_i F_i^z = \frac{1}{2}\sum_i \sum_{j(\neq i)} x_{ij}F_{ij}^z$, where $\mathbf{F}_{ij}$ is the force on ith atom due to the jth atom, the above expression for the time-dependent viscosity can be rewritten as

$$\eta_s(t) = (1/N) \sum_{i,j(j\neq i)} \sum_{l,m(l\neq m)} \langle A(\mathbf{r}_{ij}(0))A(\mathbf{r}_{lm}(t)) \rangle, \tag{165}$$

where

$$A(\mathbf{r}_{ij}) = (\rho/k_BT)^{1/2}\frac{x_{ij}z_{ij}}{2r_{ij}}v'(r_{ij})$$

$$= a(r)\cos\theta\sin\theta\cos\phi, \tag{166}$$

where $a(r) = (\rho/k_BT)^{1/2}rv'(r)/2$.

The expression for $\ddot{\eta}_s(t)$ can be written in the following form by making separate possible choice for the atomic labels [62]:

$$\ddot{\eta}_s(t) = \frac{2}{N} \sum_{i,j(j \neq i)} \langle A(\mathbf{r}_{ij}(0))\ddot{A}(\mathbf{r}_{ij}(t)) \rangle$$

$$+ \frac{4}{N} \sum_{i,j,l(l \neq j \neq i)} \langle A(\mathbf{r}_{ij}(0))\ddot{A}(\mathbf{r}_{il}(t)) \rangle$$

$$+ \frac{1}{N} \sum_{i,j,l,m(m \neq l \neq j \neq i)} \langle A(\mathbf{r}_{ij}(0))\ddot{A}(\mathbf{r}_{lm}(t)) \rangle$$

$$\equiv \ddot{\eta}_s^{(2)}(t) + \ddot{\eta}_s^{(3)}(t) + \ddot{\eta}_s^{(4)}(t). \qquad (167)$$

The first, second, and third term on the right give the contributions from the pairs, the triplets, and the quadruplets, respectively. The contribution from the pairs at $t = 0$ can be derived in the following way:

$$\ddot{\eta}_s^{(2)}(0) = \frac{2}{N} \sum_{i,j(j \neq i)} \langle A(\mathbf{r}_{ij}(0))\ddot{A}(\mathbf{r}_{ij}(0)) \rangle$$

$$= -\frac{2}{N} \sum_{i,j(j \neq i)} \langle \dot{A}(\mathbf{r}_{ij}(0))\dot{A}(\mathbf{r}_{ij}(0)) \rangle$$

$$= -\frac{2}{N} \sum_{i,j(j \neq i)} \langle [\dot{\mathbf{r}}_{ij} \cdot \nabla_{ij} A(\mathbf{r}_{ij}(0))]^2 \rangle$$

$$= -(4\rho k_B T/m) \int d\mathbf{r} [\nabla A(\mathbf{r}) \cdot \nabla A(\mathbf{r})] g(r). \qquad (168)$$

Next the expression for A is substituted in Eq. (168), and angular integration is performed. The final expression for $\ddot{\eta}_s^{(2)}(0)$ can now be written as

$$\ddot{\eta}_s^{(2)}(0) = -\frac{4\pi\rho^2}{15m} \int_0^\infty dr\, r^2 [r^2(v'')^2 + 2rv'v'' + 7(v')^2] g(r), \qquad (169)$$

where $v' = dv(r)/dr$ and $v'' = d^2 v(r)/dr^2$.

Similarly, the derivation of the triplet contribution $\ddot{\eta}_s^{(3)}(0)$ can be performed and is given by

$$\ddot{\eta}_s^{(3)}(0) = \frac{4}{N} \sum_{i,j,l(l\neq j\neq i)} \langle A(\mathbf{r}_{ij}(0))\ddot{A}(\mathbf{r}_{il}(0))\rangle$$

$$= -\frac{4}{N} \sum_{i,j,l(l\neq j\neq i)} \langle \dot{A}(\mathbf{r}_{ij}(0))\dot{A}(\mathbf{r}_{il}(t))\rangle$$

$$= -\frac{4}{N} \sum_{i,j,l(l\neq j\neq i)} \langle [\dot{\mathbf{r}}_{ij} \cdot \nabla_{ij}A(\mathbf{r}_{ij}(0))][\dot{\mathbf{r}}_{il} \cdot \nabla_{il}A(\mathbf{r}_{il}(0))]\rangle$$

$$= -\frac{4\rho^3 k_B T}{Nm} \int d\mathbf{r}_1 d\mathbf{r}_2 d\mathbf{r}_3 \nabla_{12}A(\mathbf{r}_{12}) \cdot \nabla_{13}A(\mathbf{r}_{13})g_3(\mathbf{r}_1\mathbf{r}_2\mathbf{r}_3)$$

$$= -\frac{4\rho^2 k_B T}{m} \int d\mathbf{r}ds\nabla A(\mathbf{r}) \cdot \nabla A(\mathbf{s})g_3(\mathbf{r},\mathbf{s}). \tag{170}$$

The three-particle distribution function $g_3(\mathbf{r},\mathbf{s})$ can be expressed in a series of Legendre polynomials [63]. Then expressing the Legendre polynomials in terms of spherical harmonics, we can write the expression for $g_3(\mathbf{r},\mathbf{s})$ as

$$g_3(\mathbf{r},\mathbf{s}) = g_3(r,s,\cos\Theta) = \sum_{l=0}^{\infty} Q_l(r,s)P_l(\cos\Theta)$$

$$= 4\pi \sum_{l=0}^{\infty} Q_l(r,s)(2l+1)^{-1} \sum_{m=-l}^{l} Y_{lm}^*(\Omega_r)Y_{lm}(\Omega_s), \tag{171}$$

where Θ is the angle between $\mathbf{r}$ and $\mathbf{s}$. $\nabla A(\mathbf{r})$ and $\nabla A(\mathbf{s})$ can also be expressed in terms of the spherical harmonics [64]. The above-mentioned properties when used, the expression for the triplet contribution can be simplified in the following form,

$$\ddot{\eta}_s^{(3)}(0) = -\frac{2\pi\rho^2 k_B T}{15m} \int_0^{\infty} dr\, r^2 \int_0^{\infty} ds\, s^2 \left\{ \frac{24\pi}{35} Q_3(r,s)\left[a'(r) - 2\frac{a(r)}{r}\right] \right.$$

$$\times \left[a'(s) - 2\frac{a(s)}{s}\right] + \frac{16\pi}{15}Q_1(r,s)\left[a'(r) + 3\frac{a(r)}{r}\right]\left[a'(s) + 3\frac{a(s)}{s}\right] \right\} \tag{172}$$

Now assuming that the superposition approximation for the triplet correlation function, $Q_l(r, s)$, can be written as [65]

$$Q_l(r, s) = g(r)g(s)(2l + 1)[\delta_{l0} + \frac{1}{2\pi^2 \rho} \int_0^\infty dq \, q^2 h(q) j_l(qr) j_l(qs)], \quad (173)$$

where $h(q) = S(q) - 1$ and $j_l(x)$ is the spherical Bessel function. Replacing the expression for $Q_l(r, s)$ in Eq. (172), we can write the triplet contribution as

$$\ddot{\eta}_s^{(3)}(0) = -\frac{8\rho^2}{75m} \int_0^\infty dq \, q^2 [S(q) - 1][2T_1^2(q) + 3T_2^2(q)], \quad (174)$$

where the functions $T_1(q)$ and $T_2(q)$ are defined by the following integrals:

$$T_1(q) = \int_0^\infty dr \, r^2 [rv'' + 4v'] j^x(qr) g(r), \quad (175)$$

$$T_2(q) = \int_0^\infty dr \, r^2 [rv'' - v'] j_3(qr) g(r). \quad (176)$$

The quadruplet contributions to $\ddot{\eta}_s^{(4)}(0)$ vanishes because $\ddot{\eta}_s^{(4)}(0)$ contains the ensemble averages of the scalar products of velocities of four different particles.

Thus the final expression for $\ddot{\eta}_s(0)$ can be written as

$$\ddot{\eta}_s(t = 0) = -\frac{4\pi\rho^2}{15m} \int_0^\infty dr \, r^2 [r^2 (v'')^2 + 2rv'v'' + 7(v')^2] g(r)$$
$$- \frac{8\rho^2}{75m} \int_0^\infty dq \, q^2 [S(q) - 1][2T_1^2(q) + 3T_2^2(q)]. \quad (177)$$

Note that the only approximation made in the derivation of Eq. (177) is the use of the Kirkwood superposition approximation for the triplet distribution function of the liquid [21]. In a dense liquid at low temperature (near its triple point), this is not a bad approximation [21].

B. Derivation of the Density and Current Mode Contribution

In deriving the density mode contribution to the viscosity, $\eta_{s\rho\rho}$, the formulation of Geszti is followed [39].

The starting point of the calculation is a Mori-type rephrasing of the Green–Kubo formula for the viscosity, which is given by Eq. (155). Thus the expression for viscosity can be written as

$$\eta_s = \lim_{\epsilon \to 0} \lim_{q \to 0} \frac{m^2}{q^2 V} \int_0^\infty dt (QLj^x(\mathbf{q})|\exp(iQLQt - \epsilon t)|QLj^x(\mathbf{q})). \tag{178}$$

In the above equation, $\mathbf{q}$ has been considered to have only one component that is along the z-direction. L is the Hermitian Liouville operator, $Q = 1 - P$, where P is the projector onto the dynamical quantity A^α. A^α is component of a four-vector of particle current density and particle density, defined as

$$A^\alpha = j^\alpha(\mathbf{q}) = \sum_i \frac{(p_i)^\alpha}{m} \exp(-i\mathbf{q} \cdot \mathbf{r}_i), \qquad \alpha = 1, 2, 3$$

$$A^4 = \rho(\mathbf{q}) = \sum_i \exp(-i\mathbf{q} \cdot \mathbf{r}_i) \tag{179}$$

$\alpha = 1, 2, 3$ corresponds to the $x, y,$ and z components respectively. Q is defined as

$$Q = 1 - \sum_{\mu\nu} |A^\mu)(\chi^{-1})^{\mu\nu}(A^\nu|, \tag{180}$$

where $\chi^{\mu\nu} = (A^\mu|A^\nu)$. The classical scalar product is defined by

$$(A|B) = \langle \delta A^* \delta B \rangle / k_B T, \tag{181}$$

where $\delta A = A - \langle A \rangle$.

For finding the contribution from the density term the following notation is introduced:

$$B_{\mathbf{k}}(\mathbf{q}) = Q\rho(\mathbf{k})\rho(\mathbf{q} - \mathbf{k})/N. \tag{182}$$

The contribution from the density mode is written as

$$\eta_{s\rho\rho} = \lim_{q \to 0} \frac{m^2}{q^2 V} \sum_{\mathbf{k,p}} \phi_\mathbf{k}^*(\mathbf{q}) \int_0^\infty dt \, \psi_{\mathbf{k,p}}(\mathbf{q}, t) \, \phi_\mathbf{p}(\mathbf{q}), \tag{183}$$

where the vertex function is given by

$$\phi_\mathbf{p}(\mathbf{q}) = \sum_{\mathbf{p'}} [(B|B)^{-1}]_{\mathbf{pp'}} (B_{\mathbf{p'}}(\mathbf{q})|QLj^x(\mathbf{q})) \tag{184}$$

and the propagator is written as

$$\psi_{k,p}(q,t) = (B_k(q)|\exp(iQLQt - \epsilon t)|B_p(q)).$$ (185)

From the above expressions it is evident that for the calculations of the vertex function and the propagator the knowledge of four-particle correlation functions are required. For simplification, Gaussian approximation has been assumed and the four-particle correlation function is written as the product of two two-particle correlation functions.

Replacing the expressions for $B_k(q)$ and $B_p(q)$ in Eq. (185), we can write the propagator as

$$\psi_{k,p}(q,t) \simeq \frac{1}{N^2 k_B T}(\langle\rho(-k,t)\rho(p)\rangle\langle\rho(-q+k,t)\rho(q-p)\rangle$$
$$+ \langle\rho(-k,t)\rho(q-p)\rangle\langle\rho(-q+k,t)\rho(p)\rangle)$$
$$\simeq \frac{1}{k_B T}F(k,t)F(q-k,t)(\delta_{k,p} + \delta_{k,q-p}).$$ (186)

Similarly, the expression for $(B|B)_{k,p}$ can be written where $(B|B)_{k,p} = \psi_{k,p}(q,0)$:

$$(B|B)_{k,p} = \frac{1}{k_B T}S(k)S(q-k)(\delta_{k,p} + \delta_{k,q-p}).$$ (187)

Next the calculation of the vertex function is described. The projection operator Q in Eq. (184) will introduce products of the form $(A^v|Lj^x(q))$. Considering time inversion symmetry, only $(A^4|Lj^x(q))$ will survive.

$$(\rho(q)|Lj^x(q)) = (L\rho(q)|j^x(q))$$ (188)

The operation of the Liouville operator can be replaced by time derivative; thus the term in the left-hand side of the scalar product can be written as

$$L^*\rho^*(q) = -i\frac{\partial}{\partial t}\rho^*(q).$$ (189)

Hence one obtains

$$(L\rho(q)|j^x(q)) = q(j^z(q)|j^x(q)) = 0.$$ (190)

From the above derivation it is clear that Q can be omitted from Eq. (184), and the eguation can be written as

$$(B_p(q)|QLj^x(q)) = (B_p(q)|Lj^x(q)) = (LB_p(q)|j^x(q)).$$ (191)

When substituting the expression for $B_{\mathbf{p}}(\mathbf{q})$ in the above equation, it can be shown (following similar logic) that the Q present in $B_{\mathbf{p}}(\mathbf{q})$ can be omitted. Thus Eq. (191) can be written as

$$
\begin{aligned}
(B_{\mathbf{p}}(\mathbf{q})|QLj^{x}(\mathbf{q})) = \frac{1}{Nk_{B}T} &\left[\left\langle \sum_{i,j,l} \mathbf{p} \cdot \dot{\mathbf{r}}_{i} \dot{r}_{l}^{x} \exp \mathbf{i}(\mathbf{p} \cdot \mathbf{r}_{i} + (\mathbf{q} - \mathbf{p}) \cdot \mathbf{r}_{j} - \mathbf{q} \cdot \mathbf{r}_{l}) \right\rangle \right. \\
&\left. + \left\langle \sum_{i,j,l} (\mathbf{q} - \mathbf{p}) \cdot \dot{\mathbf{r}}_{j} \, \dot{r}_{l}^{x} \exp \mathbf{i}(\mathbf{p} \cdot \mathbf{r}_{i} + (\mathbf{q} - \mathbf{p}) \cdot \mathbf{r}_{j} - \mathbf{q} \cdot \mathbf{r}_{l}) \right\rangle \right] \\
= \frac{p^{x}}{Nm} &\left[\left\langle \sum_{ij} \exp \mathbf{i}(\mathbf{p} \cdot \mathbf{r}_{ij}) \right\rangle - \left\langle \sum_{ij} \exp \mathbf{i}((\mathbf{p} - \mathbf{q}) \cdot \mathbf{r}_{ij}) \right\rangle \right] \\
= \frac{p^{x}}{m} &[S(\mathbf{p}) - S(\mathbf{p} - \mathbf{q})].
\end{aligned}
\tag{192}
$$

Now since $q \ll p$ and has only one component along the z-axis, the above expression reduces to

$$
(B_{\mathbf{p}}(\mathbf{q})|QLj^{x}(\mathbf{q})) = \frac{p^{x}}{m} q \frac{\partial S(\mathbf{p})}{\partial p^{z}}.
\tag{193}
$$

In Eq. (187), terms are taken which are leading order in $q \to 0$. Replacing it along with Eq. (184) in Eq. (193), we can write the vertex function as

$$
\phi_{\mathbf{p}}(\mathbf{q}) = q \sum_{\mathbf{p}'} (\delta_{\mathbf{p},\mathbf{p}'} + \delta_{\mathbf{p},-\mathbf{p}'})^{-1} \frac{k_{B}T}{(S(\mathbf{p}'))^{2}} \frac{p'^{x}}{m} \frac{\partial S(\mathbf{p}')}{\partial p'^{z}}.
\tag{194}
$$

The matrix inversion in the above equation can be performed easily because both $S(\mathbf{p})$ and $p^{x}\partial/\partial p^{z}$ are even in $\mathbf{p}$. Thus the final expression for the vertex function is written as

$$
\phi_{\mathbf{p}}(\mathbf{q}) = q \frac{k_{B}T}{2m} \frac{p^{x}}{(S(\mathbf{p}))^{2}} \frac{\partial S(\mathbf{p})}{\partial p^{z}}.
\tag{195}
$$

Next the expression for the vertex function [given by Eq. (195)] and that of the propagator [given by Eq. (186)] are substituted in Eq. (183), the summation is replaced as integral, angular average is performed, and variable p is renamed as q. Thus the expression for the $\eta_{s\rho\rho}$ can be written as

$$
\eta_{s\rho\rho} = k_{B}T/60\pi^{2} \int_{0}^{\infty} dq \, q^{4}[S'(q)/S(q)]^{2} \int_{0}^{\infty} dt(F(q,t)/S(q))^{2},
\tag{196}
$$

where $S'(q)$ denotes the first derivative of the static structure factor.

Note that in deriving the contribution from the density fluctuation to the total viscosity, terms of order t^2 has not been taken out. In the initial argument of separation of time scale, it was stated that contributions from terms up to order t^2 should be included only in the binary term (η_s^B), and the collective contribution term was expected to start as t^4. Thus to take out all the contributions of order t^2 from Eq. (196), the short-time dynamics has to be taken out from the propagator as has been done in case of friction. This is achieved by taking out the short-time dynamics from $(F(q,t)/S(q))^2$. Thus the corrected expression for $\eta_{s\rho\rho}$ can now be written as

$$\eta_{s\rho\rho} = k_B T/60\pi^2 \int_0^\infty dq\, q^4 [S'(q)/S(q)]^2 \int_0^\infty dt[(F(q,t)/S(q))^2$$
$$- (F_0(q,t)/S(q))^2]. \tag{197}$$

Note that although in the present derivation the subtraction of the short-time part has been performed in an *ad hoc* manner, the same expression for $\eta_{s\rho\rho}$ [given by Eq. (197)] can be obtained from a different theoretical approach [16].

In a similar way, the contribution from the transverse current correlation can be derived. $B_\mathbf{k}(\mathbf{q})$ is now written as the product of two transverse currents j^x:

$$B_\mathbf{k}(\mathbf{q}) = Q j^x(\mathbf{k}) j^x(\mathbf{q} - \mathbf{k})/N. \tag{198}$$

The contribution to the viscosity from the transverse current mode is written as

$$\eta_{sTT} = \lim_{q \to 0} \frac{m^2}{q^2 V} \sum_{\mathbf{k},\mathbf{p}} \phi_\mathbf{k}^*(\mathbf{q}) \int_0^\infty dt\, \psi_{\mathbf{k},\mathbf{p}}(\mathbf{q},t)\, \phi_\mathbf{p}(\mathbf{q}), \tag{199}$$

where the propagator can be written as

$$\psi_{\mathbf{k},\mathbf{p}}(\mathbf{q},t) \simeq \frac{k_B T}{m^2} C_{tt}(\mathbf{k},t) C_{tt}(\mathbf{q} - \mathbf{k},t)(\delta_{\mathbf{k},\mathbf{p}} + \delta_{\mathbf{k},\mathbf{q}-\mathbf{p}}); \tag{200}$$

where $C_{tt}(\mathbf{q},t)$ is the normalized transverse current autocorrelation function.

Similarly, the expression for $(B|B)_{\mathbf{k},\mathbf{p}}$ can be written where $(B|B)_{\mathbf{k},\mathbf{p}} = \psi_{\mathbf{k},\mathbf{p}}(\mathbf{q},0)$:

$$(B|B)_{\mathbf{k},\mathbf{p}} = \frac{k_B T}{m^2}(\delta_{\mathbf{k},\mathbf{p}} + \delta_{\mathbf{k},\mathbf{q}-\mathbf{p}}). \tag{201}$$

Following the same procedure, the vertex function can be calculated, and the final expression of the transverse current contribution to the viscosity is written as

$$\eta_{sTT} = k_B T m^2 / 60\pi^2 \int_0^\infty dq \, q^4 [S'(q)]^2 \int_0^\infty dt \, [C_{tt}(q,t)^2 - C_{tto}(q,t)^2], \quad (202)$$

It can be shown that the contribution from the longitudinal current goes to zero.

At high density the contribution from the transverse current terms becomes negligible since the transverse current autocorrelation function decays rapidly.

C. Similarity Between the Viscosity Expressions of Kadanoff and Swift and Those of Geszti

It can be instructive to compare the two rather different mode coupling theory expressions for the viscosity; one valid near the critical point and other near the liquid-glass transition point.

In order to do so, let us first write down the expressions for the two heat mode contributions and the density mode contribution to the zero frequency and zero wavenumber viscosity, η_{sHH} and $\eta_{s\rho\rho}$ respectively:

$$\eta_{sHH} = \frac{1}{4} \int \frac{d^3 q'}{(2\pi)^3} (q_y')^2 \frac{\left[\frac{\partial}{\partial q_x'} C_P(\mathbf{q}')\right]^2}{z_H(\mathbf{q}')[C_P(\mathbf{q}')]^2} \quad (203)$$

and

$$\eta_{s\rho\rho} = k_B T / 60\pi^2 \int_0^\infty dq \, q^4 [S'(q)/S(q)]^2 \int_0^\infty dt [(F(q,t)/S(q))^2 - (F_0(q,t)/S(q))^2]. \quad (204)$$

Although these two expressions do not immediately look similar, they essentially denote quite similar contributions to the viscosity. Let us recall that the wavenumber-dependent specific heat is expressed in terms of the correlation between entropy fluctuations. The entropy fluctuation in turn is expressed as a linear combination of the energy and density fluctuation. Now, if we neglect energy fluctuation, then the wavenumber-dependent specific heat is nothing but the correlation between the density fluctuations and thus is proportional to the static structure factor.

According to Kadanoff and Swift, the relaxation of an eigenstate is given by an exponential in time, where the relaxation time is the inverse of the eigenvalue. Thus it assumes that there is no Gaussian contribution to the correlation functions. Hence the second term in the integrand of the time integration in Eq. (204) becomes zero. In the limit of exponential decay, the heat mode contribution to the intermediate scattering function is given by

$$F(q,t) = S(q)e^{-z_H(\mathbf{q})t}. \tag{205}$$

With this form of $F(q,t)$, the time integration in Eq. (204) can be easily performed and the expression of the density mode contribution to the viscosity reduces to

$$\eta_{s\rho\rho} = k_B T / 120\pi^2 \int_0^\infty dq\, q^4 \frac{[S'(q)]^2}{[S(q)]^2 z_H(q)}. \tag{206}$$

As discussed above, $C_P(q)$ is proportional to $S(q)$ if the energy fluctuation is neglected. Thus, the two seemingly different expressions [(203) and (204)] have a similar long-time contribution.

XI. DERIVATION OF THE MCT EXPRESSION OF THE DYNAMIC STRUCTURE FACTOR

To derive the dynamic structure factor, let us once again write down the generalized relaxation equation in terms of its components and in the frequency plane,

$$zC_{\mu\nu}(\mathbf{q},z) - i\Omega_{\mu\lambda}(\mathbf{q})C_{\lambda\nu}(\mathbf{q},z) + \Gamma_{\mu\lambda}(\mathbf{q},z)C_{\lambda\nu}(\mathbf{q},z) = \tilde{C}_{\mu\nu}(\mathbf{q}). \tag{207}$$

In the above expression, summation over repeated indices is implied. The matrix elements of the frequency and memory function are given by

$$\Omega_{\nu\lambda}(\mathbf{q}) = \sum_k (LA_\nu(\mathbf{q})|A_k^*(\mathbf{q}))(A_k(\mathbf{q})|A_\lambda^*(\mathbf{q}))^{-1} \tag{208}$$

and

$$\Gamma_{\nu\lambda}(\mathbf{q},z) = \sum_k (F_\nu(\mathbf{q},z)|F_k^*(\mathbf{q}))(A_k(\mathbf{q})|A_\lambda^*(\mathbf{q}))^{-1}. \tag{209}$$

The first five components are related to the conserved hydrodynamic variables, the density, the longitudinal and transverse current, and the temperature, respectively.

To obtain an approximate expression for the density autocorrelation function, first we consider that the density fluctuation is coupled only to the longitudinal current fluctuation, and its coupling to the temperature fluctuation and other higher-order components are neglected.

With this consideration the relaxation equation will give rise to a set of coupled equations involving the time autocorrelation function of the density and the longitudinal current fluctuation, and also there will be cross terms that involve the correlation between the density fluctuation and the longitudinal current fluctuation. This set of coupled equations can be written in matrix notation, which becomes identical to that derived by Gotze from the Liouvillian resolvent matrix [3].

To write down the expression for the dynamic structure factor, we need explicit expressions for the components of the frequency matrix, memory function matrix, and the normalization matrix $\tilde{C}(\mathbf{q})$.

The definition of the classical scalar product, discussed earlier in the derivation of the viscosity, is used in the derivation of the frequency and the normalization matrix. The normalization matrix is diagonal, and its matrix elements are the following: $C_{\rho\rho} = NS(q)/k_BT$ and $C_{ll} = N/m$. The diagonal components of the frequency matrix are zero due to time inversion symmetry. The off-diagonal elements are the following: $\Omega_{\rho l} = q$ and $\Omega_{l\rho} = qk_BT/mS(q) = \langle \omega_q{}^2 \rangle/q$.

In derivation of the matrix elements of Γ, we note that the random force $F_\mu(\tau) = e^{iQL\tau}QiLA_\mu$ is zero when $A_\mu = \rho(\mathbf{q})$, since $QL\rho(\mathbf{q}) = qQj_l(\mathbf{q}) = 0$. Thus $\Gamma_{\rho\rho} = \Gamma_{\rho l} = \Gamma_{l\rho} = 0$. Γ_{ll} is nonzero and is known as the *longitudinal current relaxation kernel*.

The expression of the dynamic structure factor, $F(qz) = \frac{1}{\rho}C_{\rho\rho}(qz)$, can be written in terms of Γ_{ll} as

$$F(qz) = \frac{S(q)}{z + \dfrac{\langle \omega_q{}^2 \rangle}{z + \Gamma_{ll}(qz)}}. \tag{210}$$

The above expression has been used by Leutheusser [34] and Kirkpatrick [30] in the study of liquid–glass transition. Leutheusser [34] has derived the expression of the dynamic structure factor from the nonlinear equation of motion for a damped oscillator. In their expression they refer to the memory kernel as the dynamic longitudinal viscosity.

Equation (210) when compared with the viscoelastic model of the dynamic structure factor we can identify the memory kernel in the viscoelastic model, which is written as

$$\Gamma_{ll}(qz) = \frac{\Delta_q}{z + 1/\tau_q}. \tag{211}$$

As discussed before, the viscoelastic model is known to provide a correct description of $F(qz)$ in the intermediate density regime. Even in a supercooled liquid, it can provide correct short-time description, but fails in the long time, where the contribution from the hydrodynamic modes become important.

Thus we note that the memory kernel has a short-time and a long-time part. It is the long-time part which is not present in the viscoelastic model, becomes important in the supercooled-liquid–near-glass transition, and gives rise to the long-time tail of the dynamic structure factor.

A. Derivation of the Memory Kernel

The derivation of $\Gamma_{ll}(qz)$ can be performed in the same way as the calculation of the friction [66]:

$$\Gamma_{ll}(\mathbf{q}z) = \frac{1}{\rho k_B TmV} \int d1 \dots d2' \exp(-i\mathbf{q} \cdot \mathbf{r}_1)[\hat{\mathbf{q}} \cdot \nabla_{\mathbf{r}_1} v(\mathbf{r}_1 - \mathbf{r}_2)]G(12; 1'2'z)$$
$$\times [\hat{\mathbf{q}} \cdot \nabla_{\mathbf{r}_1'} v(\mathbf{r}_1' - \mathbf{r}_2')]\exp(i\mathbf{q} \cdot \mathbf{r}_{1'}) \qquad (212)$$

where the four-point function $G(12; 1'2', t')$ describes the correlated motion of two disturbances. It describes the time-dependent probability that two particles move from the positions (r_1', p_1') and (r_2', p_2') at t', and the same or possibly other particles are found at (r_1, p_1) and (r_2, p_2) at time t. $G(12; 1'2', t')$ also contains information on the static correlation between the solvent particles through its initial value $\tilde{G}(12; 1'2')$. z is the Laplace frequency, and $G(12; 1'2'z)$ is obtained from the usual Laplace transformation of $G(12; 1'2', t')$.

An exact equation of $G(12; 1'2'z)$ is given by [66]

$$zG(12; 1'2'z) - \int d1''d2''\Omega(12; 1''2'')G(1''2''; 1'2'z)$$
$$+ \int d1''d2''\Gamma(12; 1''2''z)G(1''2''; 1'2'z) = \tilde{G}(12; 1'2'), \quad (213)$$

where $\Omega(12; 1'2')$ in the lowest order in density is written as

$$\Omega(12; 1'2') = -[(1/m)\mathbf{p_1} \cdot \nabla_{\mathbf{r}_1} + (1/m)\mathbf{p_2} \cdot \nabla_{\mathbf{r}_2} - \nabla_{\mathbf{r}_1} v(\mathbf{r}_1 - \mathbf{r}_2) \cdot (\nabla_{\mathbf{p}_1} - \nabla_{\mathbf{p}_2})]$$
$$\times [\delta(11')\delta(22') + \delta(12')\delta(21')]. \qquad (214)$$

Now the propagator for G can symbolically be written as

$$R = (z - \Omega + \Gamma)^{-1}. \qquad (215)$$

Similarly as in the case of calculation of friction, the short-time part of the propagator, R^B, can be separated out. R^B represents a single binary collision. Then a Dyson expansion can be performed to represent the full R as a sequence of binary collisions.

In the expression of $\Gamma_{ll}(\mathbf{q}z)$ when the full Green's function is replaced by its short time part, G^B, we obtain the binary contribution of the memory kernel, $\Gamma_{ll}^B(\mathbf{q}z)$. The expression of the memory kernel with the long time part of the Green's function $(G - G^B)$ gives the recollision term. This recollision term $\Gamma_{ll}^R(\mathbf{q}z)$ can be written in terms of the T matrix and the propagator. The propagator is expressed in terms of the product of the hydrodynamical variables; and the two disturbances are assumed to propagate independently, thus considering disconnected motion. As described in the calculation of friction, the exact derivation of the T matrix is difficult and is performed only at certain limits.

The final expression for the long-time part of the memory kernel is thus given by

$$\Gamma_{ll}^R(\mathbf{q}z) = R_{\rho\rho}^l(\mathbf{q}z) - \Gamma_{ll}^B(\mathbf{q}z)R_{\rho L}^l(\mathbf{q}z) - R_{\rho L}^l(\mathbf{q}z)\Gamma_{ll}(\mathbf{q}z)$$
$$- \Gamma_{ll}^B(\mathbf{q}z)[R_{LL}^l(\mathbf{q}z) + R_{TT}^l(\mathbf{q}z)]\Gamma_{ll}(\mathbf{q}z). \qquad (216)$$

The superscript 'l' denotes the contributions from the different modes to the memory kernel of the longitudinal current. $R_{\rho\rho}^l(z)$ contains the coupling to the density. $R_{\rho L}^l(z)$ contains the coupling to one density mode and one longitudinal current mode, and R_{LL}^l and R_{TT}^l give the coupling to the longitudinal and transverse current modes, respectively. The Γ_{ll}^B term goes as t^2 and $R_{\rho\rho}^l$ term starts as t^4 and all the other terms start as t^6.

Γ_{ll}^B can be written in terms of a Gaussian function,

$$\Gamma_{ll}^B(\mathbf{q}, t) = \Gamma_{ll}^B(\mathbf{q}, t = 0) \exp[-t^2/\tau_l^2(\mathbf{q})] \qquad (217)$$

The time constant τ_l can be obtained from the second derivative of $\Gamma_{ll}^B(\mathbf{q}, t)$ at $t = 0$. The expression of $\Gamma_{ll}^B(\mathbf{q}, t = 0)$ is given by

$$\Gamma_{ll}^B(\mathbf{q}, t = 0) = \omega_0^2 + \gamma_d^l(\mathbf{q}) + \frac{q^2}{\beta m} - \langle \omega(q)^2 \rangle. \qquad (218)$$

If the full memory kernel is replaced in Eq. (210), then $\Gamma_{ll}^B(\mathbf{q}, t = 0)$ should become identical to Δ_q. However, in the expression of $\Gamma_{ll}^B(\mathbf{q}, t = 0)$ the term $(q^2/\beta m)$ comes with a prefactor 1, whereas in Δ_q it comes with a prefactor 3. Thus, a factor of $(2q^2/\beta m)$ is found to be missing in the former. As has been discussed by Balucani and Zoppi [16], this extra term arises from the coupling with the non-hydrodynamic modes, not included in our derivation of $\Gamma_{ll}^B(\mathbf{q}, t = 0)$.

There is another way to model the short-time binary part, and it can be written in the following way:

$$\Gamma_{ll}^B(\mathbf{q}, t) = \Delta_q \exp(-t^2/\tau_{\Delta_q}), \qquad (219)$$

where τ_{Δ_q} is obtained from the calculation of higher moments of the dynamic structure factor.

In the supercooled liquid, the important part of the memory kernel is its long-time part, $\Gamma_{ll}^R(\mathbf{q}, t)$. The recollision term contains the contribution from the hydrodynamic modes. As discussed by many authors [3, 30, 34], among all the hydrodynamic modes the density fluctuation is found to yield the main contribution to the memory kernel in the supercooled fluid regime.

Thus the ring collision term reduces only to the density mode contribution and is given by

$$\begin{aligned}
\Gamma_{ll}^R(\mathbf{q}, t) \simeq R_{\rho\rho}^l(\mathbf{q}, t) = \frac{\rho}{m\beta} & \int [d\mathbf{q}'/(2\pi)^3](\hat{\mathbf{q}} \cdot \mathbf{q}')[\hat{\mathbf{q}} \cdot \mathbf{q}'c(\mathbf{q}') + \hat{\mathbf{q}} \cdot (\mathbf{q} - \mathbf{q}') \\
& \times c(\mathbf{q} - \mathbf{q}')][F(\mathbf{q} - \mathbf{q}', t)F(\mathbf{q}', t) \\
& - F_0(\mathbf{q} - \mathbf{q}', t)F_0(\mathbf{q}', t)], \qquad (220)
\end{aligned}$$

This is a well-known expression obtained by many authors [3, 30, 67]. This expression can be derived when the subspace orthogonal to the hydrodynamical variables is constructed only by the bilinear products of the density fluctuation. In the language used by Kadanoff and Swift, we can say that the two density mode contribution to $\Gamma_{ll}(\mathbf{q}, t)$ is given by $R_{\rho\rho}^l$.

Note that the expression for $R_{\rho\rho}^l$ itself depends on the dynamic structure factor. Thus the calculation of $F(q, t)$ and $R_{\rho\rho}^l$ should be performed self-consistently. We shall discuss in the next section that this self-consistency provides a feedback mechanism and leads to a divergence of the dynamic structure factor at zero frequency.

XII.   THEORY OF GLASS TRANSITION

In a supercooled liquid near the glass transition temperature, the self-consistent calculation is the *only way* to explain the anomalies in different dynamical quantities. As mentioned before, the first such self-consistent calculation was done by Geszti to explain the behavior of the viscosity near the glass transition temperature. He had argued that an increase in the viscosity slows down the structural relaxation and thus the relaxation of the density mode. This in turn increases the density mode contribution to the viscosity, $\eta_{s\rho\rho}$

[39]. Thus the feedback mechanism leads to a divergence of the viscosity at a finite nonzero temperature (glass transition temperature).

Detailed study of the dynamic structure factor and related transport coefficients near the liquid–glass transition have been carried out by different authors [30, 34, 68, 69]. Although there is considerable difference in the details of these different treatments, the basic idea remains the same. The essential driving force for the glass transition is the freezing of the density fluctuation near the wavenumber $q \simeq q_m$ where the static structure factor is sharply peaked. Thus, the glass transition is not caused by a small wavenumber infrared singularity, but rather it is a phenomenon where the intermediate wavenumbers are important. As discussed before, there is already a softening of the heat mode near $q \simeq q_m$. Near the glass transition there is a general slowing down of all the dynamical quantities, and this slowing down is most effectively coupled to wavenumbers near q_m. For details we refer to the classical review article by Gotze [3].

In this section we will not discuss the glass transition in every detail. Instead, the merit of the self-consistent calculation of the dynamic structure factor and how it works as a nonlinear feedback mechanism near glass transition will be discussed here.

The expression for the dynamic structure factor has been presented in the previous section, and that in the Laplace frequency plane is given by Eq. (210). When written in the time plane, Eq. (210) provides the following equation of motion for the dynamic structure factor,

$$\ddot{\Phi}(q,t) + \langle \omega_q^2 \rangle \, \Phi(q,t) + \int_0^t dt' \Gamma(q,t-t') \dot{\Phi}(q,t') = 0, \qquad (221)$$

where Φ is the normalized density time correlation function. The time-dependent memory function, $\Gamma(q,t)$, can be rewritten in the following way:

$$\Gamma(q,t) = \Gamma^B(q,t) + \Gamma_R(q,t). \qquad (222)$$

It has been discussed in the previous section that the long-time part in the memory function gives rise to the slow long-time tail in the dynamic structure factor. In the case of a hard-sphere system the short-time part is considered to be delta-correlated in time. In a Lennard-Jones system a Gaussian approximation is assumed for the short-time part. Near the glass transition the short-time part in a Lennard-Jones system can also be approximated by a delta correlation, since the time scale of decay of $\Gamma_R(q,t)$ is very large compared to the Gaussian time scale. Thus the binary term can be written as

$$\Gamma^B(q,t) \simeq \gamma_q \delta(t). \qquad (223)$$

With the above-mentioned approximation for the short-time part of the memory function, Eq. (221) can be rewritten as

$$\ddot{\Phi}(q,t) + \gamma_q \dot{\Phi}(q,t) + \langle \omega_q^2 \rangle \Phi(q,t) + \int_0^t dt' \Gamma^R(q, t - t') \dot{\Phi}(q, t') = 0.$$

(224)

The expression for the long-time part of the memory function is given by Eq. (220):

$$\Gamma^R(\mathbf{q},t) = \frac{\rho}{m\beta} \int [d\mathbf{q}'/(2\pi)^3](\hat{\mathbf{q}} \cdot \mathbf{q}')[\hat{\mathbf{q}} \cdot \mathbf{q}'c(\mathbf{q}') + \hat{\mathbf{q}} \cdot (\mathbf{q} - \mathbf{q}')c(\mathbf{q} - \mathbf{q}')]$$
$$\times [F(\mathbf{q} - \mathbf{q}', t)F(\mathbf{q}', t) - F_0(\mathbf{q} - \mathbf{q}', t)F_0(\mathbf{q}', t)].$$

(225)

As discussed by Kirkpatrick [30], $\Gamma(q)$ can be replaced by $\Gamma(q_m)$ since there is a marked softening near this wavenumber. The maximum contribution from Eq. (225) in the long time comes when both the dynamic structure factors are evaluated near q_m. The Gaussian part of the dynamic structure factor can also be neglected in the asymptotic limit. Thus the long-time part of the memory function now contains the vertex function and a bilinear product of the dynamic structure factors, all evaluated at or near q_m. To make the analysis simpler, the wavenumber dependence of all the quantities are not written explicitly.

With the above-mentioned simplifications, the equation of motion for the density time autocorrelation function reduces to the following form of a nonlinear equation of motion for a damped oscillator:

$$\ddot{\Phi}(t) + \gamma \dot{\Phi}(t) + \langle \omega_q^2 \rangle \Phi(t) + 4\lambda \langle \omega_q^2 \rangle \int_0^t d\tau \Phi^2(\tau) \dot{\Phi}(t - \tau) = 0.$$

(226)

The above equation is an integrodifferential equation that has an unusual structure. Here $\langle \omega_q^2 \rangle^{1/2}$ is the frequency of the free oscillator and γ is the damping constant. The fourth term on the left-hand side of Eq. (226) has the form of the memory kernel, and its strength is controlled by the dimensionless coupling constant λ which contains the contribution from the vertex function.

The rest of the analysis is now carried out, using the formalism presented by Leutheusser [34]. Let us introduce a Fourier–Laplace transform or one-sided Fourier transform,

$$\Phi(z) = \mathcal{L}\{\Phi(t)\} = i \int_0^\infty dt e^{izt}\Phi(t), \qquad \text{Im } z > 0$$

(227)

Note that here z is not the usual Laplace frequency (used in rest of the review), but it is a Fourier frequency. Equation (226) can be written in the Fourier–Laplace space (z) as

$$\Phi(z) = -\cfrac{1}{z - \cfrac{\langle \omega_q^2 \rangle}{z + \eta_l(z)}}, \qquad (228)$$

where

$$\eta_l(z) = i\gamma + 4\lambda \langle \omega_q^2 \rangle \mathcal{L}\{\Phi^2(t)\}. \qquad (229)$$

Equation (228) is the normalized density correlation function in the Fourier frequency plane and has the same structure as Eq. (210), which is the density correlation function in the Laplace frequency plane. $\eta_l(z)$ in Eq. (228) is the memory function in the Fourier frequency plane and can be identified as the dynamical longitudinal frequency. Equation (228) provides the expression of the density correlation function in terms of the longitudinal viscosity. On the other hand, η_l itself is dependent on the density correlation function [Eq. (229)]. Thus the density correlation function should be calculated self-consistently. To make the analysis simpler, the frequency and the time are scaled by $\langle \omega_q^2 \rangle^{1/2}$ and $\langle \omega_q^2 \rangle^{-1/2}$, respectively. As the initial guess for η_l, the coupling constant λ is considered to be weak. Thus η_l in zeroth order is

$$\eta_l(z) = i\gamma. \qquad (230)$$

When Eq. (230) is replaced in Eq. (228), the density correlation function is found to have two simple poles,

$$\Phi(z) = -\frac{a_1}{z + iv_1} - \frac{a_2}{z + iv_2}, \qquad (231)$$

where $v_{1/2} = [\gamma \mp (\gamma^2 - 1)^{1/2}]/2$ and $a_{1/2} = [1 \pm \gamma/(\gamma^2 - 1)^{1/2}]/2$. When Eq. (231) is substituted in Eq. (229), the longitudinal viscosity is given to the first order in λ by the following expression:

$$\eta_l(z) = i\gamma - \cfrac{4\lambda}{z - \cfrac{2}{z + i\gamma - \cfrac{2}{z + 2i\gamma}}}. \qquad (232)$$

Thus the zero-frequency longitudinal viscosity in the first order in λ is given by

$$\eta_l = \gamma + 2\lambda(\gamma + 1/\gamma). \tag{233}$$

When Eq. (233) is compared with Eq. (230), the zero-frequency value of the longitudinal viscosity in first order is found to be larger than its zeroth-order value. This suggests that in every loop of the self-consistent calculation the zero-frequency longitudinal viscosity will increase, which might lead to a divergence of the zero-frequency value of $\eta_l(z)$ and $\Phi(z)$.

It is *not* possible to understand the nature of the divergence just by doing a self-consistent calculation since that implies an infinite loop calculation. An alternative way is to make an ansatz for $\Phi(z)$ and examine its validity. The following ansatz for $\Phi(z)$ was assumed by Leutheusser:

$$\Phi(z) = -\frac{f}{z} + (1 - f)\,\Phi_v(z). \tag{234}$$

In the ansatz, $\Phi(z)$ is written as a combination of a zero-frequency pole contribution with weight f and a remaining part $\Phi_v(z)$ with weight $(1 - f)$, where $\Phi_v(t = 0) = 1$.

Then Eq. (229) can be rewritten and is found to have a similar structure:

$$\eta_l(z) = -4\lambda f^2/z + \eta_{lv}(z), \tag{235}$$

where

$$\eta_{lv}(z) = i\gamma + 8\lambda f(1 - f)\Phi_v(z) + 4\lambda(1 - f)^2/\mathcal{L}\{\Phi_v^2(z)\}. \tag{236}$$

When Eq. (235) is substituted in Eq. (228) and the expression is compared with the ansatz given by Eq. (234), the weight factor f is given by

$$f = (1 + \sqrt{1 - 1/\lambda})/2. \tag{237}$$

Equation (237) shows that when the coupling constant is larger than the critical value $\lambda_c = 1$ the ansatz [Eq. (234)] leads to an acceptable solution. This implies that for $\lambda < 1$, density fluctuations decay to zero for a long time but for $\lambda \geq 1$ they decay to a finite value f. The value of f increases from $f = 1/2$ for $\lambda = 1$ to $f = 1$ for $\lambda \to \infty$. Thus the spectrum of density fluctuation exhibits a delta function peak at zero frequency, with strength f which is the characteristic of a glassy phase. Thus in the glass phase the translational motion is frozen in and the vibrational motion around the arrested position is described by $\Phi_v(z)$.

The above analysis shows that on approaching the glass transition, the slowing down of the density fluctuations is controlled by the increasing longitudinal viscosity, which in turn is coupled via a nonlinear feedback mechanism to the slowly decaying fluctuations. This leads to a divergence of the structural relaxation time at a certain critical coupling constant, λ_c.

As shown by Leutheusser, the above analysis can be extended to show that at glass transition the density fluctuations decay with a long-time power law $\Phi(t) \sim t^{-\alpha}$ with $\alpha = 0.395$. As one approaches the transition, the viscosity is predicted to diverge as $\epsilon^{-\mu}$ and $\epsilon^{-\mu'}$ below and above the transition, respectively. $\epsilon = |1 - \lambda/\lambda_c|$, $\mu = (1 + \alpha)/2\alpha$, and $\mu' = \mu - 1$, where $\mu \simeq 1.7$–2. It is shown by Kirkpatrick [30] that the diffusion coefficient near glass transition goes to zero as ϵ^{μ}.

XIII. RELATION BETWEEN FRICTION AND VISCOSITY IN THE NORMAL LIQUID

In this section the studies of the relation between the friction (ζ) on a tagged solute and the viscosity (η_s) of the medium is presented for neat liquids in the normal regime. The well-known Stokes relation is often used to connect the friction (ζ) on a spherical molecule with the viscosity (η_s) of the medium and is given by

$$m\zeta = C\pi\eta_s R, \qquad (238)$$

where m is the mass of the solute, C is the hydrodynamic boundary condition, and R is the radius of the solute. This connection between friction and viscosity goes beyond the ordinary Stokes relation; even the generalized hydrodynamics describes the frequency (ω)-dependent friction in terms of frequency-dependent viscosity [23].

While the hydrodynamic theory always predicts this near equivalence of the friction and the viscosity, microscopic theories seem to provide a rather different picture. In the mode coupling theory (MCT), the friction on a tagged molecule is expressed in terms of contributions from the binary, density, and transverse current modes. The latter can of course be expressed in terms of viscosity. However, in a neat liquid the friction coefficient is primarily determined *not* by the transverse current mode but rather by the binary collision and the density fluctuation terms [59]. Thus for neat liquids there is no *a priori* reason for such an intimate relation between the friction and viscosity to hold.

A further motivation of this study comes from the following observations. Many chemical dynamic processes, such as nonpolar solvation dynamics [70], can be described in terms of the frequency-dependent viscosity because viscoelastic responses are required to understand the processes involving the rate of change in shape or size of the molecules in liquids [57]. Note that it is the frequency-dependent viscosity which is readily accessible experimentally, whereas the frequency-dependent friction is a theoretical entity. Another place where a knowledge of this interrelationship between $\zeta(\omega)$ and $\eta_s(\omega)$ is required is in understanding the viscosity dependence of activated processes in viscous liquids; this is a subject of much current interest [71]. In the elegant calculation of the frequency-dependent friction by Zwanzig and Bixon [23], the frequency dependence of the viscosity was assumed to be given by the following Maxwell relation:

$$\eta_s(\omega) = \frac{\eta_s}{1 + i\omega\tau_s},$$
(239)

where τ_s is the viscoelastic relaxation time, given by $\tau_s = \eta_s/G_\infty$, where G_∞ is the infinite frequency shear modulus and $\eta_s = \eta_s(\omega = 0)$. From the above expression it is clear that this model assumes only one time scale. On the other hand, recent experimental [72, 73] and computer simulation studies [48, 74] have amply demonstrated that the solvent response is bimodal with at least two widely different time scales describing the response.

Mode coupling theory provides the following rationale for the known validity of the Stokes relation between the zero frequency friction and the viscosity. According to MCT, both these quantities are primarily determined by the static and dynamic structure factors of the solvent. Hence both vary similarly with density and temperature. This calls into question the justification of the use of the generalized hydrodynamics for molecular processes. The question gathers further relevance from the fact that the time (t) correlation function determining friction (the force–force) and that determining viscosity (the stress–stress) are microscopically different.

An attempt has been made to answer the following questions. What is the relation between $\eta_s(t)$ and $\zeta(t)$ at short times? Does the ratio between the two retain a Stokes-like value at all times? And how does the relation behave as a function of frequency? The analysis seems to suggest that if one includes only the binary interaction in the calculation of the time scale of the short-time dynamics, both viscosity and friction exhibits nearly the same time scale. When the triplet dynamics is included, both the responses become slower with the viscosity being affected more than the friction. The time scale of both the responses are of the order of 100 fs. It is shown that both the frequency-dependent viscosity and the friction exhibit a clear bimodal dynamics.

The validity of the Stokes relation has also been investigated from the microscopic point of view, and the following surprising result is obtained. Individually and separately, the ratio of both the bare (binary dominated) and the mode coupling contributions to the friction and the viscosity follows a Stokes-like relation. Contrary to the hydrodynamic picture, it was found that in the case of neat liquids at high density, it is more appropriate to think that the viscosity is being controlled by the diffusion or the friction. This is because in this regime the viscosity is primarily determined by the structural relaxation of the surrounding liquid, which in turn is determined by the diffusion.

A. Time Dependence of Friction and Viscosity

The calculational methods for the friction and the viscosity have been presented in Sections IX and X, respectively. Here we will not describe them but just state the results.

The comparative study between the time-dependent friction and the viscosity at $\rho^* = 0.844$ and $T^* = 0.728$ is depicted in Fig. 3. In this figure both the viscosity and the friction have been normalized to unity at $t = 0$ by their respective initial values. This figure has several interesting features. Both friction and viscosity exhibit a pronounced ultrafast Gaussian decay which accounts for almost 90% of the total relaxation. The Gaussian time constants

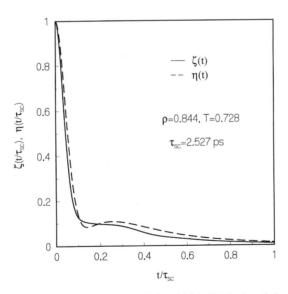

Figure 3. The time dependence of the friction $\zeta(t)$ (solid line) and the viscosity $\eta_s(t)$ (dashed line), for a Lennard-Jones liquid near its triple point ($\rho^* = 0.844$ and $T^* = 0.728$). The friction and the viscosity are normalized by their initial values to facilitate comparison of the dynamics. The time is scaled by the usual dimensionless time, $\tau_{sc} = (m\sigma^2/k_BT)^{0.5}$. For more details see the text.

are equal to 130 fs for the friction and 160 fs for the viscosity. The second interesting aspect is that both quantities exhibit slow long-time decay and are also comparable.

It is worthwhile to discuss the relative contributions of the binary and the three-particle correlations to the initial decay. If the triplet correlation is neglected, then the values of the Gaussian time constants are equal to 89 fs and 93 fs for the friction and the viscosity, respectively. Thus, the triplet correlation slows down the decay of viscosity more than that of the friction. The greater effect of the triplet correlation is in accord with the more collective nature of the viscosity. This point also highlights the difference between the viscosity and the friction. As already discussed, the Kirkwood superposition approximation has been used for the triplet correlation function to keep the problem tractable. This introduces an error which, however, may not be very significant for an argon-like system at triple point.

B. Frequency Dependence of Viscosity: Comparison with Maxwell Relation

Figure 4 depicts the imaginary part of the frequency-dependent viscosity which clearly demonstrates the bimodality of the viscoelastic response. In the same figure the prediction from the Maxwell's relation have also been plotted. In the latter the relaxation time τ_s is calculated by the well-known

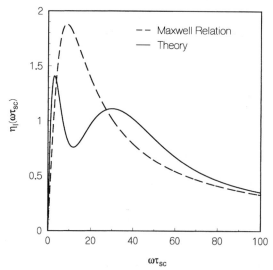

Figure 4. The imaginary part of the calculated viscosity is plotted as a function of the Fourier frequency at the triple point (solid line). Also shown is the prediction of the Maxwell viscoelastic model (dashed line), given by Eq. (239). The viscosity is scaled by $\sigma^2/\sqrt{(mk_BT)}$ and the frequency is scaled by τ_{sc}^{-1}, where $\tau_{sc} = [m\sigma^2/k_BT]^{1/2}$. For more details see the text.

expression, $\tau_s = \eta_s(z = 0)/G_\infty$, where both $\eta_s(z = 0)$ and G_∞ are calculated using MCT. It shows that the Maxwell relation produces only one peak at low frequency and provides inadequate description at higher frequencies. Another important aspect of this graph is that the simple Maxwell relation fails to describe adequately even the low frequency peak.

C. Stokes Relation Revisited

As emphasized before, the hydrodynamic derivation (based on the contribution of the current mode alone [75]) of the relation between the friction and the viscosity has no validity in the case of neat liquids (where the tagged molecule is one of the solvent molecules). On the other hand, the experiments [76], the computer simulations [77], and the MCT calculations presented here all show that the ratio of friction to viscosity at high density almost always lies between 4π and 6π even for a neat liquid. Therefore, it is imperative to analyze the cause of apparent validity of the Stokes relation in greater depth.

In the following a semiquantitative argument is presented on the recovery of the hydrodynamic boundary condition from microscopic considerations.

An analysis of the relevant integrals shows that the dominant contribution of the density mode to the viscosity and the friction comes from intermediate length scale ($8 \geq q\sigma \geq 3$). That is, more than 90% of the contribution comes from a region surrounding the sharp first peak of the static structure factor—that is, around $q\sigma = 2\pi$. At these values of the wavenumber, the dynamic structure factor is well-determined by the following simple mean field expression:

$$F(q,t) = S(q)\exp\left(\frac{-Dq_m^2 t}{S(q_m)}\right), \tag{240}$$

where q_m is the position of the first peak of the static structure factor. Here D is the self-diffusion coefficient that is determined from the velocity autocorrelation function after the frequency-dependent friction and the mean square displacement are calculated self-consistently. This self-consistency scheme is presented in Section IX. The diffusion coefficient thus calculated is then used in the above expression to provide a correct intermediate scattering function to be used in the calculation of the viscosity.

Further simplification can be made by using a simple prescription for the wavenumber dependence of the structure factor, as shown by Balucani et al. [78]. The above prescription provides fairly accurate values for the zero-frequency friction and the viscosity.

Therefore, it is clear from the above discussion that the collective contribution to the viscosity is dominated by the *structural relaxation* which in turn is determined by the rate of diffusion. Thus, it is more appropriate to consider the viscosity of the medium as being determined by the diffusion. This is of course a matter of perspective.

There is, however, an even more interesting aspect. Using the results of Balucani [79], it can be shown that the initial value of the viscosity and the friction are related approximately by

$$\frac{m\zeta(t=0)}{\eta_s(t=0)\mathcal{R}} \approx \frac{20}{\rho^*}. \tag{241}$$

For $\rho^* = 0.844$ and $T^* = 0.72$, Eq. (241) gives a value of the ratio equal to 23.6966.

It is already found that the decay of the normalized viscosity is slightly slower than that of the friction and the ratio of the time constants is 160/124. Thus, the ratio of the contribution from the bare part to the zero-frequency friction to that of the viscosity is equal to $23.6966 \times 124/160$, which is equal to 18.364. Therefore, the ratio of the bare part of the zero-frequency friction to that of viscosity is nearly identical to 6π.

It is to be noted that in the above discussion although the numerical values of the prefactor is close to 6π, it does not in any way imply the stick boundary condition. The above calculation is based only on microscopic considerations; on the other hand, the boundary condition can only be obtained by studying the somewhat macroscopic velocity profile of the solvent. Thus, the main point here is that in the high-density liquid regime, the ratio of the friction to the viscosity attains a constant value independent of the density and the temperature.

It is now interesting to discuss the simulated values of this ratio at high density. For an argon system near the triple point at $\rho = 0.021\,\text{Å}^{-3}$ and $T = 86.5\text{K}$ the ratio is 4.7π. At $\rho = 0.021\,\text{Å}^{-3}$ and $T = 95\text{K}$ the ratio is 5.1π. In computing the above ratios the friction is obtained from the Einstein relation using the known value of the diffusion coefficient [77]. It is perhaps fair to allow an uncertainty of 5–10% in the determination of this ratio by either theory or simulation.

XIV. RELATION BETWEEN FRICTION AND VISCOSITY IN THE SUPERCOOLED LIQUID

In the previous section we have discussed the relation between the time- and frequency-dependent friction and viscosity in the normal liquid regime. The study in this section is motivated by the recent experimental (see Refs. 80–87) and computer simulation studies [13, 14, 88] of diffusion of a tagged particle in the supercooled liquid where the tagged particle has nearly the same size as the solvent molecules. These studies often find that although the friction/diffusion can be connected to the viscosity in the normal liquid regime, there is an apparent decoupling between them in the supercooled regime (see Refs. 13, 14, and 80–88). The measured diffusion coefficient becomes increasingly larger than the value predicted by the Stokes–Einstein (SE) relation (see Refs.13, 14, and 80–88), which connects the diffusion of a solute to the viscosity of the medium and is given by

$$D = \frac{k_B T}{C \pi \eta_s R}, \tag{242}$$

where D is the diffusion coefficient of the solute, k_B is Boltzmann's constant, and T is the absolute temperature. Thus according to the SE relation, D should follow the temperature dependence of T/η_s. In separate studies, Sillescu and co-workers [82] and Cicerone and Ediger [84] have found that the translational diffusion in supercooled o-terphenyl (OTP) has significantly weaker temperature dependence than that predicted by the SE relation. Cicerone and Ediger have reported that for probes which are of the same size as OTP molecules, the translational diffusion near glass transition occurs two orders of magnitude faster than that expected based on rotation times [84].

Conceptually, one can envisage several different microscopic scenario for the observed decoupling. The first scenario is that as the liquid is supercooled, the hopping mode of diffusion becomes dominant over the conventional hydrodynamic mode. The fact that hopping can indeed take place in a supercooled liquid has been confirmed by molecular dynamics simulations [13, 14], although it is yet to be observed in experiments. If the hopping indeed becomes the dominant mechanism of mass transport in the supercooled liquid, then one can find a way to explain the decoupling because the hopping can couple only to the high-frequency motion of the solvent. Another currently popular conjecture is the existence of dynamic inhomogeneity in the systems where different regions are assumed to have rather different relaxation times. While the diffusion is fast through fluid-like regimes, it is slow in the frozen, solid-like regions. To explain several experimental studies, Ediger and co-workers have suggested the existence of such spatial inhomogeneity in liquids near glass transition [81, 84–87]. Cicerone and Ediger [85] have used a photobleaching technique to observe the rotational dynamics of dilute probe molecules in supercooled OTP. They have shown that the nonexponential rotational relaxation of the probe molecules are at least partly due to the presence of spatial heterogeneity in the host dynamics. These heterogeneities are long-lived near glass transition. In a separate study [84] these authors have used this concept of spatial inhomogeneity to explain the enhanced diffusion of probes of same size as the OTP molecules, near glass transition.

Thus the existence of such inhomogeneity seems to be necessary to understand the decoupling of the self-diffusion from the viscosity for solutes having the same size as that of the solvent. The existence of such inhomogeneity has also been suggested by Tarjus and Kivelson [89] from their computer simulation studies.

It is found from the present analysis that the straightforward application of the mode coupling theory fails to account for the decoupling in a neat liquid. In order to explain this decoupling, an extended MCT has been developed which takes into account the inhomogeneity of the medium. The extended semiphenomenological theory seems to explain the decoupling of the same size solute rather well. The above results are in excellent agreement with the experimental results of Heuberger and Sillescu [83].

A. Expression for the Dynamic Structure Factor in the Supercooled Regime

In Section XI we discussed the calculational method of the dynamic structure factor in the supercooled regime. We also discussed that the memory function Γ_{ll} needs to be calculated self-consistently with the dynamic structure factor itself. Near the glass transition, the dynamic structure factor is expected to diverge. This leads to an infinite loop numerically formidable calculation.

Here we present a different prescription to calculate the dynamic structure factor or the intermediate scattering function in the supercooled regime. This is a quantitative approach based on the basic result of the mode coupling theory. The effect of the mode coupling term in the intermediate scattering function is written in a simpler way by the following expression:

$$F_{MC}(q,t) = F(q,t) + F_{sing}(q,t). \tag{243}$$

The normal part, $F(q, t)$, is still given by the viscoelastic model [Eq. (143)] while the singular part, $F_{sing}(q, t)$, is given by the mode-coupling theory. Note that it is this singular part which at glass transition becomes a nondecaying time-independent but q-dependent term which is related to the Debye–Waller factor of the glass. The singular term is therefore approximated by the following form [90]:

$$F_{sing}(q,t) = A_q \exp((-t/\tau_{MC}(q))^\beta) \tag{244}$$

with $\tau_{MC}(q) = cS(q)/D_0 q^2$, where D_0 is the diffusion coefficient of the solvent. This form of the singular part of the structure factor has been successful in fitting the intermediate scattering function obtained from the neutron scattering experiments [90]. The prefactor A_q is known to be weakly wavenumber-dependent; in this work it is assumed to be equal to $0.3S(q)$ [90]. This is, however, not a serious approximation. The value of the exponent (β) is experimentally found to be close to 0.58. In the calculation of $\tau_{MC}(q)$, the constant c is chosen in such a way that the whole $F_{MC}(q,t)$ represents the behavior of $F(q,t)$ in normal liquid regime.

The most important prediction of the mode coupling theory is the temperature or the density dependence of the relaxation time, $\tau_{MC}(q)$. MCT predicts that this relaxation time grows as a power law as the glass transition is approached (from the supercooled liquid side). This is because the diffusion coefficient D_0 of the liquid goes to zero in the following fashion:

$$D_0(T) = D_0(T_0)(T - T_g)^\gamma \qquad (245)$$

or

$$D_0(\rho) = D_0(\rho_0)(\rho_g - \rho)^\gamma, \qquad (246)$$

where γ is an exponent predicted to have a value equal to 1.8 for Lennard-Jones-like systems [90]. T_0 and ρ_0 are the temperature and density below which the mode coupling effects are negligible.

As already mentioned, many of the predictions of the mode coupling theory, such as the power-law divergence of the relaxation time and the nonexponential relaxation functions, have been verified experimentally. This is not to say that the theory does not have difficulties in explaining experimental results at glass transition [91]. One important limitation at present is that the theory fails to predict the correct transition temperature (or density). Another criticism of the theory is its failure to satisfactorily describe the emergence of hopping transport at low temperatures near the glass transition. The primary concern here is to study the relaxations at temperatures (or densities) away from the glass transition point itself. For this purpose, the use of Eqs. 244 and 246 is reasonable because they describe the emergence of slow density relaxation properly.

In determining the dynamic structure factor, the value of $D_0(T_0)$ is needed to be specified. Here the decoupling is studied as a function of the change of density. A noticeable long-time tail in density relaxation appears only very near to the glass transition line. This makes the choice of T_0 or ρ_0 rather easy. It is found that, for the reduced temperature, $T^* = 0.8$, the reduced density ρ_0^* is 0.91.

It should be stressed at this point that the construction of the dynamic structure factor outlined above is based on the basic results of the mode coupling theory.

B. Effect of the Dynamic Structure Factor
on Friction and Viscosity

In the supercooled liquid regime it is the structural relaxation which contributes the most to the friction and viscosity. Thus we can neglect the contribution from the transverse and longitudinal current.

The calculation of the friction and the viscosity are performed in a similar way as has been discussed in Sections IX and X, respectively. The only difference is that instead of the viscoelastic expression for $F(q, t)$, Eq. (243) has been used.

With the above-described formulation of the dynamic structure factor the viscosity and the friction are calculated in the supercooled liquid. It is found that long-time tail of the dynamic structure factor affects both the viscosity and the friction, leading to their divergence near the glass transition. Thus for a homogeneous liquid the friction and viscosity are coupled even in the supercooled liquid regime. The calculated viscosity is plotted against density in Fig. 5, which shows a divergence near the glass transition density. The divergence of viscosity in supercooled liquids near the glass transition has also

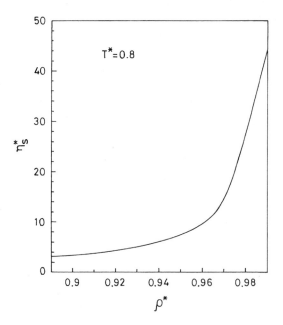

Figure 5. The theoretical result for viscosity (η) as a function of reduced density (ρ^*) near the glass transition point, at reduced temperature $T^* = 0.8$. ρ^* has been varied from 0.89 to 0.99, where the glass transition density, ρ_g^*, is 1.0. The plot shows that the viscosity diverges near the glass transition point. For more details see the text.

been studied by other authors using a different theoretical model. Dattagupta and Turski have shown that the local orientational ordering leads to liquid–solid freezing [92]. Later Dattagupta showed that the mismatch in the orientation of two locally ordered neighboring region in supercooled liquid causes an internal stress which leads to a large enhancement of viscosity [93].

C. Effect of Inhomogeneity on Decoupling of Diffusion / Friction from Viscosity

In the previous section it was stated that a straightforward application of the MCT cannot explain the decoupling of diffusion/friction from viscosity in the supercooled liquid. As discussed before, there are two explanations suggested for this decoupling in neat liquids. One of them explains diffusion in supercooled liquid in terms of a hopping mechanism, whereas the other explains it in terms of inhomogeneity in the solvent. As argued by Bagchi [94] earlier, the hopping can decouple from the bulk viscosity because the former is intrinsically a high-frequency phenomenon. Such hopping mechanisms of mass transport have indeed been observed for binary mixtures of spherical atoms only very near to the glass transition temperature [13, 14]. On the other hand, Sillescu and coworkers [83] have observed the decoupling rather far from the glass transition temperature. So the only other proposed explanation is in terms of the random distribution of quenched inhomogeneity in supercooled liquids. The implication of this hypothesis is discussed below.

In the inhomogeneous model it is assumed that in the deeply supercooled liquid, there exist regions with different degrees of fluidity [89]. While all these regions contribute with respective weights in determining the viscosity, the diffusion is determined more by the fluid regions [84]. At the heart of this inhomogeneity hypothesis lies the size of a typical domain. One may make an estimate of the length of the domain by comparing the rotational relaxation times with the tracer diffusion studies. Cicerone et al. [86] suggest that such a domain, if it exists, must be less than 2.5 nm. Tarjus and Kivelson [89] have recently argued that such a size is reasonable. The emergence of such length was predicted previously by Kirkpatrick [30]. A semiquantitative calculation of the diffusion considering an inhomogeneous solvent is discussed here. It is shown that such a decoupling can indeed be understood by combining the inhomogeneity hypothesis with MCT.

Domains having different degrees of fluidity may appear as a consequence of density fluctuations at intermediate length scales. Note that in an ergodic system the probability of fluctuation of the wavenumber-dependent density, $\rho_{\mathbf{q}}$, is given by the following well-known expression:

$$P(\rho_{\mathbf{q}}) \sim \exp\left(-\frac{\rho_{\mathbf{q}}\rho_{-\mathbf{q}}}{S(\mathbf{q})}\right). \tag{247}$$

In a supercooled liquid the static structure factor $S(\mathbf{q})$ is nearly zero for small wavenumbers (because of very low compressibility); hence density fluctuation can take place only at intermediate wavenumbers where $S(\mathbf{q})$ is large. For example, a fluctuation at $q\sigma = 1$ would imply a length scale approximately comparable to $2\pi\sigma$, which is of the same order as estimated by Cicerone et al. [86].

Let us consider that the solvent has nonoverlapping domains that have distinct static and dynamic characteristics. The diffusion in such a system can be considered as a random walk in a random environment formed by the independently relaxing domain [95]. Hence the diffusions in three dimensions is the weighted average of the diffusion in all these different domains [84]. The average diffusion D can then be written as

$$D = \sum_{i=1}^{N} P_i D_i, \tag{248}$$

where D_i is the local diffusion coefficient in domain i and P_i is the probability distribution of the domain i. The calculation of the diffusion in these different noninteracting domains can be performed within MCT by calculating the friction in each of these domains. This is reasonable because the friction is determined primarily by the local effects and can thus be calculated independently for each domain. In terms of averaging over the friction in different domains the expression of diffusion can be written as

$$D = \frac{k_B T}{m} \sum_{i=1}^{N} P_i \frac{1}{\zeta_i} = \frac{k_B T}{m} \left\langle \frac{1}{\zeta} \right\rangle, \tag{249}$$

where N is the number of nonoverlapping, distinct domains and ζ_i is the friction in the domain i.

As mentioned earlier, the most obvious characteristic of these domains is their density. Near the glass transition, even a small difference in density makes a large difference in the dynamic properties. In the theoretical discussions of glassy liquids, the most important microscopic dynamic quantity is the intermediate scattering function $F(q,t)$, which is related to the dynamic structure factor $S(q,\omega)$ by a Fourier transform. $F(q,t)$ can be drastically different for a solid-like and a liquid-like domain. As shown by the calculations presented in the previous section, a very slow long-time tail appears in the intermediate scattering function as the glass transition is approached; such a tail is absent for the fluid-like domain. In fact, this slow long-time tail becomes nondecaying at glass transition to give rise to the elastic peak in

$S(q, \omega)$. As stated earlier, such a slow tail in $F(q, t)$ can significantly increase the friction. In calculating the friction, the effect of the inhomogeneity on the static and dynamic correlations is to be considered only for intermediate and long wavenumbers. The small wavevector limit probes the collective dynamics of the solvent, and hence it is the average friction which contributes in this region. It is fair to assume that $q\sigma \simeq 1$ separates small from intermediate wavenumber regime.

Let us now turn to the calculation of viscosity in an inhomogeneous solvent. This calculation is different from (almost inverse of) diffusion. The viscosity of the solvent is the weighted average of the local viscosity in the different domains. Hence it can be written as

$$\eta_s = \frac{1}{N} \sum_{i=1}^{N} P_i \eta_{si}, \qquad (250)$$

where η_{si} is the viscosity of the domain i. For each domain the local viscosity is calculated considering the dynamic inhomogeneity of that domain. In a similar way as in the case of friction, a particular domain is characterized primarily by the dynamic correlation $F(q, t)$. While calculating the viscosity the integration over the wavenumber is again divided into two separate regions as done in the case of friction. Again, the effect of inhomogeneity is considered for intermediate and long wavenumbers, whereas in the case of small wavenumbers it is the average viscosity which contributes.

As discussed, it is most appropriate to focus attention directly on the intermediate scattering function $F(q, t)$, which is to be considered as the key random variable. If $P[F(q, t)]$ is the normalized probability distribution of the functional $F(q, t)$, then the modified MCT expression for friction takes the following form:

$$\left\langle \frac{1}{\zeta} \right\rangle = \int d[F(q, t)] P[F(q, t)] \frac{1}{\zeta_B + R_{\rho\rho}[F(q, t)]}. \qquad (251)$$

The major contribution to $R_{\rho\rho}$ comes from the intermediate wavenumber—that is, from immediate neighbors. In this wavenumber regime, $F(q, t)$ will have the characteristic of either the solid- or the liquid-like domain. Thus the calculated friction will have only two distinct values, that of the solid-like domain and that of the liquid-like domain. The expression of the self-diffusion is now given by

$$D = \frac{k_B T}{m} \left(\frac{\mathbf{A}}{\zeta_s} + \frac{1 - \mathbf{A}}{\zeta_l} \right), \qquad (252)$$

where ζ_s is the friction in the solid-like domains and ζ_l is the friction in the liquid-like domains. **A** is the probability of having solid-like domains and $(1 - \mathbf{A})$ is the probability of having liquid-like domains. Note that Eq. (252) represents the quenched two-step model of Zwanzig [95].

Similarly, because of the existence of two distinct domains, the viscosity can be written as

$$\eta_s = \mathbf{A}\eta_{ss} + (1 - \mathbf{A})\eta_{sl}, \tag{253}$$

where η_{ss} is the viscosity in the solid-like region and η_{sl} is the viscosity in the liquid-like region. Both are calculated in the same procedure as discussed in Section X but with different $F(q, t)$ values.

The calculated self-consistent MCT values of the viscosity and the friction for the solid-like region are found to be about two orders of magnitude larger than those of the same for the liquid-like domain, and the expressions (252) and (253) reduce to

$$D \approx \frac{k_B T}{2m\zeta_l} \tag{254}$$

and

$$\eta_s \approx \frac{\eta_{ss}}{2}, \tag{255}$$

which gives a large decoupling of diffusion from viscosity, by about two orders of magnitude. Hence even if the average viscosity of the solvent diverges, the measured diffusion coefficient need not become zero, because its value is primarily determined by the friction in the liquid-like regions. It should be stated here that this analysis essentially quantifies the physical picture of decoupling proposed by Cicerone and Ediger [84].

D.   Comments on the Above Analysis

1. There are two assumptions at the heart of the inhomogeneity hypothesis. First, of course, is the existence of the domains with different dynamic properties. Second, the dynamic properties of the domain are sensitive to small variations of temperature or density, because only small fluctuations in density are allowed at such high density. Such sensitivity to small change in density (or temperature) can happen only in the supercooled liquid, not too far from the glass transition temperature.

2. The analysis presented here requires the existence of distinct domains of certain dimension to explain the decoupling. What is crucial is the

bifurcation of the inhomogeneous medium into homogeneous regimes which can sustain liquid-like or solid-like dynamics. Since the dimension of these domains are predicted not to be large, it is not necessary to have an emergence of any macroscopic length.

XV. POWER LAW MASS DEPENDENCE OF DIFFUSION

For future theoretical developments in the field of transport properties of binary and higher-order mixtures, the simplest case seems to study the influence of the variation in mass of one of the species on the transport properties. Without a full understanding of this pure mass effect on the transport properties, it is not possible to analyze the effect of the translational–rotational coupling in real molecules. Toward this goal the simplest system that can be considered is a binary mixture at infinite solute dilution where the effect of the solute–solvent mass ratio on the solute diffusion can be studied.

Although there has not been much theoretical work other than a quantitative study by Hynes et al [58], there are some computer simulation studies of the mass dependence of diffusion which provide valuable insight to this problem (see Refs. 96–105). Alder et al. [96, 97] have studied the mass dependence of a solute diffusion at an infinite solute dilution in binary isotopic hard-sphere mixtures. The mass effect and its influence on the concentration dependence of the self-diffusion coefficient in a binary isotopic Lennard-Jones mixture up to solute–solvent mass ratio 5 was studied by Ebbsjo et al. [98]. Later on, Bearman and Jolly [99, 100] studied the mass dependence of diffusion in binary mixtures by varying the solute–solvent mass ratio from 1 to 16, and recently Kerl and Willeke [101] have reported a study for binary and ternary isotopic mixtures. Also, by varying the size of the tagged molecule the mass dependence of diffusion for a binary Lennard-Jones mixture has been studied by Ould-Kaddour and Barrat by performing MD simulations [102]. There have also been some experimental studies of mass diffusion [106–109].

The above-mentioned computer simulation and experimental studies have addressed various aspects of mass dependence, but they all show that the self-diffusion coefficient of a tagged molecule exhibits a weak mass dependence, especially for solutes with size comparable to or larger than the size of the solvent molecules. Sometimes this mass dependence can be fitted to a power law, with a small exponent less than 0.1 [99]. This weak mass dependence has often been considered as supportive of the hydrodynamic picture. In hydrodynamics the diffusion of a solute is conventionally described by the well-known Stokes–Einstein (SE) relation, which predicts that the diffusion is totally independent of the mass of the solute. Kinetic theory, on the other hand, predicts a completely opposite picture. For example, the Enskog theory predicts a square root mass dependence, as given by the expression

$$D_E = \frac{3}{8\rho\sigma^2 g(\sigma)} \sqrt{\frac{k_B T}{2\pi\mu}}, \tag{256}$$

where μ is the reduced mass of the solute–solvent pair, σ is the diameter of the solvent, and $g(\sigma)$ is the value of the radial distribution function at contact. Equation (256) predicts too strong a mass dependence which is not observed in computer simulation (see from Refs. 96–105) and experimental studies [106].

According to the SE relation, the product $D\eta$ should remain constant for systems having particles of same size and studied at the same temperature. Recent studies [102, 104, 105] have found that the SE relation does not hold when the mass of the particles are changed. Walser et al. [104] have performed MD simulations of water molecules with different mass and different molecular mass distribution. They have shown that although the viscosity increases and the diffusion decreases with mass, the product of the two does not remain constant. They have found that the product $D\eta$ is not correlated with the molecular mass, but it is correlated for those systems with the same mass distribution. Thus, while the mass dependence is not as strong as predicted by the kinetic theory, it is also not totally negligible.

Thus neither the kinetic theory nor the hydrodynamic theory can explain the correct mass dependence of diffusion. Clearly, the hydrodynamic and the kinetic theories describe two opposite limits of diffusion. While the first one assumes the validity of the Navier–Stokes hydrodynamics at the molecular length scales, the second one tends to describe diffusion only in terms of binary collisional dynamics. While hydrodynamics assumes that the diffusion occurs via the coupling of the solute velocity with only the collective transverse current mode of the solvent, the Enskog kinetic theory neglects coupling of the solute motion to all the hydrodynamic modes. In both these pictures the diffusion due to the structural relaxation of the surrounding solvent is totally neglected. The more recent mode coupling theory (MCT) seems to interpolate between the two limits and takes into account the contributions of the structural relaxation.

A mode coupling theoretical study seems to provide an accurate description of the dependence of the diffusion coefficient of the solute on mass. A power law mass dependence of the diffusion is obtained which is in good agreement with the simulation results of Bearman and Jolly [99]. The value of the exponent is found to be equal to 0.099. The reason for the weak power law dependence is that the mass dependence enters largely through the binary term whose contribution is small in dense liquids. In addition, the contribution of the density term moves in the opposite direction when the mass is increased, thus further weakening the effects of the binary term.

A.   Theoretical Formulation

The system studied here consists of one solute molecule of mass M and the N solvent molecules, each of mass m. The pair potential for the solvent–solvent pair and the solute–solvent pair is assumed to be given by the simple Lennard-Jones 12-6 potential.

The mode coupling expression for the frequency-dependent friction as presented in Section IX is given by

$$\frac{1}{\zeta(t)} = \frac{1}{\zeta^B(t) + R_{\rho\rho}(t)} + R_{TT}(t), \qquad (257)$$

where $\zeta^B(t)$ is the binary part of the friction, $R_{\rho\rho}(t)$ is the friction due to the coupling of the solute motion to the collective density mode of the solvent, and $R_{TT}(t)$ is the contribution to the diffusion (inverse of friction) from the transverse current mode of the solvent.

The expression of the binary friction $\zeta^B(t)$ for different solute–solvent mass ratio is given by

$$\zeta^B(t) = \omega_{012}^2 \exp(-t^2/\tau_\zeta^2), \qquad (258)$$

where ω_{012} is now the Einstein frequency of the solute in presence of the solvent and is given by

$$\omega_{012}^2 = \frac{\rho}{3M} \int d\mathbf{r} g(r) \nabla^2 v(r), \qquad (259)$$

here $g(r)$ is the radial distribution function.

In Eq. (258), the relaxation time τ_ζ is determined from the second derivative of $\zeta^B(t)$ at $t = 0$ and is given exactly by

$$\omega_{012}^2/\tau_\zeta^2 = (\rho/6M\mu) \int d\mathbf{r} (\nabla^\alpha \nabla^\beta v(\mathbf{r})) g(r) (\nabla^\alpha \nabla^\beta v(\mathbf{r}))$$
$$+ (1/6\rho) \int [d\mathbf{q}/(2\pi)^3] \gamma_d^{\alpha\beta}(\mathbf{q})(S(q) - 1)\gamma_d^{\alpha\beta}(\mathbf{q}), \qquad (260)$$

where summation over repeated indices is implied. μ is the reduced mass of the solute–solvent pair. Here $S(q)$ is the static structure factor. The expression for $\gamma_d^{\alpha\beta}(\mathbf{q})$ is written as a combination of the distinct parts of the second moments of the longitudinal and transverse current correlation functions $\gamma_d^l(\mathbf{q})$ and $\gamma_d^t(\mathbf{q})$, respectively:

$$\gamma_d^{\alpha\beta}(\mathbf{q}) = -(\rho/M)\int d\mathbf{r}\exp(-i\mathbf{q}\cdot\mathbf{r})g(\mathbf{r})\nabla^\alpha\nabla^\beta v(\mathbf{r})$$
$$= \hat{q}^\alpha\hat{q}^\beta\gamma_d^l(\mathbf{q}) + (\delta_{\alpha\beta} - \hat{q}^\alpha\hat{q}^\beta)\gamma_d^t(\mathbf{q}), \qquad (261)$$

where $\gamma_d^l(\mathbf{q}) = \gamma_d^{zz}(\mathbf{q})$ and $\gamma_d^t(\mathbf{q}) = \gamma_d^{xx}(\mathbf{q})$.

The expression for $R_{\rho\rho}(t)$ for different solute–solvent mass ratio can be written as

$$R_{\rho\rho}(t) = \frac{\rho k_B T}{M}\int [d\mathbf{q}'/(2\pi)^3](\hat{q}\cdot\hat{q}')^2 q'^2[c(q')]^2[F^s(q',t)F(q',t)$$
$$- F_0^s(q',t)F_0(q',t)]. \qquad (262)$$

Similarly, the expression for $R_{TT}(t)$ is given by

$$R_{TT}(t) = \frac{1}{\rho}\int [d\mathbf{q}'/(2\pi)^3][1 - (\hat{q}\cdot\hat{q}')^2][\gamma_d^t(q')]^2\omega_{012}^{-4}[F^s(q',t)C_{tt}(q',t)$$
$$- F_0(q',t)C_{tt0}(q',t)]. \qquad (263)$$

The solute dynamic variables required to calculate the density and the current contribution to the friction are the inertial part of the self-intermediate scattering function, $F_0^s(q,t)$, given by

$$F_0^s(q,t) = \exp\left(-\frac{k_B T}{M}\frac{q^2 t^2}{2}\right) \qquad (264)$$

and the self-intermediate structure factor, $F^s(q,t)$. Assuming a Gaussian approximation, the expression for $F^s(q,t)$ can be written as

$$F^s(q,t) = \exp\left(-\frac{q^2\langle\Delta r^2(t)\rangle}{6}\right), \qquad (265)$$

where $\langle\Delta r^2(t)\rangle$ is the mean square displacement (MSD) which is calculated self-consistently with the frequency-dependent friction through the iterative scheme described in Section IX.

The expression for the solvent dynamical variables in Eqs. (262) and (263) remains the same as presented in Section IX.

Note that in the above expressions, the mass of the solute enters in a complex fashion. First, it enters in the binary friction—even here the contribution is more complex than what was envisaged in the Enskog theory. The mass also enters in the collective contributions.

B. Results

The mass dependence of the solute diffusion has been studied at $\rho^* = 0.844$ and $T^* = 0.728$. The diffusion coefficient of the solute is found to have a weak mass dependence. The diffusion is found to decrease as the mass of the solute is increased. In Fig. 6, both the binary and density term contribution to the total friction is plotted against the mass ratio. It is found that the current term contribution remains small and almost unaltered over the whole range of solute-solvent mass ratio studied here.

The same plot also shows that the binary part of the friction increases slowly and monotonically with the solute mass. On the other hand, the density term is first found to decrease for solutes almost twice as massive as the solvent and then it increases with the mass of the solute. The reason behind this initial decrease of the density term with the mass of the solute is the following. The maximum contribution from the density term to the total friction is around $q\sigma = 2\pi$. Now at this wavenumber, the time scale of the short-time collective motion of the solvent ($F_0(q, t)$) is larger than the time scale of the

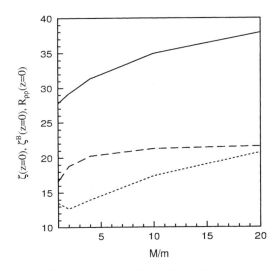

Figure 6. The total friction (represented by solid line), the binary contribution to the friction (represented by long dashed line), and the density contribution to the friction (represented by the short dashed line) plotted against the solute–solvent mass ratio at $\rho^* = 0.844$ and $T^* = 0.728$. The friction is scaled by τ_{sc}^{-1}. For more details see the text.

inertial motion of the solute of the same mass. As the mass of the solute is increased, the time scale of its inertial motion increases and thus $F_0(q,t)F_0^s(q,t)$ increases until the time scale of the inertial motion of the solute becomes larger than the time scale of the short-time collective dynamics of the solvent. It is found that until the solute–solvent mass ratio is below 2, the inertial time scale of the solute remains smaller and $F_0(q,t)F_0^s(q,t)$ increases with the mass of the solute. Now the increase in the value of the product, $F_0(q,t)F_0^s(q,t)$, decreases the contribution from the density term. Thus the contribution from the density term is found to initially decrease with the solute–solvent mass ratio and then increase with it. Though the density term decreases initially, the total friction is found to always increase with the mass of the solute. The initial increase is a little slower due to the opposite effect of the solute mass on the density and the binary term.

The most interesting result suggested by this theoretical investigation is the power law dependence of the solute diffusion on mass as has also been observed in computer simulation studies [99]. The power law dependence is clearly manifested in Fig. 7, where $\ln(D_1/D_2)$ is plotted against $\ln(M/m)$. Here D_1 is the diffusion of the solvent and D_2 is the diffusion of the solute. The slope of the line is found to be 0.099. This seems to suggest a weak mass

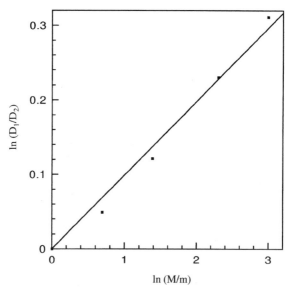

Figure 7. Plot of $\ln(D_1/D_2)$ versus $\ln(M/m)$ at $\rho^* = 0.844$ and $T^* = 0.728$, where D_1 is the self-diffusion of the solvent and D_2 is that of the solute. M and m are the masses of the solute and the solvent, respectively. The squares are the calculated values, and the solid line is the linear fit. The slope of the straight line is 0.099. The plot shows a power law mass dependence of the solute diffusion. The slope of the plot suggests that this mass dependence is weak.

dependence of the solute diffusion, which is in agreement with the MD simulation results.

This mass dependence study can be extended to investigate the dependence of the relaxation processes in proteins on the solvent viscosity. Experimentally, this has been studied by changing the molecular composition of the solvent [110, 111], thus changing the solute–solvent mass ratio. Hence the theoretical analog of this experimental study will be to investigate the dependence of the relaxation processes in proteins on solute–solvent mass ratio.

Another important problem in this area is the much stronger mass dependence of the viscosity, observed in simulations [104]. The same effect is observed between ordinary and heavy water; an understanding of this behavior is still awaited. Because the difference between water and heavy water can at least partly be modeled by using different interaction energy parameter, ϵ, one can attribute this anomaly partly to the dynamics and partly to the statics. Also note that to investigate the mass dependence of viscosity we need to change the mass of the solvent while keeping the solute mass fixed, a problem that has not been addressed here. Since the MCT expressions are asymmetric with respect to the solute and solvent mass, change in the solvent mass might have a stronger effect leading to a stronger mass dependence of the viscosity.

XVI. SIZE DEPENDENCE OF DIFFUSION

Diffusion of small solute particles (atoms, molecules) in a dense liquid of larger particles is an important but ill-understood problem of condensed matter physics and chemistry. In this case one does not expect the Stokes–Einstein (SE) relation between the diffusion coefficient D of the tagged particle of radius R and the viscosity η_s of the medium to be valid. Indeed, experiments [83, 112–115] have repeatedly shown that in this limit SE relation (with slip boundary condition) significantly *underestimates* the diffusion coefficient. The conventional SE relation is $D = C'k_BT/R\eta_s$, where k_BT is the Boltzmann constant times the absolute temperature and C' is a numerical constant determined by the hydrodynamic boundary condition. To explain the enhanced diffusion, sometimes an empirical modification of the SE relation of the form

$$D = \text{const}/\eta_s^\alpha \tag{266}$$

is used [83, 114]. The value of the exponent α is typically $\simeq 2/3$. This fractional viscosity dependence of D is often referred to as the microviscosity effect, which implies that the viscosity around a small solute is rather different from that of the bulk due to size effects. On the other hand, Zwanzig and Harrison [116] proposed that it is more meaningful to discuss the experimental results in terms of an *effective* hydrodynamic radius which is determined, among many factors, by the solute–solvent size ratio. However, neither the fractional viscosity dependence of D nor the origin of the effective hydrodynamic radius is well-understood.

Although detailed microscopic calculations of the problem mentioned above are not available, there exist several computer simulation studies [102, 117], which also find the anomalous enhanced diffusion, even for simple model potentials such as the Lennard-Jones. The physical origin of the enhanced diffusion is not clear from the simulations. The enhancement can be as large as 50% over the hydrodynamic value. What is even more surprising is that the simulated diffusion constant becomes smaller than the hydrodynamic prediction for very small solutes, with sizes less than one-fifteenth of the solvent molecules. These results have defied a microscopic explanation.

A microscopic calculation of the size-dependent diffusion is presented here. The calculation is based on the well-known mode coupling theoretical approach. The theoretical calculation is shown to give an excellent agreement with the simulation results and provides a physical interpretation of the enhanced diffusion. It is found that the enhanced diffusion of smaller solutes arises from the decoupling of the solute motion from the density mode of the solvent.

An analysis of the limiting expressions for all the three contributions to the friction/diffusion is performed considering the size of the solute to be large and keeping all other parameters unchanged. This analysis shows that the final expression for the zero-frequency friction reduces to a form given by Hynes et al. [58] and by Mehaffey and Cukier [8]. The contribution from the density mode becomes zero as the microscopic structure of the solvent becomes unimportant to a large solute. It is also shown that in the limit of large solute radius it is analytically possible to recover a Stokes–Einstein-like relation.

An early crossover to the hydrodynamics takes place when the solute–solvent interaction energy is changed along with the solute–solvent size ratio. This early crossover is found to be due to the nonvanishing contribution from the density mode for increased solute–solvent interaction. This density mode contribution leads to a faster decrease of the microscopic contribution to the diffusion and thus an early crossover to the hydrodynamic picture.

A. Theoretical Formulation

Let us consider a single-tagged solute particle among the solvent molecules. $v_{12}(r)$ denotes the interaction pair potential between the solute and a solvent molecule while $v(r)$ denotes the same for a pair of solvent molecules. Since the theory is developed for a spherically symmetrical potential, both $v_{12}(r)$ and $v(r)$ are assumed to be given by the Lennard-Jones potential.

The LJ diameter is denoted by σ_1 and σ_2 for the solvent and the solute, respectively. $\mathcal{R}^{-1} = \sigma_2/\sigma_1$ is the solute-to-solvent size ratio. ϵ_1 and ϵ_2 are the energy scales for the solvent–solvent and the solute–solute pair interaction, respectively. The expression for the solute–solvent interaction pair potential is given by

$$v_{12}(r) = 4\epsilon_{12}\left[\left(\frac{\sigma_{12}}{r}\right)^{12} - \left(\frac{\sigma_{12}}{r}\right)^6\right], \tag{267}$$

where $\sigma_{12} = (\sigma_1 + \sigma_2)/2$ and $\epsilon_{12} = \sqrt{\epsilon_1\epsilon_2}$. Both the solute and the solvent are considered to have the same mass 'm'.

The expression for the solvent–solvent interaction pair potential is given by

$$v(r) = 4\epsilon_1\left[\left(\frac{\sigma_{11}}{r}\right)^{12} - \left(\frac{\sigma_{11}}{r}\right)^6\right]. \tag{268}$$

Since the goal is to study the friction on a solute which is different in size from the solvent, in the mode coupling expressions of friction the terms representing the coupling between the solute and the solvent are calculated using the solute–solvent interaction potential. Thus the binary terms $c_{12}(q)$ and γ'_{d12} are all calculated from $v_{12}(r)$, and all the other solvent static and dynamical quantities are calculated from $v(r)$.

The liquid is characterized by the reduced number density $\rho^* = \rho\sigma_1$ and the reduced temperature $T^* = k_B T/\epsilon_1$.

B. Anomalous Diffusion of Small Solutes

This section presents the study of the diffusion of solutes smaller than the size of the solvent in the normal liquid regime. A detailed calculation of the size dependence of the diffusion coefficient has been carried out. The solute–solvent size ratio is varied from $1/20$ to 1.

Note that the binary HMSA [60] scheme gives the solute–solvent radial distribution function only in a limited range of solute–solvent size ratio. It fails to provide a proper description for such a large variation in size. Thus, here the solute–solvent radial distribution function has been calculated by employing the well-known Weeks–Chandler–Anderson (WCA) perturbation scheme [118], which requires the solution of the Percus–Yevick equation for the binary mixtures [119].

The results are compared with the existing computer simulation studies.

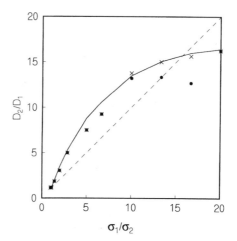

Figure 8. The ratio of the self-diffusion coefficient of the solute (D_2) to that of the solvent molecules (D_1) plotted as a function of the solvent–solute size ratio (σ_1/σ_2) for equal mass. The solid line represents the values calculated from the present mode coupling theory. The filled circles and the crosses represent the computer–simulated [102] and the modified computer-simulated values, respectively. For comparison we have also shown the results predicted by the Stokes–Einstein relation (represented by the dashed line). Here the range of density studied is $\rho^*(=\rho\sigma^3) = 0.85$–$0.92$ at $T^*(=k_BT/\epsilon) = 0.75$.

The study is performed at reduced temperature $T^* = 0.75$ and reduced density $\rho^* = 0.844$–0.92. This is precisely the system studied in computer simulations [102]. The variation of the self-diffusion coefficient with the solute size is shown in Fig. 8, where the size of the solute molecule has been varied from 1 to 1/20 times that of the solvent molecule. In the same figure the computer-simulated values [102] are also plotted for comparison with the calculated results. The calculated results are in good agreement with the computer simulations. Both the theoretical results and the computer simulation studies show an enhanced diffusion for size ratios $\mathcal{R}(\mathcal{R} = \sigma_1/\sigma_2)$ between 1.5 and 15. This is due to the sharp decoupling of the solute dynamics from the solvent density mode.

For neat liquids the contribution from the current term is found to be much less at high densities [59]. Thus there is a decoupling of the solute motion from the current mode of the solvent even for dense neat liquids. For smaller solutes the current term contribution further reduces and in addition there takes place this decoupling of the solute motion from the solvent density mode that gives rise to the enhanced diffusion of smaller solutes.

The decoupling scenario can be clearly envisaged from. Fig. 9, which shows the time dependence of (a) the self-intermediate scattering function

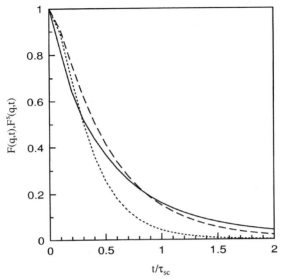

Figure 9. Time dependencies of the single-particle and the collective intermediate scattering functions compared for two different solute sizes at a particular wavenumber $q^* = 6.001$ at reduced temperature $T^* = 0.75$ and in the *normal* density regime ($\rho^* = 0.89$). The solid line represents the collective intermediate scattering function. The long-dashed line is the single-particle intermediate scattering function for solute–solvent size ratio 1.0 and the short-dashed line is for solute-solvent size ratio 0.5. The plots show that the decoupling of the solute motion from the solvent dynamics increases as the solute size is decreased. The time is scaled by τ_{sc}, where $\tau_{sc} = [m\sigma^2/k_BT]^{1/2}$.

for solute–solvent size ratio 0.5 and (b) the intermediate scattering function of the solvent for a given wavenumber (chosen to be equal to $6.0/\sigma$) in the normal density regime. For comparison we have also plotted in the same graph the self-intermediate scattering function for solute–solvent size ratio 1. The plots show that for size ratio 1 the solute motion is fully coupled to the solvent dynamics, whereas when the solute–solvent size ratio is 0.5 there is a disparity in the time scale of the solute and the solvent dynamics. The time scale of the solute motion is much smaller than the time scale in which the solvent dynamics takes place. This reduces the contribution of the density mode of the solvent to the solute friction/diffusion.

Another factor that contributes to the decoupling is the two particle direct correlation function. The product $c_{12}(q)F(q,t)$ defines the modified structure of the solvent probed by the solute. The value of the direct correlation function is less for smaller solutes at all wavevectors. The smaller the value of the two-particle direct correlation function, the lesser will be the contribution of the density mode to the total friction.

This sharp decoupling also explains the saturation of the diffusion for $\mathcal{R} > 10$. For $\mathcal{R} > 10$, due to almost full decoupling of the solute motion from the solvent dynamics there is hardly any contribution from the density mode and the friction is determined by the binary part. Thus in the time scale at which the solute dynamics takes place, the solvent remains nearly static. Now once the solute becomes very small, it does not feel the static structure of the solvent around it and can diffuse through the interparticle distances. Thus further decrease in the size of the solute does not decrease the binary friction leading to the near saturation of the diffusion when plotted against the solute–solvent size ratio.

Another surprising result obtained both in the computer simulation studies and in the theoretical analysis is the smaller value of the diffusion coefficient than that predicted by the SE relation for $\mathcal{R} > 15$. The SE relation predicts that the friction is proportional to $\mathcal{R}^{-1}$ whereas this present microscopic calculation shows a near constant value of the friction for $\mathcal{R} > 15$. This leads to a smaller value of the calculated diffusion coefficient than that predicted by the SE relation.

Note that there is an apparent disagreement between the theory and simulation for $\mathcal{R} = 12-18$. While the theory gives a smooth curve, there is a dip in the simulation result [102]. In the theoretical calculation the system is studied in the limit of zero solute concentration. Thus the size of the solute should not influence the diffusion of the bulk solvent. It should remain constant for a particular density and temperature as the solute size is varied. The simulated system is made comparable with the theoretically studied system by fixing the value of the diffusion of a solvent molecule in the presence of other solvent molecules, $D_1 = 0.024$ for $\rho^* = 0.92$. This leads to a smooth curve shown in Fig. 8. The agreement between theory and simulations is now excellent.

Thus it is found that there is a decoupling of tracer diffusion from viscosity in the normal liquid regime. Though there is a large decoupling of the motion of the tracer particle from the solvent dynamics, surprisingly the decoupling leads to only a small enhancement of the diffusion. This is because in this region both the friction and the viscosity are dominated by the short time-part.

Decoupling of small solutes from the solvent viscosity becomes stronger in the supercooled liquid, as glass transition is approached [83]. There can be two reasons for this stronger decoupling.

1. The long-time tail in the solvent dynamic structure factor is responsible for the large value of the viscosity in supercooled liquid. The solute dynamics for smaller solutes are faster than that of the solvent. Thus, the solute dynamics is decoupled from this long-time tail of the solvent dynamic structure factor.

2. Since a smaller solute is coupled more strongly to its immediate neighbors, the solvent inhomogeneity (discussed in Section XIV) has a stronger effect on the solute dynamics.

C. Analysis of the Limiting Expressions for the Terms Contributing to the Friction/Diffusion of a Large Solute

Let us first rewrite the expression for the friction in terms of the binary friction and the contributions from different hydrodynamic modes:

$$\frac{1}{\zeta(z)} = \frac{1}{\zeta^B(z) + R_{\rho\rho}(z)} + R_{TT}(z). \tag{269}$$

The expression for the diffusion can thus be written as

$$D = D_{\text{micro}} + D_{\text{hydro}}, \tag{270}$$

where

$$D_{\text{micro}} = \frac{1}{\zeta^B(z=0) + R_{\rho\rho}(z=0)} \quad \text{and} \quad D_{\text{hydro}} = R_{TT}(z=0).$$

An analysis of the limiting expressions for the three different terms contributing to the friction/diffusion is presented, considering the solute to be large compared to the solvent. When the solute becomes very large, physically one expects the solvent to appear as a continuum medium to the solute. Thus the microscopic details of the solvent becomes unimportant.

It is found that as the solute size is increased, keeping all other parameters fixed, the peak in the solute–solvent radial distribution function slowly disappears and approaches the value 1. This implies that the probability of a solvent particle, provided that there is a solute at the origin, is same everywhere. The solute–solvent static structure factor $S_{12}(q)$, which can be obtained from $g_{12}(r)$, will also have no structure and will have a uniform value; that is, $S_{12}(q) = 1$ for all wavenumbers.

For a theoretical analysis the radial distribution function is approximated to be given by the following step function:

$$
\begin{aligned}
g_{12}(r) &= 0, & r &< r_{\min} \\
&= 1, & r &\geq r_{\min},
\end{aligned} \tag{271}
$$

where $r_{\min} = 2^{1/6}\sigma_{12}$. Other than the size, all other parameters of the solute are considered to be the same as the solvent (i.e., $\epsilon_2 = \epsilon_1$ and $M = m$).

1. Binary Contribution

With the above-mentioned radial distribution function, expression for the Einstein frequency after performing the angular integration Eq. (114) reduces to

$$\omega_{012}^2 = \frac{4\epsilon_1 \rho}{3m} \int_{r_{min}}^{\infty} dr \left(132 \left(\frac{\sigma_{12}}{r} \right)^{12} - 30 \left(\frac{\sigma_{12}}{r} \right)^6 \right). \tag{272}$$

Once the integration is performed, it can be easily shown that $\omega_{012}^2 \propto \sigma_{12}$.

It has been already discussed (see Section XIII) that the contribution from the three-particle contribution to the binary time constant of the friction is small; thus for simplicity the analysis for τ_ζ is performed considering the contribution only from the two-particle term. Thus the second term in Eq. (120) is neglected and the angular integration in the first term is performed. This reduces Eq. (120) to the following form:

$$\frac{\omega_{012}^2}{\tau_\zeta^2} = \frac{64\pi\epsilon_1^2 \rho}{m^2} \int_{r_{min}}^{\infty} dr \left[A + \frac{B}{3} + \frac{C}{3} \right], \tag{273}$$

where $A, B,$ and C are given by the following expressions:

$$A = \left[144 \frac{\sigma_{12}^{24}}{r^{26}} + 36 \frac{\sigma_{12}^{12}}{r^{14}} - 144 \frac{\sigma_{12}^{18}}{r^{20}} \right],$$

$$B = \left[3168 \frac{\sigma_{12}^{18}}{r^{20}} - 576 \frac{\sigma_{12}^{12}}{r^{14}} - 4032 \frac{\sigma_{12}^{24}}{r^{26}} \right],$$

$$C = \left[28224 \frac{\sigma_{12}^{24}}{r^{26}} + 2304 \frac{\sigma_{12}^{12}}{r^{14}} - 16128 \frac{\sigma_{12}^{18}}{r^{20}} \right], \tag{274}$$

When the integration is performed, it is found that $\omega_{012}^2/\tau_\zeta^2 \propto \sigma_{12}^{-1}$; thus $\tau_\zeta \propto \sigma_{12}$. From Eq. (115) the binary friction at zero frequency can be obtained as

$$\zeta^B(z=0) = \omega_{012}^2 \tau_\zeta \frac{\sqrt{\pi}}{2} = C_1 \sigma_{12}^2, \tag{275}$$

where C_1 is a constant. Note that the binary friction is proportional to the square of the solute–solvent diameter, which is in accord with the Enskog expression for the friction. At the limit of large solute radius, we have $\sigma_{12} \simeq \sigma_2/2$; thus the binary friction becomes proportional to the square of the solute radius.

2.   Density Mode Contribution

Since $S_{12}(q) = 1$ for all the wavenumbers, using Ornstein–Zernike relation it can be shown that $c_{12}(q) = 0$ for all q.

This implies that the density mode does not have any contribution. Physically, this can be explained in the following way. Since the solvent appears as a continuous medium to the solute, the microscopic structure of the solvent is ignored, leading to a absence of the short-range correlations. It can also be argued that the time scale at which the density fluctuation takes place is very small compared to the time scale of the solute motion. Thus the solute is ignorant to the processes happening in such a short time scale and can only sense the average value of the density.

3.   Transverse Current Mode Contribution

The analysis of the transverse current mode contribution [given by Eq. (135)] is performed in the asymptotic limit because this term is known to contribute in the long time. The short time part of the propagator (second term in the integrand) can be neglected in the long time. Since the solute is very large, it has a long time scale. Thus $F^s(q, t)$ has a much slower time dependence than that of the current autocorrelation function and can approximately be written as $F^s(q, t) = F^s(q, t = 0) = 1$. With the above-mentioned approximations, Eq. (135) in the frequency plane at $z = 0$ reduces to

$$R_{TT}(z = 0) = \frac{1}{3\pi^2\rho} \int_0^\infty dq\, q^2 [\gamma^t{}_{d12}(q)]^2 \omega_{012}^{-4} \int_0^\infty dt C_{tt}(q, t). \qquad (276)$$

One can now approximate the current autocorrelation function in the diffusive limit. When the time integration is performed, the above expression reduces to

$$R_{TT}(z = 0) = \frac{m}{3\pi^2\eta_s} \int_0^\infty dq [\gamma^t{}_{d12}(q)]^2 \omega_{012}^{-4} \qquad (277)$$

Next, for simplicity, an approximation for the vertex function is considered. It has been shown by Balucani that $\gamma^t_{d12}(q)$ can approximately be given by [16]

$$\gamma^t_{d12}(q) \simeq \omega_{012}^2 [j_0(qr_{min}) + j_2(qr_{min})], \qquad (278)$$

where j_0 and j_2 are the spherical Bessel functions of order zero and two, respectively. When the above expression of the vertex function is substituted in

Eq. (276), the expression for $R_{TT}(z = 0)$ reduces to

$$R_{TT}(z = 0) \simeq \frac{m}{3\pi^2\eta_s} \int_0^\infty dq \left[j_0(qr_{min}) + j_2(qr_{min}) \right]^2$$

$$= \frac{m}{3\pi^2\eta_s r_{min}} \int_0^\infty dx \left[j_0(x) + j_2(x) \right]^2, \qquad (279)$$

where $x = qr_{min}$. Thus it implies that the contribution from the transverse current term to the diffusion is proportional to the inverse of the solute–solvent diameter, which is in accord with the Stokes–Einstein relation. The analysis of the current term is true for any solute size as no approximation for $g_{12}(r)$ has been made.

Note that even without the approximation of the vertex function [given by Eq. (278)], it is possible to demonstrate that $R_{TT}(z = 0) \propto \sigma_{12}^{-1}$, but the analysis becomes cumbersome. Thus the approximation for the vertex function is not necessary, but it is made to simplify the analysis.

4.   Expression for Friction/Diffusion

The final expression for the zero-frequency friction for a large solute is given by

$$\frac{1}{\zeta(z = 0)} = \frac{1}{C_1\sigma_{12}^2} + \frac{m}{C_2\eta_s\sigma_{12}} \qquad (280)$$

and the diffusion is given by

$$D = \frac{k_BT}{C_1 m\sigma_{12}^2} + \frac{k_BT}{C_2\eta_s\sigma_{12}}, \qquad (281)$$

where C_2 is another constant. Note that the expression for the friction is similar to the form given by Hynes et al. [58] and also by Mehaffey and Cukier [8]. One more interesting feature is that the finite solvent size correction to the Stokes relation given by Hynes et al. is automatically incorporated in this formulation due to the presence of r_{min} in the expression for R_{TT}.

When $\sigma_2 \gg \sigma_1$ since the first term is proportional to σ_2^{-2} and the second one to σ_2^{-1}, the diffusion is primarily given by the second term. Thus the hydrodynamic regime can be recovered.

D.   Early Crossover to the Hydrodynamics: Effect of Solute–Solvent Interaction

A numerical calculation of the friction/diffusion by only changing the size of the solute, predicts a crossover to the current mode dominated regime beyond

TABLE I
The Calculated Value of the Diffusion, D_{cal} and the Simulated
Value [60], D_{sim}, for Xe in Ne [a]

σ_2/σ_1	M/m	ϵ_2/ϵ_1	D_{cal}	D_{sim}
1.44	6.5	6.09	0.01979	0.018 ± 0.004

[a] The ratio of the different parameters are also given. The
diffusion values are given in units of $\sqrt{k_B T \sigma^2 / m}$.

solute–solvent size ratio about 6. Conventionally, this crossover is expected
to happen earlier. It can be argued that for solutes larger than the solvent the
interaction energy between the solute–solvent pair is usually larger than that
between the solvent–solvent pair.

The diffusion of solutes has been studied by increasing its size, its mass,
and the interaction energy parameter. To check the validity of the MCT
scheme with the change of all the three parameters (the size, the mass, and
the interaction energy) of the solute, the diffusion of Xe in Ne has been studied where computer-simulated values are available [60]. The ratios of the
different solute–solvent parameters and calculated and simulated values of
diffusion are presented in Table I.

The agreement between the calculated and simulated values of diffusion
suggests that the MCT scheme works well for binary system at infinite solute
dilution. It should be mentioned here that the good agreement is only possible
when the change of interaction energy is also considered along with the size.

To study the crossover to the hydrodynamics, the size ratio, the mass ratio,
and also epsilon ratio is varied by considering different solute–solvent systems such as Xe in Ne, Kr in He, Xe in He, CCl_4 in He, and so on. The Lennard-Jones parameters for these systems are obtained from the standard
literature [120, 121]. The results obtained are presented in the Table II.

TABLE II
The Calculated values of D_{micro} and D_{hydro} for Different
Solute–Solvent Size Ratio [a]

σ_2/σ_1	M/m	ϵ_2/ϵ_1	D_{micro}	D_{hydro}
1.44	6.50	6.09	0.0133	0.00663
1.48	20.93	16.05	0.00878	0.00669
1.74	32.80	22.45	0.006496	0.00678

[a] The mass ratio and the ratio of the interaction energy are also
changed accordingly. The diffusion values are given in units of
$\sqrt{k_B T \sigma^2 / m}$.

From the numbers presented in Table II, it is found that D_{micro} decreases very fast as the solute size is increased; on the other hand, D_{hydro} increases slowly with the increase in the solute size. For solute–solvent size ratio 1.48, D_{micro} is bigger than D_{hydro}. When σ_2/σ_1 is about 1.74, D_{hydro} becomes just bigger than D_{micro}, suggesting that already a crossover to the current mode has taken place, although at this stage the diffusion takes place via both channels. For $\sigma_2/\sigma_1 = 4.0$, D_{hydro} is found to be almost 5 times larger than D_{micro}, which implies that a sharp crossover to the current mode-dominated region has taken place.

The reason for the early crossover can be understood from the following discussion. When the interaction energy between the solute and the solvent is increased, the peak of the radial distribution function does not disappear. Thus $c_{12}(q) \neq 0$ for all wavenumbers. Hence the density mode contribution does not become zero as happens in the case where the size of the solute is only increased. Hence D_{micro}, along with the binary term, also contains the contribution from the density mode. This results in faster decrease of D_{micro}, leading to an early crossover.

The above analysis suggests that incorporation of the change in mass and most importantly the change in the interaction energy parameter with the change in size of the solute can cause an early crossover to the current-mode-dominated region. It should be pointed out here that the values of the solute–solvent size ratio at which the crossover takes place is not universal but system-dependent. It very much depends on how the interaction energy of the solute–solvent potential changes with the solute size.

XVII. SUBQUADRATIC QUANTUM NUMBER DEPENDENCE OF THE OVERTONE VIBRATIONAL DEPHASING IN MOLECULAR LIQUIDS

This is a subject of great interest to the physical chemistry/chemical physics community. In this section we discuss how the mode coupling theory calculation of the friction on atoms connected by a chemical bond can lead to a better understanding of vibrational dephasing in dense liquids. In particular, the application of MCT is shown to provide a possible explanation of an old ill-understood problem.

Vibrational dephasing provides us with a powerful method to probe the interaction of a chemical bond with the surrounding medium. Over the years, many experimental techniques have been developed to study dephasing of bonds in many molecular systems at various temperatures, pressures, and concentrations [122–124]. One popular experimental technique is the isotropic Raman lineshape. The other methods involve a coherent excitation of the vibration with a laser pulse and monitoring the decay of the phase coherence, after a delay time, by anti-Stokes Raman scattering measurements at the phase-matched angle. However, only the fundamental and the lower overtones of a given vibrational mode can be studied by these techniques.

Recent developments in ultrafast higher-order nonlinear spectroscopic techniques like resonance Raman and IR echo experiments have now made the dephasing of higher quantum levels accessible. Experiments involving such new techniques have revealed interesting subquadratic observations of the dephasing rate as a function of quantum number n [124, 126] (as against the theoretically predicted quadratic dependence on n) in some systems. For example, in systems like $CDCl_3$ and CD_3I, the ratio of the dephasing rates between the overtone and the fundamentals of the C–D stretching mode are 2.1 and 2.4, respectively [126], as against a theoretically expected value of 4 in both the systems. This is surprising because in liquids the frequency time correlation function is expected to be short and in the motionally narrowed limit. Overtone transition studies that have been attempted earlier [127, 128] have reported similar subquadratic quantum number dependence, although these observations involved a considerable amount of uncertainty because of low signal-to-noise factors of the bandwidth measurements.

Among various theories of vibrational dephasing in liquids, the works of Madden and Lynden-Bell [129], Oxtoby [125], Fischer and Laubereau [130], and Schweizer and Chandler [123] are certainly the important ones. These have investigated the connection between pure dephasing and vibrational lineshapes. Fischer and Laubereau [130] studied the effects of collinear collision of a diatomic with an atom on vibrational dephasing. The elegant work of Oxtoby was based on Kubo's well-known stochastic theory of line shapes [131]. Subsequently, Schweizer and Chandler developed a unified molecular theory that explored the effects of the vibration–rotation coupling on dephasing and also the role of repulsive and attractive forces separately [123]. By assuming a distinct separation of time scales between the latter two, they could analyze their individual contributions separately.

However, none of the above-mentioned theories can satisfactorily explain the observed nonquadratic quantum number dependence of the dephasing rate. That the situation is somewhat complex can be understood from the following analysis. The average dephasing time $\langle \tau_v \rangle$ is given by [125]

$$\langle \tau_v \rangle = \int_0^\infty dt \langle Q(t)Q(0) \rangle, \tag{282}$$

where $Q(t)$ is the time-dependent normal coordinate whose dephasing is being studied and $\langle Q(t)Q(0) \rangle$ is the relevant time correlation function. The Fourier transform of this function is the observable in isotropic Raman experiments. The quadratic dependence of the dephasing rate $\langle \tau_v \rangle^{-1}$ on the quantum number n arises when $\langle Q(t)Q(0) \rangle$ decays exponentially as

$$\langle Q(t)Q(0) \rangle = \exp(-n^2 t/\tau). \tag{283}$$

The above form is well known and is the one usually assumed in the Kubo–Oxtoby theory.

Note that in a dense liquid the decay of the vibrational correlation function need *not* be single exponential as assumed in Eq. (283). It can be biphasic with a Gaussian decay at short times followed by an exponential decay at longer times [123, 124]. Now if the decay of $\langle Q(t)Q(0) \rangle$ is fully Gaussian as in the following form,

$$\langle Q(t)Q(0) \rangle = \exp(-n^2 t^2/\tau^2), \tag{284}$$

then the dependence of $\langle \tau_v \rangle^{-1}$ on n is *linear*. Thus, the quantum number dependence will critically depend on the relative magnitudes of the two components of the biphasic decay. It was earlier believed that in dense liquids the Gaussian component was negligible. For example, in Oxtoby's work, the force–force correlation function (required to determine $\langle Q(t)Q(0) \rangle$) which is related to friction is assumed to be delta-correlated [125]. Microscopic calculation of the frequency-dependent friction obtained from the mode coupling theory (MCT) [132] and computer simulations [74] have clearly demonstrated that the force–force correlation function has a rich structure and is certainly not delta-correlated. The main point here is that although $\langle Q(t)Q(0) \rangle$ will always show quadratic n dependence (in the exponential), the *experimental observables*, like $\langle \tau_v \rangle$ or the width of the Raman lineshape function, can show subquadratic n dependence. The microscopic calculations show that this indeed happens.

A particular advantage of MCT is that it separates the dynamics probed by the normal coordinate into two parts: (1) the initial ultrafast part that arises from the binary, uncorrelated, short-range interaction of the normal coordinate Q with the surrounding solvent molecules and (2) the slower part that is coupled to the collective density fluctuations of the solvent. Thus, one can indeed imagine the first part as a homogeneous contribution while imagining the second slow part as an inhomogeneous one. However, such a classification is not required here. The present separation is also different from the one suggested by Schweizer and Chandler. In their analysis, it is the attractive part that couples to the density fluctuations. It is easy to show from the MCT theory that while the attractive forces make dominant contribution to the binary part, both the attractive and the repulsive parts contribute to the slower relaxation.

In order to investigate the quantum number dependence of vibrational dephasing, an analysis was done on two systems: C–I stretching mode in neat-CH_3I and C–H mode in neat-$CHCl_3$ systems. The C–I and C–H frequencies are widely different ($525\,cm^{-1}$ and $3020\,cm^{-1}$, respectively) and so also their anharmonic constants. Yet, they both lead to a subquadratic quantum number dependence. The time-dependent friction on the normal coordinate is found to have the universal nonexponential characteristics in both systems—a distinct inertial Gaussian part followed by a slower almost-exponential part.

The overtone dephasing rates are found to be substantially subquadratic in n dependence and show good qualitative agreement with the experimental observations and also computer simulation studies [133]. For higher levels ($n \geq 4$), a linear dependence on n was obtained. This linear dependence of the dephasing rate on the quantum number n has a simple physical explanation. As n increases, the dephasing becomes faster and is determined mainly by the initial fast dynamics of the liquid. This naturally leads to a linear dependence on n for large n.

A. Theory of Vibrational Line Shape

We shall now briefly review the Kubo–Oxtoby theory of vibrational lineshape. The starting point for most theories of vibrational dephasing is the stochastic theory of lineshape first developed by Kubo [131]. This theory gives a simple expression for the broadened isotropic Raman line shape ($I(\omega)$) in terms of the Fourier transform of the normal coordinate time correlation function by

$$I(\omega) = \int_0^\infty dt \, \exp(i\omega t)\langle Q(t)Q(0)\rangle. \qquad (285)$$

The experimental observables are either the lineshape function $I(\omega)$, as in the classical experiments, or the normal coordinate time correlation function, $\langle Q(0)Q(t)\rangle$, as in the time domain experiments of Tominaga and Yoshihara [126]. The normal coordinate time correlation is related to the frequency modulation time correlation function by

$$\langle Q(t)Q(0)\rangle = \text{Re} \, \exp(i\omega_v t)\left\langle \exp\left[i\int_0^t dt' \Delta\omega(t')\right]\right\rangle, \qquad (286)$$

where ω_v is the vibrational frequency and $\hbar\Delta\omega(t) = V_{nn}(t) - V_{00}(t)$ is the fluctuation in energy between vibrational levels of 0 and n, where n represents the quantum level of the overtone transitions; V_{nn} is the Hamiltonian matrix element of the coupling of the vibrational mode to the solvent bath.

$\Delta\omega(t)$ is, therefore, the instantaneous shift in the vibrational frequency due to interactions with the solvent molecules.

A cumulant expansion [134] of Eq. (286) gives

$$\langle Q(t)Q(0)\rangle = \text{Re } \exp(i\omega_v t + i\langle\Delta\omega\rangle t)\exp\left[-\int_0^\infty dt1 \int_0^{t1} dt2\langle\Delta\omega(t1)\Delta\omega(t2)\rangle\right].$$

$$(287)$$

The double integral in Eq. (287) is rewritten in the following form:

$$\langle Q(t)Q(0)\rangle = \text{Re } \exp(i\omega_v t + i\langle\Delta\omega\rangle t)\exp\left[-\int_0^t dt'(t - t')\langle\Delta\omega(t')\Delta\omega(0)\rangle\right],$$

$$(288)$$

where $\langle\Delta\omega(t')\Delta\omega(0)\rangle$ is the frequency–time correlation function. Assuming a weak coupling of the vibration to the solvent bath and extending the integration limit to infinity, the following expression is obtained:

$$\langle Q(t)Q(0)\rangle \simeq \exp(-t/\tau_v), \qquad (289)$$

where

$$\tau_v^{-1} = \int_0^\infty dt'\langle\Delta\omega(t')\Delta\omega(0)\rangle. \qquad (290)$$

When Eq. (289) is substituted in Eq. (285), one finds a Lorentzian line shape where the half-width at half-maximum (HWHM) is τ_v^{-1}. Thus by assuming an exponential decay, as in Eq. (289), τ_v can be obtained directly from either Eq. (282) or Eq. (290). However, $\langle Q(0)Q(t)\rangle$ may be nonexponential, as opposed to exponential as assumed in Oxtoby's work, and, as shown later, may give rise to an overall subquadratic overtone dependence of the rate.

The main contributions to the frequency–time correlation function are assumed to be, as in the earlier works [123, 124], from the vibration–rotation coupling and the repulsive and attractive parts of the solvent–solute interactions. In several theories, the (faster) repulsive and the (slower) attractive contributions are assumed to be of widely different time scales and are treated separately. However, this may not be true in real liquids because the solvent dynamic interactions cover a wide range of time scales and there could be a considerable overlap of their contributions. The vibration–rotation coupling contribution takes place in a very short time scale; and by neglecting the cross-correlation between this mechanism and the atom–atom forces, they could be treated as two separate contributions. Therefore, the frequency-time correlation function can be written as

$$\langle \Delta\omega(t)\Delta\omega(0)\rangle = \langle \Delta\omega_{VR}(t)\Delta\omega_{VR}(0)\rangle + \langle \Delta\omega_{RA}(t)\Delta\omega_{RA}(0)\rangle, \quad (291)$$

where subscripts VR and RA represent the vibration–rotation and repulsive–attractive contributions to the dephasing. The total lineshape is, therefore, assumed to be a product of the above two contributions.

The frequency–time correlation function is dependent on the frequency and the force constants of the vibrational mode whose dephasing is being considered. They are determined by fitting the vibrational bond energies to a Morse potential of the following form:

$$V(r) = D_e[1 - \exp[-\beta(r - r_e)]]^2, \quad (292)$$

where r is the positional coordinate, r_e is the equilibrium bond length, and β and D_e are the Morse potential fit parameters. The potential can be expanded about the equilibrium position $r = r_e$ using the Maclaurin series expansion as

$$V(r) = V(r = r_e) + (dV/dr)_{(r=r_e)}(r - r_e) + (d^2V/dr^2)_{(r=r_e)}(r - r_e)^2 + \cdots \quad (293)$$

and so Eq. (292) can be rewritten as

$$V(r) = \tfrac{1}{2}(2D_e\beta^2)(r - r_e)^2 + \tfrac{1}{6}(-6D_e\beta^3)(r - r_e)^3, \quad (294)$$

neglecting the higher-order terms. Let $x = r - r_e$ represent the displacement from the equilibrium position r_e. The potential function of the vibrational diatomic is of the following usual form:

$$H_{vib} = 1/2\mu\omega_v^2 x^2 + 1/6fx^3 \quad (295)$$

where H_{vib} represents the hamiltonian for the vibrational mode. Assuming Q to be the normal coordinate and using the relation $Q = \mu^{1/2}x$, the above equation for the vibration of an anharmonic oscillator can now be rewritten in the form

$$H_Q = K_{11}Q^2 + K_{111}Q^3. \quad (296)$$

The harmonic force constant K_{11} and the anharmonicity parameter K_{111} are obtained by comparing Eq. (294) or Eq. (295) with Eq. (296).

For deriving an expression for the frequency–time correlation function the formulation of Oxtoby is followed. If V is the anharmonic oscillator–medium interaction, then by expanding V in the vibrational coordinate Q using Taylor's series we obtain

$$\hbar \Delta \omega_n(t) = (Q_{nn} - Q_{00}) \left(\frac{\partial V}{\partial Q} \right)_{Q=0} (t) + \frac{1}{2} (Q_{nn}^2 - Q_{00}^2) \left(\frac{\partial^2 V}{\partial^2 Q} \right)_{Q=0} (t) + \cdots$$

(297)

In the estimation of $\Delta \omega_n(t)$, only the first two terms are considered, neglecting the higher-order terms. $(Q_{nn} - Q_{00})$ and $(Q_{nn}^2 - Q_{00}^2)$ are the quantum mechanical expectation values of the anharmonic oscillator. They can be calculated using perturbation theory and is given by

$$Q_{nn} - Q_{00} = \frac{3n\hbar(-K_{111})}{\omega_v^3}$$

(298)

$$Q_{nn}^2 - Q_{00}^2 = \frac{n\hbar}{\omega_v}.$$

(299)

If the vibrations were harmonic, the first term would not be present, but in fact anharmonicities are large enough.

The derivatives appearing in Eq. (297) can be rewritten in terms of the atomic displacement ϵ_i, if the potentials are taken to be atom-additive functions of relative atomic separations $V(|r_i - r_j|)$. The derivatives are given as [135]

$$\frac{\partial V}{\partial Q} = \sum_i \left(\frac{\partial V}{\partial Q} \cdot \frac{\partial \epsilon_i}{\partial Q} \right) = -\sum_i F_i \cdot \left(\frac{\partial \epsilon_i}{\partial Q} \right),$$

(300)

where i represents the atom and F_i represents the force on it. Also if the potential has a strong repulsive part, then $(\partial^2 V / \partial Q^2)$ can be approximated as

$$\frac{\partial^2 V}{\partial Q^2} \simeq -\frac{1}{L} \sum_i F_i \cdot \left(\frac{\partial \epsilon_i}{\partial Q} \right) \left| \left(\frac{\partial \epsilon_i}{\partial Q} \right) \right|,$$

(301)

where L is defined as the characteristic potential range.

The atomic displacement ϵ_i is related to the normal coordinate Q through

$$\epsilon_i = \frac{l_{ik} Q_k}{m_i^{1/2}},$$

(302)

where i represents the atom, k labels the normal mode Q of interest, and $\mathbf{l_{ik}} = l_{ik}u_{ik}$ represents a characteristic vector along the normal mode Q_k. For a diatomic molecule,

$$\mathbf{l_{ik}} = (m_i/\mu)^{1/2}\gamma_i, \tag{303}$$

where

$$\gamma_i = \frac{m_i}{m_i + m_j}. \tag{304}$$

An important ingredient of Oxtoby's work was the decomposition of the force on the normal coordinate, $-(\partial V/\partial Q)$, in terms of the force on the atoms involved. Assuming that the forces on the different atoms of the diatomic are uncorrelated and that the area of contact of each atom with the solvent is a half-sphere, Oxtoby derived the following expression for the frequency–time correlation function:

$$\langle \Delta\omega_{RA}(t)\Delta\omega_{RA}(0)\rangle = \frac{n^2}{2}\sum_i \left[\frac{3(-K_{111})l_{ik}}{\omega_v^3 m_i^{1/2}} + \frac{l_{ik}^2}{2\omega_v Lm_i}\right]^2 \langle \mathbf{F_i(t)F_i(0)}\rangle, \tag{305}$$

where $\langle \mathbf{F_i(t)F_i(0)}\rangle$ represents the force–force correlation function (or the friction $\zeta(t)$) on the atom i moving along the direction of vibration. The presence of n^2 in the above equation is the reason why the vibrational dephasing rate is usually assumed to exhibit the quadratic n dependence. However, while $\langle \Delta\omega(t)\Delta\omega(0)\rangle$ may have an n^2 dependence, the average dephasing rate $(\langle \tau_v\rangle^{-1})$ can show subquadratic dependence because $\langle Q(t)Q(0)\rangle$ is largely Gaussian at short times.

Following Oxtoby's method, the bond friction here is calculated in terms of the friction (or the force–force time correlation function) on the individual solute atoms connected by the bond. The calculation of the friction on a solute atom in a medium of solvent spheres is performed following the method described in Section IX.

B. Relative Role of the Attractive and Repulsive Forces

In their important work, Schweizer and Chandler [123] assumed a separation of time scales between the attractive and the repulsive forces. Based on this argument, they could analyze the respective roles of the attractive and the repulsive contributions and were the first to point out that the vibrational dephasing is largely controlled by the attractive forces. MCT theory, on the other hand, is based on the separation of time scales between the binary collision and the repeated recollisions. For Lennard-Jones potential, the contribution to the Einstein frequency and hence the binary part arises primarily from the attractive part of the potential. On the other hand, the $R_{\rho\rho}(t)$ is determined by both the attractive and repulsive contributions, although it is the latter that renders it the dominant contribution. A rough idea of the relative roles of the attractive and repulsive forces can, therefore, be obtained by comparing the contributions of the binary ($\zeta^B(t)$) and the density ($R_{\rho\rho}(t)$) parts. Interestingly, it is the binary part which makes the major contribution at short times, and therefore the attractive part becomes increasingly more important as n increases. The important role of the attractive part implied by MCT in dephasing is in complete agreement with the conclusions arrived in Schweizer and Chandler's [123] work where it was explained using a separation of the attractive and the repulsive contributions although in a manner different from the procedure adopted within the MCT framework. The overall implications of MCT are also in agreement with the suggestion by Myers and Markel [124] that the dephasing could be dominated by the attractive interactions or that the repulsive interactions could have a rather long effective correlation time. Note that MCT is able to treat the attractive and repulsive contributions together and in a consistent manner.

C. Vibration–Rotation Coupling

As described by Schweizer and Chandler [123] and also by Myers and Markel [124], the effective potential for the vibration–rotation centrifugal coupling is the rotational kinetic energy which is given by

$$V_{VR} = J^2/2I, \tag{306}$$

where J is the angular momentum, $I = \mu r^2$ is the moment of inertia, and μ is the reduced mass. Expanding this as a Taylor series around the equilibrium bond length, r_e, results in [124]

$$V_{VR} = \frac{J^2}{2I_m r_e}\left[\frac{r_e}{2} - \Delta r + \frac{3(\Delta r)^2}{2r_e} - \cdots\right], \tag{307}$$

where I_m is now the moment of inertia at r_e, and Δr gives the displacement from equilibrium. The centrifugal contribution to the broadening of the line shape is then given by

$$
\begin{aligned}
\langle \Delta\omega_{VR}(t')\Delta\omega_{VR}(0)\rangle_{n,0} &= \langle[V_{VR,nn}(t) - V_{VR,00}(0)][V_{VR,nn}(t) - V_{VR,00}(0)]\rangle/\hbar^2 \\
&= (\Delta R/\hbar I_m r_e)^2\langle\Delta J^2(t)\Delta J^2(0)\rangle, \tag{308}
\end{aligned}
$$

where

$$\Delta R = (Q_{nn} - Q_{00}) - 3(Q_{nn}^2 - Q_{00}^2)/2r_e. \qquad (309)$$

Q_{nn} and Q_{nn}^2 are the expectation values of the bond length displacement and its square, $(r - r_e)$ and $(r - r_e)^2$, in the nth vibrational level of the isolated molecule. These can be easily calculated using quantum perturbation theory. Following Schweizer and Chandler [123], $\langle \Delta J^2(t) \Delta J^2(0) \rangle$ can be written as

$$\langle \Delta J^2(t) \Delta J^2(0) \rangle = \langle J^2(t) J^2(0) \rangle - \langle J^2 \rangle^2; \qquad (310)$$

and then rewriting the correlation function $\langle J^2(t) J^2(0) \rangle$ as

$$\langle J^2(t) J^2(0) \rangle = 0.5 \langle J^4 \rangle [1 + \langle \mathbf{J} \cdot \mathbf{J}(t) \rangle^2 / \langle J^2 \rangle^2] \qquad (311)$$

and, finally, using $\langle J^4 \rangle = 2 \langle J^2 \rangle^2$ and $\langle J^2 \rangle = 2Ik_BT$, one obtains

$$\langle \Delta \omega_{VR}(t') \Delta_{VR} \omega(0) \rangle_{n,0} = (2k_BT \Delta R / \hbar r_e)^2 [J_{\mathrm{corr}}(t)]^2, \qquad (312)$$

where J_{corr} is the normalized angular momentum autocorrelation function, $\langle \mathbf{J} \cdot \mathbf{J}(t) \rangle^2 / \langle \mathbf{J}^2 \rangle$. This is available experimentally from anisotropic Raman lineshape data for some systems.

Note that the angular momentum–momentum correlation function related to the vibrational–rotation friction too is highly nonexponential. This again could significantly alter the n^2 dependence of the rate. However, for the systems studied here, the contribution of vibration–rotation coupling is negligible.

D. Results and Discussion

The formulation presented earlier is applied to two cases: the C–I stretching vibration of CH_3I and the C–H stretching of $CHCl_3$. The following simplification of these polyatomic systems have been assumed to calculate the friction. First, a polyatomic solvent system is reduced to a monatomic spherical system. The effective diameters for this reduction have been estimated from fits to experimentally obtained temperature and density data particular to the system [121, 123]. Second, the polyatomic solute is replaced by the diatomic comprising the bond. So, the CH_3–I and C–H vibrations could be treated as diatomics dissolved in a solvent for simplicity [123, 125]. The main attempt is, therefore, to estimate quantitatively the time-dependent friction on the CH_3 and I solute spheres in a medium of CH_3I pseudospheres. Similarly, C and H solute spheres in a medium of $CHCl_3$ pseudospheres. As the aim here is to understand the quantum number (n) dependence of the overtone dephasing qualitatively, the above approximation is not expected to be serious.

1. Time-Dependent Friction

The solute–solvent and the solvent–solvent interaction potentials are assumed to be given by Lennard-Jones potential. The Lennard-Jones interaction parameters for dissimilar solvent (i) and solute (j) spheres are estimated from those of the interaction of similar spheres through the combining rule $\epsilon_{ij} = (\epsilon_{ii}\epsilon_{jj})^{1/2}$ and $\sigma_{ij} = (\sigma_{ii} + \sigma_{jj})/2$ [121, 124].

The liquid is characterized by the reduced atomic density $\rho\sigma_{ii}^3$, and the reduced temperature is given by $T^* = k_B T/\epsilon_{ii}$. The time-dependent friction is calculated following the procedure presented in Section IX.

The Lennard-Jones parameters for the CH_3 system are taken as $\epsilon/k_B = 146.6K$ and $\sigma = 3.85\,\text{Å}$, for the I system as $\epsilon/k_B = 400K$ and $\sigma = 4\,\text{Å}$ [124]. For CH_3I, these LJ parameters are $\epsilon/k_B = 467$ and $\sigma = 4.6\,\text{Å}$ [121]. The reduced atomic density $\rho\sigma^3$ (or ρ^*) for CH_3I is 0.918 at a temperature of 303 K [121, 123]. Here ρ is the number density.

In the case of $CHCl_3$, the LJ parameters for the C system are taken as $\epsilon/k_B = 100.7K$ and $\sigma = 3.214\,\text{Å}$ [136], for the H system they are taken as $\epsilon/k_B = 37K$ and $\sigma = 2.93\,\text{Å}$ [137] and for the $CHCl_3$ system they are taken as $\epsilon/k_B = 494K$ and $\sigma = 5.045\,\text{Å}$ [121]. The reduced density $\rho\sigma^3$ value of 0.819 corresponding to 303 K has been used [121, 123].

The time-dependent friction profiles ($\langle F(t)F(0)\rangle$ versus t plots) that have been obtained using MCT for CH_3 and I are shown in Figs. 10 and 11, respectively. In both cases, the friction on the atom shows a *strong bimodal response* in $\langle F(t)F(0)\rangle$ profile—that is, Gaussian behavior in the initial time scale followed by a slowly relaxing component. There is even a rise in friction in the intermediate time scale (see Fig. 11). This arises from the coupling of the solute motion to the collective density relaxation of the solvent. As mentioned earlier, the Gaussian component arises from the binary collisions and the slower part arises from correlated recollisions. However, *the frictions on CH_3 and I are found to be quite different*. It is much higher in the case of I than in that of CH_3. This is expected because although the LJ diameters of these CH_3 and I spheres are nearly equal, their individual masses are considerably different ($CH_3 = 15\,g/mol$ and $I = 126\,g/mol$). Because of this mass disparity, the positional coordinate Q corresponding to the equilibrium point would be much closer to the iodine atom (I). As a result, the iodine atom is much more static than CH_3. The large value of the friction for I also arises from its large ϵ value as against the small value for CH_3, because the friction on a molecule is found to increase with ϵ [138]. Therefore, CH_3 is more involved in collisions, is more free to move because it is smaller and lighter, and thus the friction on it is substantially reduced. Also, the contribution of the heavy atom gets reduced because of the presence of the mass term in the denominator of the prefactor term corresponding to each atom in

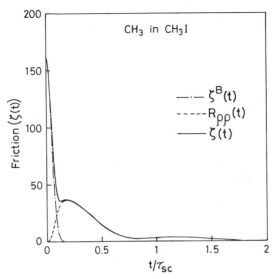

Figure 10. The calculated total friction ($\zeta(t)$) as a function of time, along with the relative contributions to it from the binary ($\zeta^B(t)$) and the density relaxation $R_{\rho\rho}(t)$ terms for the system CH_3 in CH_3I. The reduced temperature $T^*(=k_BT/\epsilon)$ is 1.158 and the reduced density ρ^* for CH_3I is 0.918. The time-dependent frictions are scaled by τ_{sc}^{-2}, where $\tau_{sc} = [m_i\sigma_j^2/k_BT]^{1/2} \simeq 1.1$ ps. i and j represent the solute atom and the solvent atom, respectively. The plot shows a clear Gaussian component in the initial time scale for the binary part $\zeta^B(t)$ and slower damped oscillatory behavior for the $R_{\rho\rho}(t)$ part.

Eq. 305. Thus, the effective friction on the vibrational coordinate becomes less than the earlier estimates [123, 125] of the same.

The same arguments should hold well for C and H solutes in the case of $CHCl_3$, because their individual masses are considerably different (C = 12 g/mol and H = 1.008 g/mol), which means that H is more free to move than C and therefore the solvent influence on H is likely to have a dominant role in the determination of the C–H bond friction. Furthermore, C is actually shielded by the presence of three Cl atoms, a factor that has not been considered here. However, this is not expected to be serious because the dephasing is more sensitive to the friction dynamics of the H atom. The time-dependent friction profiles for H and C show similar strong bimodal behavior as in the case of CH_3 and I systems.

2. Vibration–Rotation Contribution

The vibration–rotation contribution has been computed from the experimentally reported value of τ_J [124] for CH_3I while for $CHCl_3$ the small moment of inertia associated with the proton stretch renders the vibration–rotation

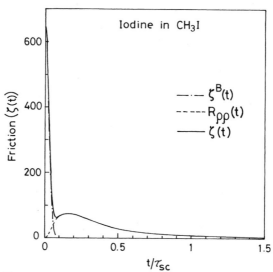

Figure 11. The calculated time-dependent total friction ($\zeta(t)$) and its relative contributions from the binary term ($\zeta^B(t)$) and the density relaxation term $R_{\rho\rho}(t)$ for the iodine atom in CH$_3$I. The reduced temperature $T^* (= k_B T/\epsilon)$ is 0.363 and the reduced density ρ^* for CH$_3$I is 0.918. The time-dependent friction is scaled by τ_{sc}^{-2}, where $\tau_{sc} = [m_i \sigma_j^2 / k_B T]^{1/2} \simeq 3.3$ ps. The friction is found to be much higher for Iodine than CH$_3$. This figure has been taken from Ref. 133.

contribution negligible. The angular momentum correlation data used for the calculation of the vibration–rotation coupling reported by Myers and Markel [124] for the case of CH$_3$I has been used here. They have approximated the experimental J_{orr} to the following form:

$$J_{corr}(t) = [1 - (2\omega_J t/3\pi)]^2 \cos(\omega_J t) \quad \text{for } t \leq 3\pi 2/\omega_J, \quad (313)$$

$$= 0 \quad \text{for } t \leq 3\pi 2/\omega_J, \quad (314)$$

where $\omega_J = 10 \, \text{ps}^{-1}$. The correlation function decay is highly nonexponential with a significant negative region and is over within 0.5 ps. This, along with the strongly Gaussian binary collisional contributions, is mainly responsible for the highly non-exponential $\langle Q(t)Q(0) \rangle$ profile observed in the initial time scale (see Fig. 12).

3. Dephasing Rates

The calculated $\ln\langle Q(t)Q(0) \rangle$ for the first three quantum levels for the CH$_3$I case is shown in Fig. 12. The overtone normal coordinate time correlation function, $\langle Q(t)Q(0) \rangle$, is obtained from the frequency-modulation time corre-

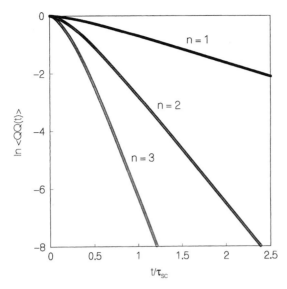

Figure 12. Theoretically obtained plots of $\ln \langle Q(t)Q(0)\rangle$ versus t (where t is scaled by τ_{sc}^{-2}, $\tau_{sc} = [m_{CH_3}\sigma_{CH_3I}^2/k_BT]^{1/2} \simeq 1.1$ ps) for the first three quantum levels ($n = 1,2,3$) of the CH_3–I mode in CH_3I from the friction estimates (shown in Figs. 10 and 11) and the vibration–rotation contribution. The equilibrium CH_3–I bond length was set to $r_e = 2.14$ Å. The results show an increasing Gaussian behavior in the short-time scale with increasing quantum number n. This figure has been taken from Ref. 133.

lation function, $\langle \Delta\omega(t)\Delta\omega(0)\rangle$, using Eq. (288). The values of other parameters that are required in the calculation of $\langle \Delta\omega(t)\Delta\omega(0)\rangle$ using Eq. (305) are obtained from the values reported earlier in Oxtoby's work [125] and is presented in Table III.

The overtone dephasing rates are found to be substantially subquadratic (close to $3n$ in the case of CH_3I and $1.5n$ in the case of $CHCl_3$) toward higher quantum levels (see Fig. 13). These results show good qualitative agreement with the experimental observations that have been reported for the C–I stretching of CH_3I in hexane [124] and C–D stretching of CD_3I [126]. The subquadratic quantum number dependence of the vibrational overtone

TABLE III
The Values of the Parameters for C–H and C–I Modes Taken
from Ref. 125

	ω_v (cm^{-1})	L ($\times 10^{-9}$ cm)	$-K_{111}$ (10^{47}g$^{-1/2}$cm^{-1}s^{-2})
CH_3–I	525	2.69	1.72
C–H	3020	2.92	236

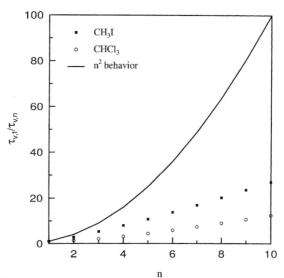

Figure 13. Ratios of vibrational dephasing time (τ_{v1}/τ_{vn}) plotted as a function of the quantum number for C–I stretching (525 cm^{-1}) in neat CH$_3$I and for C–H stretching (3020 cm^{-1}) in CHCl$_3$. The results show highly subquadratic dependence in both the cases. This figure has been taken from Ref. 133.

dephasing in the case of CH$_3$I is in good agreement with the computer simulation studies [133]. In particular, the dephasing times obtained for the 1 level of C–I stretching mode of CH$_3$I and the C–H mode in CHCl$_3$ is about 2.6 ps and 1.4 ps, respectively, which are in good agreement with the experimental observations. The results obtained for the dephasing times from the fundamental overtone are compared with that predicted by the hydrodynamic theory [125] and the experimental results [139–141] in Table IV. In view of the approximations involved in the modeling, the above results are clearly

TABLE IV

Dephasing Times Predicted from Fundamental Overtone of the C–I Stretching Mode in CH$_3$I and C–H Stretching Mode in CHCL$_3$ Using Mode Coupling Theory (MCT), the Hydrodynamics Theory (HT) [125], and the Experimental (Exptl) values [139–141]

	ω_v (cm^{-1})	$\tau_v 1$ (MCT) (ps)	τ_{v1} (HT) (ps)	τ_{v1} (Exptl) (ps)
C–I	525	2.63	1.2	2.3 [139]
C–H	3020	1.40	0.27	1.35 [140, 141]

fortuitous. However, one can at least believe that the present theory can reproduce the experimental results semiquantitatively.

The subquadratic n dependence clearly arises from the nonexponential component of $\langle Q(t)Q(0)\rangle$ (shown in Fig. 12) in the initial time scale which increases with increase in the quantum number n, which strongly reflects the presence of the Gaussian components of binary friction. This dominant Gaussian behavior is responsible for the nearly linear n dependence in the higher quantum levels.

XVIII. ISOMERIZATION DYNAMICS IN VISCOUS LIQUIDS: A STUDY OF THE VISCOSITY EFFECTS VIA MCT

Isomerization reactions, which involve rotation of a bulky group around a body-fixed axis, often play a fundamental role in chemical and biological processes both in solution and in organized assemblies. In nature, isomerization reactions are found to occur over diverse time scales, ranging from nanosecond for high barrier reactions to a few tens of femtoseconds for barrierless reactions. The study presented in this section is performed for high barrier isomerization reactions in viscous liquids, with an aim to understand the origin of the *fractional viscosity dependence of the rate*. The traditional model of barrier crossing employs a bistable potential where a sizeable activation barrier separates the initial state (reactant) from the final state (product). In such a picture, the crucial step in the reaction is the crossing of the reactant over the barrier to the product region. If the dynamic solvent effects are neglected, then the rate of the barrier crossing is given by the well-known transition state theory (TST). The validity of the TST could be limited in the condensed phases where the reactive motion is often coupled strongly to the environment. The most important and often discussed effect is the reduction of the reaction rate by the solvent frictional forces experienced during the barrier crossing. Over the decades many theoretical (see Refs. 142–153) as well as experimental (see Refs. 154–163]) investigations have been carried out to understand this effect. The rate of barrier crossing as given by the well-known Kramers' theory [148] which is based on the stochastic Langevin equation and is actually the continuum limit of the usual gas phase harmonic TST [150] can be written as

$$k^{Kra} = \frac{\omega_R}{2\pi\omega_b} \left[\left(\frac{\zeta^2}{4} + \omega_b^2 \right)^{1/2} - \zeta/2 \right] \exp(-E_b/k_BT), \qquad (315)$$

where E_b is the height of the barrier, ζ is the zero frequency friction, ω_R is the frequency of motion in the reactant well, and ω_b is the barrier frequency. k_B is the Boltzmann constant and T is the absolute temperature of the system. The above expression provides a simple dependence of rate on friction (ζ) and predicts that in the limit of high friction, the rate should vary inversely with friction. If one further assumes that this friction is proportional to the zero frequency shear viscosity (η) (via the Stokes–Einstein relation with a proper boundary condition), then one predicts that at high viscosity the rate should be inversely proportional to the solvent viscosity. This limiting result is known as the Smoluchowski limit of Kramers' theory and can be easily tested in experiments.

Many experiments (see Refs. 154–160) have shown that Kramers' theory fails to describe the viscosity dependence of rate in isomerization reactions. This is especially the case where the barrier frequency (ω_b) giving the curvature at the barrier top is large. In this case both experimental (see Refs. 154–160) and simulation studies [147, 153] find a rate that decreases with viscosity at a rate *much slower* than that predicted by Kramers' theory. In fact, at high viscosities, it is often found that the rate could be fitted to a form given by

$$k^*_{\text{iso}} = k_{\text{iso}} \exp(E_b/k_B T) = A\eta^{-\alpha}, \tag{316}$$

where A is a viscosity-independent constant. The values of the exponent α lie in the range $1 \geq \alpha \geq 0$. It appears that the higher the barrier frequency, the lower the value of the exponent α. Of course, one recovers the TST result when the exponent is equal to zero.

An elegant explanation for the unusual viscosity dependence was provided by the non-Markovian rate theory (NMRT) of Grote and Hynes [149] which incorporates the idea of frequency dependence of the friction. According to this theory the friction experienced by the reactive motion is *not* the zero frequency macroscopic friction (related to viscosity) but the friction at a finite frequency which itself depends on the barrier curvature. The rate is obtained by a self-consistent calculation involving the frequency-dependent friction.

The situation is far more complex for reactions in high viscous liquids. The frequency-dependent friction, $\hat{\zeta}(z)$ [in the case of Fourier frequency-dependent friction $\zeta(\omega)$], is clearly bimodal in nature. The high-frequency response describes the short time, primarily binary dynamics, while the low-frequency part comes from the *collective* that is, the long-time dynamics. There are some activated reactions, where the barrier is very sharp (i.e., the barrier frequency ω_b is $\geq 100\,\text{cm}^{-1}$). In these reactions, the dynamics is governed only through the ultrafast component of the total solvent response and the reaction rate is completely decoupled from the solvent viscosity. This gives rise to the well-known TST result. On the other hand, soft barriers (i.e., ω_b in the range of $5-10\,\mathrm{cm}^{-1}$) probe mainly the long-time dynamics and the reaction rate can be strongly coupled to the solvent viscosity, giving rise to inverse viscosity dependence. Many isomerization reactions fall in the intermediate regime (i.e., ω_b in the range of $10-100\,\mathrm{cm}^{-1}$). In these cases the reaction dynamics probes both the short-time and long-time dynamics of the liquid, and the reaction rate can be coupled to the viscosity with fractional viscosity dependence. The fractional viscosity dependence behavior decreases with increasing the barrier frequency (ω_b) and ultimately one recovers the well-known TST result in the limit of large ω_b.

In a recent study by Biswas and Bagchi [164], this problem was addressed where the high-frequency frictional response of the liquid was modeled by the Enskog friction. This provides an upper limit of the friction at high frequencies. Thus, *while the generalized hydrodynamic friction [23, 165] underestimates the friction at high frequency, the Enskog limit overestimates the friction in this limit.* For continuous potentials, the friction should be a rapidly decreasing function of frequency, a feature *totally absent* in the Enskog approximation of the friction. It was, of course, not clear to what extent the rate calculation could be affected by this approximation. This is also related to the question of the *sensitivity of the rate to the solute–solvent interaction potential.* It is worth mentioning that use of the generalized hydrodynamic description underestimates the friction at large viscosity. This is because in the usual application the viscoelastic relaxation time τ_s is assumed to be proportional to the viscosity (η) [166].

In this section a detailed investigation of the rate is presented by using a fully microscopic calculation of the friction which refrains from approximating the short-time response by the Enskog form. A similar calculation has been carried out for viscosity. As the short-time friction is expected to be a sensitive function of the interatomic potential, the comparison between the present calculation for continuous potential and the previous one by Biswas and Bagchi [164] could provide valuable insight into the problem.

A detailed study of the viscosity dependence of the rate has been carried out for a large number of thermodynamic state points. A subsequent analysis reveals that over a large variation of viscosity, the rates can indeed be fitted well to Eq. (316) and the exponent α is found to depend strongly on the barrier frequency (ω_b).

In the limit of very large viscosity, such as the one observed near the glass transition temperature, it is expected that rate of isomerization will ultimately go to zero. It is shown here that in this limit the barrier crossing dynamics itself becomes irrelevant and the Grote–Hynes theory continues to give a rate close to the transition theory result. However, there is no paradox or difficulty here. The existing theories already predict an interpolation scheme that can explain the crossover to inverse viscosity dependence of the rate in the limit of very large viscosity. This also provides an understanding of the expected quenching of the isomerization rate at very large viscosities.

A. Grote–Hynes Theory

The rate theory of Grote and Hynes [149] included the non-Markovian (memory) effects by considering the following generalized Langevin equation (GLE) for the dynamics along the reaction coordinate:

$$\mu \frac{dv}{dt} = F(x) - \int_0^t d\tau \zeta(\tau) v(t - \tau) + R(t), \tag{317}$$

where μ is the effective mass and v represents the velocity along the reaction coordinate, $F(x)$ is the systematic force arising from the potential in the barrier region, $\zeta(t)$ is the time-dependent friction, and $R(t)$ is the random force from solvent assumed to be Gaussian. $F(x)$ is assumed to arise from a static potential which is an inverted parabola, so that

$$F(x) = \mu \omega_b^2 x. \tag{318}$$

In general, both $\zeta(t)$ and $R(t)$ appearing in Eq. (317) arise from microscopic motions of heat-bath (solvent) modes interacting with reaction coordinate [167–169]. For isomerization reactions in solvents, both of them arise from the microscopic motions of the solvent molecules interacting with the isomerizing moiety. So, they must be related to each other. They are related by the following relation (known as fluctuation–dissipation theorem):

$$\zeta(t) = \beta \langle R(0)R(t) \rangle, \tag{319}$$

where β is the inverse of the Boltzmann constant (k_B) times the absolute temperature (T) and $\langle \cdots \rangle$ represents the statistical average over heat-bath modes at temperature T. By using the probability distribution from the generalized Fokker–Planck equation, Grote and Hynes [149] obtained, after some rather lengthy analysis, the following simple and elegant expression for the rate constant, k^{GH},

$$k^{GH} = k^{TST}(\lambda_r/\omega_b), \tag{320}$$

where k^{TST} is the transition state rate constant given by

$$k^{TST} = \frac{\omega_R}{2\pi} \exp(-E_b/k_B T). \tag{321}$$

Here, ω_R is the frequency of motion in the reactant well, and E_b is the height of the transition-state barrier. λ_r is the effective barrier frequency with which the reactant molecule passes, by diffusive Brownian motions through the barrier region and is given by the following self-consistent relation

$$\lambda_r = \frac{\omega_b^2}{\lambda_r + \hat{\zeta}(\lambda_r)/\mu}, \tag{322}$$

where $\hat{\zeta}(\lambda_r)$ is the Laplace transform of the time-dependent friction:

$$\hat{\zeta}(\lambda_r) = \int_0^\infty dt\, e^{-\lambda_r t} \zeta(t). \tag{323}$$

Equation (320) predicts the TST result for very weak friction ($\lambda_r \simeq \omega_b$) and predicts the Kramers' result for low barrier frequency (i.e., $\omega_b \to 0$) so that $\hat{\zeta}(\lambda_r)$ can be replaced by $\hat{\zeta}(0)$ in Eq. (322). If the barrier frequency is large ($\omega_b \geq 10^{13}\,\mathrm{s}^{-1}$) and the friction is not negligible ($\hat{\zeta}(0)/\mu \simeq \omega_b$), then the situation is not so straightforward. In this regime, which often turns out to be the relevant one experimentally, the effective friction $\hat{\zeta}(\lambda_r)$ can be quite small even if the zero frequency (i.e., the macroscopic) friction (proportional to viscosity) is very large. The non-Markovian effects can play a very important role in this regime.

It should be noted here that the Grote–Hynes rate expression [Eq. (320)] reduces to the Kramers' result for a delta function friction assumption:

$$\zeta(t) = \zeta \delta(t). \tag{324}$$

It is also important to note that for a system coupled linearly to a finite discrete set of harmonic oscillators, the rate of escape over the barrier is described by the TST [150] and the TST rate is exactly given by Eq. (320). In the limit of continuum of oscillators, the Kramers–Grote–Hynes result is regained [Eq. (320)] [150].

In order to apply the Grote–Hynes formula to realistic cases, the frequency-dependent friction is required which is calculated from the mode coupling theory (MCT) presented in Section IX.

B. Frequency-Dependent Friction

The total frequency-dependent friction calculated from the MCT, $\hat{\zeta}(z)$, is plotted against the Laplace frequency (z) in Fig. 14. In the same figure the Enskog friction ζ_E and the binary contribution $\zeta^B(z)$ are also shown. Note here that in the high-frequency regime the frequency-dependent total friction is much less than the Enskog friction and is dominated entirely by the binary

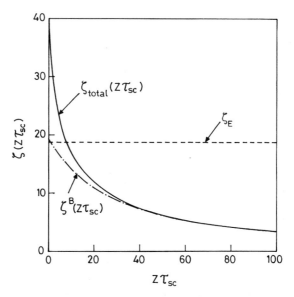

Figure 14. The frequency-dependent total friction $\hat{\zeta}_{total}(z)$ (solid line) and the binary friction $\zeta^B(z)$ (dashed–dot line) plotted as a function of Laplace frequency (z). For comparison, the calculated Enskog friction ζ_E is also shown (dashed line). The calculation has been performed for $\rho^* = 0.85$ and $T^* = 0.85$. The frequency-dependent friction, the Enskog friction, and the frequency are scaled by τ_{sc}^{-1}. This figure has been taken from Ref. 170.

contribution. Therefore, the use of the Enskog friction in barrier crossing dynamics, for systems with continuous potentials, would always overestimate the solvent-induced impedance on reaction rate.

C. Barrier Crossing Rate

The isomerization rate is calculated using the Grote–Hynes formula, given by Eqs. (320) and (322). The frequency-dependent friction ($\hat{\zeta}(z)$) and viscosity (η) has been obtained from the mode coupling theory presented in Section IX. For convenience the rate is expressed in terms of the dimensionless quantity κ in the following form:

$$\kappa = k^{GH}/k^{TST} = \lambda_r/\omega_b. \tag{325}$$

Now to find λ_r iteratively, Eq. (322) can be recast as follows:

$$\kappa^2 + \frac{\kappa\hat{\zeta}(\omega_b\kappa)}{\mu\omega_b} - 1 = 0. \tag{326}$$

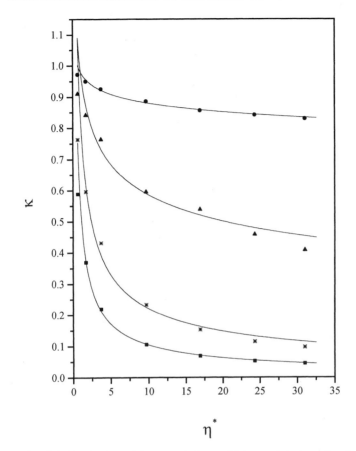

Figure 15. Calculated values of the transmission coefficient κ plotted as a function of the solvent viscosity η^* for four barrier frequencies ω_b at $T^* = 0.85$. The squares denote the calculated results for $\omega_b = 3 \times 10^{12} \text{ s}^{-1}$, the asterisks denote results for $\omega_b = 5 \times 10^{12} \text{ s}^{-1}$, the triangles denote results for $\omega_b = 10^{13} \text{ s}^{-1}$ and the circles denote results for $\omega_b = 2 \times 10^{13} \text{ s}^{-1}$. The solid lines are the best-fit curves with exponents: $\alpha \simeq 0.72$ for $\omega_b = 3 \times 10^{12} \text{ s}^{-1}$, $\alpha \simeq 0.58$ for $\omega_b = 5 \times 10^{12} \text{ s}^{-1}$, $\alpha \simeq 0.22$ for $\omega_b = 10^{13} \text{ s}^{-1}$, and $\alpha \simeq 0.045$ for $\omega_b = 2 \times 10^{13} \text{s}^{-1}$. Note here that the barrier crossing rate becomes completely decoupled from the viscosity of the solvent at $\omega_b = 2 \times 10^{13} \text{s}^{-1}$. The transmission coefficient κ is obtained by using Eq. (326). Note here that the viscosity is calculated using the procedure given in Section X and is scaled by $\sigma^2 / \sqrt{m k_B T}$, and ω_b is scaled by τ_{sc}^{-1}. For discussion, see the text. This figure has been taken from Ref. 170.

Thus κ is the transmission coefficient that measures deviation of the Grote–Hynes rate k^{GH} from the transition state theory rate k^{TST}.

κ has been calculated for a large number of thermodynamic state points (T^*, ρ^*) by varying the solvent density (ρ^*) widely (ρ^* is varied from 0.6

TABLE V

The Value of the Exponent α Obtained from the Best-Fit
Curves (Fig. 15) for Four Different Barrier Frequencies ω_b, at
$T^* = 0.85$

ω_b	α
$3 \times 10^{12} \, s^{-1}$	0.72
$5 \times 10^{12} \, s^{-1}$	0.58
$10^{13} \, s^{-1}$	0.22
$2 \times 10^{13} \, s^{-1}$	0.045

to 1.05) at a fixed reduced temperature, $T^* = 0.85$, for four different barrier frequency (ω_b) values. These four different barrier frequencies are $3 \times 10^{12} \, s^{-1}, 5 \times 10^{12} \, s^{-1}, 10^{13} \, s^{-1}$, and $2 \times 10^{13} \, s^{-1}$. The calculated rates have been plotted as a function of solvent viscosity (η) for all these four barrier frequencies, ω_b (Fig. 15). For all these frequencies, the rate can be fitted very well by fractional viscosity dependence, with exponent (α) in the range of $1 \geq \alpha > 0$. The value of the exponent obtained from the best-fit curves for all four barrier frequencies are presented in Table V.

As can be seen from the numbers, the exponent α is clearly a function of barrier frequency (ω_b) and its value is decreasing with increase in ω_b. For $\omega_b = 2 \times 10^{13} \, s^{-1}$, its value almost goes to zero ($\alpha < 0.05$), which clearly indicates that beyond this frequency the barrier crossing rate is entirely decoupled from solvent viscosity so that one recovers the well-known TST result that neglects the dynamic solvent effects.

In Fig. 16, the values of the exponent α are plotted against the barrier frequency; this is clearly a kind of master curve that summarizes the essence of much of the work reported here. When the rate is calculated from Grote–Hynes theory, this curve depends only weakly on temperature.

There have been several other theoretical studies by different authors [144, 145, 156, 163] where the frequency-dependent friction was modeled by using the modified version of the generalized hydrodynamic expression [23, 165]. These theories failed to reproduce the experimental results at certain limits. Barbara and coworkers attributed this failure of the G–H theory to the non-availability of a reliable frequency-dependent friction and called for the use of a friction better than the hydrodynamic friction.

It is useful to note here that the value of α is determined by the Grote–Hynes formula, which to some extent limits the sensitivity of the rate to viscosity (η). On the other hand, if the binary part of the friction is replaced by the Enskog friction, then larger values of α are obtained [164]. This is because the Enskog friction is directly proportional to $g(\sigma)$, which increases with increasing density. The same correlations also lead to an increase in solvent

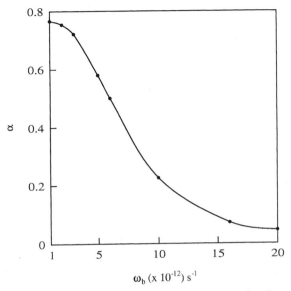

Figure 16. Calculated values of the exponent α plotted as a function of the barrier frequency ω_b. The solid circles are the values of α obtained for different barrier frequencies. The solid line is simply an aid to the human eye. The calculations have been performed by varying the reduced density ρ^* at a constant reduced temperature $T^* = 0.85$ for a particular value of ω_b. For further details, see the text. This figure has been taken from Ref. 170.

viscosity. The above results also emphasize the sensitivity of the rate to the intermolecular potential! Recently a detailed mode coupling theory analysis of isomerization dynamics has been carried out by Murarka et al. [170].

D. Can Diffusion Be the Rate-Determining Step?

The solvents used by Sumi and coworkers [162] in their experimental studies are all high viscous liquids ($\eta \gg 10\,\mathrm{P}$). Diffusion in such systems is naturally very slow. For example, the translational diffusion (D) of a phenyl group in a liquid with viscosity $\eta = 10\,\mathrm{P}$ is about $8.9 \times 10^{-9}\mathrm{cm}^2\,\mathrm{s}^{-1}$ at 300 K. Since isomerization reactions considered here involve large-amplitude spatial motion of bulky groups, one certainly needs to find out the time required by the isomerizing moiety to traverse the distance, by diffusion, from the reactant minimum to that of the product. This time (τ_d) is given by the following well-known relation $\tau_d = \Delta s^2/6D$. The distance Δs is obtained from simple geometric arguments and is given by $\Delta s = r\Delta\theta$, where r is the distance between the body-fixed axis and the center of the sphere of hydrodynamic radius d and $\Delta\theta$ is the change in the twisting angle during the isomerization reaction.

TABLE VI

The Diffusion Rate k_d at $\eta = 10\,P$ and the Barrier Crossing
Rate $k_b \simeq k^{TST}$ at 300 K and 200 K for tS[a]

Temperature (K)	k_d (s^{-1})	$k_b \simeq k^{TST}$ (s^{-1})
300	3.45×10^7	3.5×10^{10}
200	2.3×10^7	2.5×10^9

[a] k^{TST} is calculated for an isolated tS molecule in gas phase
where $E_b \simeq 3.5$ kcal/mol.

Let us now consider the photoinduced isomerization reaction of tS. For this isomerization reaction, $\Delta\theta$ in the excited-state energy surface is $90°$ and Δs is approximately equal to 3.92×10^{-8} cm. At $\eta = 10\,P$, the time required (τ_d) to travel this distance is 2.9×10^{-8} s at 300 K and 4.35×10^{-8} s at 200 K. The diffusion rate is given by $k_d = 1/\tau_d$. In order to understand at what condition the diffusion itself becomes the rate-determining step, the values of the diffusion rate and the barrier crossing rate for TS is presented in Table VI at two different temperatures. If the barrier frequency (ω_b) is 10^{13} s^{-1}, then at this large viscosity the Grote–Hynes theory would predict an *almost* complete decoupling of the barrier crossing rate from the viscosity. That is, k^{GH} is comparable to k^{TST} at this high barrier frequency. Thus the barrier crossing rate can be assumed to be given by k^{TST}. k^{TST} has been calculated for the isolated tS molecule in the gas phase, where the height of the activation barrier is small ($E_b \simeq 3.5$ kcal/mol).

From the numbers presented in Table VI it is clear that while both k_d, and k^{TST} are temperature-dependent, the dependence of k_d is certainly much weaker. However, the recent full Monte Carlo and molecular dynamics simulation study of Gershinsky and Pollak [171] on the viscosity dependence of tS isomerization rate in liquid ethane shows that the height of the barrier (E_b) increases systematically as the solvent density is increased by varying the external pressure. It should be noted here that their study is restricted to ethane solvent and to very low viscosity, up to 0.59 cP [161] (highest pressure studied is 600 MPa). Correspondingly, E_b increases systematically from 3.5 kcal/mol to 4.3 kcal/mol, which leads to a decrease in k^{TST} with density. If the barrier height is kept invariant with pressure (i.e., density), then at high viscosity of 10 P and at 300 K we have $k_b \gg k_d$. Therefore, the rate will be determined by diffusion alone and could be inversely proportional to η. But as the barrier height increases with pressure as suggested by Gershinsky and Pollak [171], the reverse may take place at this high viscosity and the rate may be determined by the barrier crossing dynamics.

Let us next describe the ground-state Z/E isomerization of 4-(dimethylamino)-4$'$-nitroazobenzene (DNAB). This system was studied by Sumi and

TABLE VII

The Diffusion Rate k_d at $\eta = 10^5$ P and the Barrier Crossing
Rate $k_b \simeq k^{TST}$ at 300 K and 200 K for DNAB[a]

Temperature (K)	k_d (s^{-1})	$k_b \simeq k^{TST}$ (s^{-1})
300	8×10^2	2.65×10^3
200	5.5×10^2	0.1

[a] k^{TST} is calculated at $\omega_R = 10^{13}$ s^{-1} where $E_b \simeq 12$ kcal/mol.

coworkers [162]. They varied the viscosity over six orders of magnitude by varying the external pressure. The activation barrier for this reaction is rather high ($E_b \simeq 12$ kcal/mol). Therefore, k^{TST} is very small compared to that for low barrier reaction. Because of this very low value of k^{TST}, the diffusion rate (k_d) for this reaction needs to be calculated only at very high viscosity ($\eta = 10^5$ P) in order to capture the crossover. This will also help in understanding the high-pressure experimental results of Sumi and coworkers [162]. For this ground-state isomerization reaction, $\Delta\theta$ in the ground-state energy surface is 180°. The diffusion rate and the barrier crossing rate for this system are presented in Table VII. The numbers suggest that for DNAB isomerization reaction at 300 K as $k_b > k_d$, the diffusion is the rate-determining step.

This analysis (assuming invariance of E_b with density) shows that for DNAB and *trans*-stilbene isomerization reactions, the crossover from the barrier-crossing-dominated to the diffusion-controlled regime occurs at different viscosities. The magnitude of the viscosity at which this crossover occurs is strongly temperature-dependent. Clearly, the crossover depends on the systems being studied as it is governed by the barrier height (E_b). Thus for *trans*-stilbene the crossover occurs at lower viscosity, than that in DNAB. Note that this crossover can be studied experimentally only if viscosity is varied by changing the external pressure at a constant temperature.

In order to complete the above analysis, one needs to solve the full non-Markovian Langevin equation (NMLE) with the frequency-dependent friction for highly viscous liquids to obtain the rate. This requires extensive numerical solution because now the barrier crossing dynamics and the diffusion cannot be treated separately. However, one may still write phenomenologically the rate as [172],

$$\frac{1}{k} = \frac{1}{k_b} + \frac{1}{k_d}, \tag{327}$$

where $1/k_d$ gives the time required for diffusion from reactant to the product surface (note that the factor 2 is absorbed in k_d itself). Such a description can be useful at intermediate to large viscosity. For large ω_b, k_b is only weakly dependent on η. However, k_d always remains inversely proportional to η. Thus, at small η, $k_b \ll k_d$ and the rate is determined by k_b only. It is in this regime that one expects fractional viscosity dependence. At very large viscosities, on the other hand, one may get back a stronger viscosity dependence. An equation like (327) has already been proposed by Sumi and coworkers [167] for isomerization in viscous liquids, although from an entirely different model. The analysis presented here seems to provide an understanding of the eventual quenching of the isomerization rate at very high viscosities.

XIX. TIME-DEPENDENT DIFFUSION IN TWO-DIMENSIONAL LENNARD-JONES FLUIDS

Diffusion in two-dimensional (2-D) systems is a subject not only of great academic interest but also of practical relevance. Many important chemical and biological processes occur in systems that are best described in two dimensions. However, theoretical study of such dynamic processes face the following paradoxical situation. The different theoretical studies of diffusion in 2-D systems lead to two very different conclusions. Earlier mode coupling theoretical analysis predicts that due to the presence of the t^{-1} tail in the velocity autocorrelation function the diffusion coefficient *diverges* in two dimensions [57]. On the other hand, the kinetic theory finds that it is the collision operator which diverges [57]. Since the diffusion coefficient is roughly the inverse of the collision operator, the kinetic theory predicts a *zero* diffusion coefficient in 2-D systems [57]. As discussed by Pomeau and Resibois [57], both these methods have certain inconsistencies. In the mode coupling theory approach the existence of diffusion coefficient was assumed in the beginning, and then it was shown that diffusion coefficient diverges. On the other hand, in the kinetic theory approach it was assumed that the ring collision term is small compared to the Boltzmann collision operator, and then it was concluded that the ring collision term is infinite. Thus in both theoretical approaches the initial assumptions and final conclusions are not consistent with each other.

Even conclusions drawn from the different computer simulation studies also do not seem to be consistent. The classic computer simulation study of Alder and Wainwright was performed at low density [173]. In this study they have shown that the long-time diffusion coefficient diverges in 2-D systems due to the existence of persistent hydrodynamic flows. On the other hand, some recent molecular dynamics simulations of 2-D systems have reported estimates of the self-diffusion coefficient [174]. In particular, it appeared from these simulations that a diffusion coefficient *might*, after all, exist at higher densities due to the absence of the persistent hydrodynamic flows that were found to be present in the low density studies. Thus neither the theoretical studies nor the computer simulation results lead to a definite conclusion regarding the existence of diffusion coefficient in two dimension.

In order to understand the above questions/paradoxes, a mode coupling theoretical (MCT) analysis of *time-dependent* diffusion for two-dimensional systems has been performed. The study is motivated by the success of the MCT in describing the diffusion in 3-D systems. The main concern in this study is to extend the MCT for 2-D systems and study the diffusion in a Lennard-Jones fluid. An attempt has also been made to answer the anomaly in the computer simulation studies.

The present mode coupling theoretical approach does not presume the existence of the diffusion coefficient, thus avoiding the inconsistencies that were present in the earlier studies [57]. According to the present calculation, it is found that the diffusion coefficient diverges in the long time. It is also found that although the long-time diffusion coefficient (D) diverges due to the long-wavelength hydrodynamic fluctuations, one can still exactly define and accurately calculate the time-dependent diffusion $(D(t))$.

The saturation of $D(t)$ in the long time implies that a finite value of diffusion coefficient exists (which has been reported in certain computer simulation studies [174]). Such a saturation in $D(t)$ can also be achieved from MCT calculation by artificially introducing a lower wavenumber cutoff. This suggests that the saturation in $D(t)$ is a system size effect and arises due to suppression of long-wavelength fluctuations. The agreement between MCT prediction and the simulation results [175] is satisfactory. The rise in $D(t)$ with time in the long time can be fitted to a logarithmic form. It has also been shown that the rise in $D(t)$ is sharper at lower density, and this has been attributed to the earlier emergence of the current term at lower density.

A. Theoretical Formulation

All the particles with the same mass m, and diameter σ are defined to interact through the pairwise additive LJ potential,

$$v(r) = 4\epsilon \left[\left(\frac{\sigma}{r} \right)^{12} - \left(\frac{\sigma}{r} \right)^{6} \right], \tag{328}$$

where ϵ is the interaction parameter. The system is characterized by the reduced number density $\rho^* = \rho\sigma^2$ and reduced temperature $T^* = k_B T/\epsilon$, where k_B is the Boltzmann constant. The units of mass, length, and time are m, σ, and $\tau_{sc} = \sigma\sqrt{m/\epsilon}$, respectively. The time-dependent diffusion $D(t)$ is scaled by σ^2/τ_{sc}.

Among the earlier theoretical formulations of the diffusion in 2-D systems, certainly the kinetic theory and the mode–mode coupling theory approaches are the important ones. As mentioned before, the kinetic theory approach predicts that the collision operator diverges and thus the diffusion coefficient vanishes in 2-D systems [57]. There are various approaches in the mode coupling theory [9, 57, 176], and all these approaches have predicted that in three dimensions the VACF has a $t^{-3/2}$ tail. When these approaches are extended in two dimensions, they show the presence of t^{-1} tail in the VACF, which leads to a divergence of the diffusion in the long time. A brief discussion on the previous theories are presented here.

1. Kinetic Theory Approach

The VACF is given by

$$C_v(t) = \langle v_{1x}(t)v_{1x}(0)\rangle, \tag{329}$$

where $v_{1x}(t)$ denotes the x-component of the velocity of particle 1 at time t. The above expression can be explicitly written as [57],

$$C_v(t) = \lim_A \int dr^N dp^N v_{1x} \exp[-iL_N t] v_{1x} \rho_N^{eq}, \tag{330}$$

where A denotes the area, L_N is the Liouville operator, and ρ_N^{eq} denotes the canonical equilibrium distribution. After some steps of algebra the above expression in the frequency domain can be written as [57]

$$C_v(z) = \int d^2 p_1 v_{1x} \frac{1}{-iz - \tilde{C}_0(p_1; z)} v_{1x} \phi^{eq}(p_1), \tag{331}$$

where $\phi^{eq}(p_1)$ denotes the Maxwellian distribution

$$\phi^{eq}(p_1) = \frac{1}{(2\pi m k_B T)} \exp(-p_1^2/2m k_B T). \tag{332}$$

In Eq. (331), $\tilde{C}_0(p_1; z)$ is the frequency-dependent collision operator given by

$$\tilde{C}_0(p_1; z) = \int_0^\infty dt e^{izt} \left[-\lim_A \int dr^N dp^{N-1} \right.$$
$$\left. \times [L_N \exp(-i(1 - \hat{P}_N)L_N t)(1 - \hat{P}_N)L_N \frac{\rho_N^{eq}}{\phi^{eq}(p_1)}] \right]. \tag{333}$$

In Eq. (6), $\hat{P}_N$ is the projection operator that gives the projection along p_1 and is defined as

$$\hat{P}_N = \frac{\rho_N^{eq}}{\phi^{eq}(p_1)} \int dr^N dp^{N-1}. \tag{334}$$

Now a density expansion of the collision operator in Eq. (331) can be performed and the first few terms in this expansion are [57]

$$\tilde{C}_0(p_1;z) = \rho \tilde{C}_0^{(0)}(p_1;z) + \rho^2 \tilde{C}_0^{(1)}(p_1;z) + \cdots . \qquad (335)$$

In the above expression, $\tilde{C}_0^{(0)}(p_1;z)$ is the finite frequency generalization of the Boltzmann–Lorentz collision operator. $\tilde{C}_0^{(1)}(p_1;z)$ can be described by the finite frequency generalization of the Choh–Uhlenbeck collision operator. [57]. This operator describes the dynamical correlations created by the collisions between three particles. Using the above-mentioned description the expression of $\tilde{C}_0^{(1)}(p_1;z)$ can be shown to be written as [57]

$$\tilde{C}_0^{(1)}(p_1;z) \simeq_{z \to 0} \ln z \qquad (336)$$

Thus replacing the expression for $\tilde{C}_0^{(1)}(p_1;z)$ in Eq. (331), we find that the $C_v(z)$ goes to zero in the limit $z \to 0$. Now the diffusion coefficient is given by

$$D = \frac{1}{2} \int_0^\infty d\tau C_v(\tau) = \frac{1}{2} C_v(z=0). \qquad (337)$$

Thus as $\tilde{C}_0^{(1)}(p_1;z)$ diverges in the limit $z \to 0$ the diffusion coefficient becomes zero.

The inconsistency present in this approach is that in the density expansion of the collision operator it was assumed that $\tilde{C}_0^{(1)}(p_1;z)$ is small compared to $\tilde{C}_0^{(0)}(p_1;z)$ but finally it has been shown that the former diverges. Thus the assumption made during the density expansion is not correct, and such an expansion cannot be performed.

2. Mode Coupling Theory Approach

There have been various approaches in the mode coupling theory [9, 37, 57, 176]. All these theories have exhibited the presence of $t^{-3/2}$ of the velocity autocorrelation function in the asymptotic limit in three dimensions. Extending each of these theories for studies in two dimensions we can show that the velocity autocorrelation function has t^{-1} tail in the asymptotic limit. Since the diffusion coefficient is related to $C_V(t)$ through Eq. (337), it can be shown that D diverges in the long time due to the presence of this t^{-1} tail in the VACF.

Here the presence of this t^{-1} in the VACF is shown, and the inconsistency present in the theory has been discussed. For this study, the MCT approach that has been presented in Section IX for 3-D studies has been extended in the case of two dimensions.

The mode coupling theory expressions for the calculation of the friction in two-dimensions can be obtained in a similar way as has been described in Section IX for a neat liquid in three dimensions. In the mode coupling theory the total friction is decomposed into a short-time part and a long-time part. The long-time part that describes the correlated recollisions is obtained by expanding the total friction in the basis set of the eigenfunctions of the Liouville operator—that is, the density, the longitudinal current mode, and the transverse current mode. It is found that the density, the transverse current mode, and the short-time part of the friction makes major contribution to the total friction. Thus the frequency-dependent friction is given by the following expression:

$$\frac{1}{\zeta(z)} = \frac{1}{\zeta^B(z) + R_{\rho\rho}(z)} + R_{TT}(z). \tag{338}$$

In the above expression, $\zeta^B(z)$ is the binary part of the friction, and $R_{\rho\rho}(z)$ gives the coupling of the solute motion to the density modes of the solvent through two-particle direct correlation function. $R_{TT}(z)$ gives the coupling to the transverse current through the transverse vertex function. $R_{\rho\rho}(z)$ and $R_{TT}(z)$ are obtained through Laplace transformation of $R_{\rho\rho}(t)$ and $R_{TT}(t)$, respectively. In two dimensions, the expressions for $R_{\rho\rho}(t)$, $R_{TT}(t)$, and $\zeta^B(t)$ are given by

$$R_{\rho\rho}(t) = \frac{\rho k_B T}{m} \int [d\mathbf{q}'/(2\pi)^2](\hat{q} \cdot \hat{q}')^2 q'^2 [c_{12}(q')]^2 [F^s(q',t)F(q',t) \\ - F_0^s(q',t)F_0(q',t)], \tag{339}$$

$$R_{TT}(t) = \frac{1}{\rho} \int [d\mathbf{q}'/(2\pi)^2][1 - (\hat{q} \cdot \hat{q}')^2][\gamma_d^t(q')]^2 \omega_0^{-4} [F^s(q',t)C_{tt}(q',t) \\ - F_0^s(q',t)C_{tt0}(q',t)]. \tag{340}$$

$$\zeta^B(t) = \omega_0^2 \exp(-t^2/\tau_\zeta^2), \tag{341}$$

where ω_0 is the well-known Einstein frequency [21, 22], now in two-dimensions

$$\omega_0^2 = \frac{\rho}{2m} \int d\mathbf{r} g(r) \nabla^2 v(r), \tag{342}$$

here $g(r)$ is the radial distribution function that has been obtained from the simulation [175].

In Eq. (341), the relaxation time τ_ζ is determined from the second derivative of $\zeta^B(t)$ at $t = 0$ and is given exactly by

$$\omega_0^2/\tau_\zeta^2 = (\rho/2m^2) \int d\mathbf{r}(\nabla^\alpha \nabla^\beta v(\mathbf{r}))g(\mathbf{r})(\nabla^\alpha \nabla^\beta v(\mathbf{r}))$$

$$+ (1/4\rho) \int [d\mathbf{q}/(2\pi)^2] \gamma_d^{\alpha\beta}(\mathbf{q})(S(q) - 1)\gamma_d^{\alpha\beta}(\mathbf{q}), \qquad (343)$$

where summation over repeated indices is implied. The expression for $\gamma_d^{\alpha\beta}(\mathbf{q})$ is written as a combination of the distinct parts of the second moments of the longitudinal and transverse current correlation functions $\gamma_d^l(\mathbf{q})$ and $\gamma_d^t(\mathbf{q})$, respectively, given by

$$\gamma_d^l(\mathbf{q}) = -\frac{\rho}{m} \int d\mathbf{r} \, \exp(-i\mathbf{q} \cdot \mathbf{r})g(r)\frac{\delta^2}{\delta y^2}v(r) \qquad (344)$$

and

$$\gamma_d^t(\mathbf{q}) = -\frac{\rho}{m} \int d\mathbf{r} \, \exp(-i\mathbf{q} \cdot \mathbf{r})g(r)\frac{\delta^2}{\delta x^2}v(r). \qquad (345)$$

To calculate $R_{\rho\rho}(t)$, the two-particle direct correlation function $c_{12}(q)$ is required which is obtained from the nearly analytical expression given by Baus and Colot for a 2-D system [177]. The static structure factor $S(q)$ has been calculated from the two-particle direct correlation function through the well-known Ornstein–Zernike relation [21].

The other dynamic variables required to calculate $R_{\rho\rho}(t)$ and $R_{TT}(t)$ are the dynamic structure factor of the solvent, $F(q, t)$, the inertial part of the dynamic structure factor, $F_0(q, t)$, the transverse current autocorrelation function of the solvent, $C_{tt}(q, t)$, the inertial part of the same, $C_{tt0}(q, t)$, the self-dynamic structure factor of the solute, $F^s(q, t)$, and the inertial part of the self-dynamic structure factor of the solute, $F_0^s(q, t)$. The expressions for all the above-mentioned dynamic quantities are similar to that given in Section IX but in two dimensions.

From the studies in three dimensions it is known that in the asymptotic limit it is the current term which makes the dominant contribution. The asymptotic limit of the R_{TT} term [given by Eq. (340)] is calculated assuming $F^s(q, t)$ and $C_{tt}(q, t)$ in the diffusive limit. In this limit the expression for $R_{TT}(t)$ can be written as

$$R_{TT}(t) = \frac{1}{8\rho}\left[\pi\left(D + \frac{\eta}{\rho m}\right)t\right]^{-1}, \qquad (346)$$

where D is the diffusion coefficient of the solute and η is the viscosity of the medium. The presence of t^{-1} in $R_{TT}(t)$ gives rise to the t^{-1} tail in the VACF, which leads to the divergence of the diffusion coefficient.

The inconsistency in this approach and all other mode coupling theoretical approaches [9, 37, 57, 176] is that a finite diffusion coefficient has been assumed to define the diffusive behaviour of the self-dynamic structure factor, and then it has been concluded that this diffusion coefficient itself diverges.

3.   Present Theoretical Formulation

In the present theoretical approach this inconsistency is avoided by not assuming any diffusion coefficient in defining the self-dynamic structure factor.

The self-dynamic structure factor of the solute at all time is calculated from the mean square displacement (MSD) using the following definition:

$$F^s(q,t) = \exp\left(-\frac{q^2\langle\Delta r^2(t)\rangle}{4}\right). \tag{347}$$

The MSD is obtained from the VACF through the following expression:

$$\langle\Delta r^2(t)\rangle = 2\int_0^t d\tau\, C_v(\tau)(t-\tau). \tag{348}$$

$C_v(t)$ is obtained by Laplace inverting $C_v(z)$, which is related to the frequency-dependent friction by the following generalized Einstein relation:

$$C_v(z) = \frac{k_B T}{m(z + \zeta(z))}. \tag{349}$$

Thus the frequency-dependent friction is calculated self-consistently with the MSD.

The self-consistency is implemented through the following iterative scheme. First, the VACF is obtained from Eq. (349) by replacing the total frequency-dependent friction, $\zeta(z)$, by its binary part, $\zeta^B(z)$. The VACF thus obtained is used to calculate the MSD through Eq. (348). Now this MSD is used to calculate $R_{\rho\rho}(t)$ and $R_{TT}(t)$ and thus $\zeta(z)$. This total friction is then used to calculate the new VACF, which again is used to determine MSD and thus $\zeta(z)$. This iterative process is continued until the VACF obtained from two consecutive steps overlap.

Once the VACF is obtained self-consistently, the time-dependent diffusion, $D(t)$, is calculated from the following expression [178]:

$$D(t) = \frac{1}{2}\int_0^t d\tau\, C_v(\tau). \tag{350}$$

It is to be noted that the present formulation will still give the t^{-1} tail in $R_{TT}(t)$ and thus in the VACF, but in Eq. (346) the diffusion coefficient is now replaced by the time-dependent diffusion $D(t)$.

Thus the present formulation does not have the inconsistency that was present in the earlier approaches.

B. Results and Discussions

The numerical calculations are done both at low and high density at $\rho^* = 0.6$ and $\rho^* = 0.7932$ and at $T^* = 0.7$. The high-density calculations are performed to investigate if there is any existence of diffusion coefficient at high density as claimed by the simulation studies [174].

1. Velocity Autocorrelation Function

The VACF calculated at $\rho^* = 0.6$ and at $T^* = 0.7$ is plotted in Fig. 17 . The long-time tail present in the VACF is clearly demonstrated in the figure. As discussed before, it is the presence of this long-time t^{-1} tail in the VACF which gives rise to the divergence of the diffusion in the long time. It is found that the presence of the t^{-1} tail in the VACF calculated at high density is not

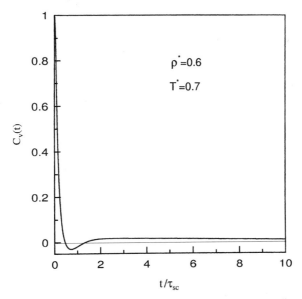

Figure 17. Normalized velocity autocorrelation function of a two-dimensional fluid plotted against reduced time. The plot shows the presence of the t^{-1} tail in $C_v(t)$ at long time. The plot is at $\rho^* = 0.6$ and $T^* = 0.7$. The time is scaled by $\tau_{sc} = \sqrt{m\sigma^2/\epsilon}$. This figure has been taken from Ref. 188.

that prominent. This is because of slower emergence of the current mode at high density. This point will be discussed later.

2. Time-Dependent Diffusion

Although the expressions for the diffusion in two-dimensions looks similar to that in the case of three dimensions, their behavior in these two dimensions are entirely different. While in three dimensions the time-dependent diffusion converges to a constant value in the long time, in two dimensions the diffusion coefficient diverges with time. The time-dependent diffusion at $\rho^* = 0.6$ and at $T^* = 0.7$ is plotted in Fig. 18. The $D(t)$ shows a divergent behavior that is expected due to the presence of the long-time t^{-1} tail present in the VACF which is shown in Fig. 17.

The time-dependent diffusion calculated at higher density, at $\rho^* = 0.7932$, and at $T^* = 0.7$ is plotted in Fig. 19. The plot shows that the diffusion at higher density also does not saturate to a finite value but increases with time. In the same figure the $D(t)$ values obtained from the simulated VACF and MSD [175] have also been plotted. The agreement between MCT and the simulations is satisfactory.

To understand the long-time behavior of time-dependent diffusion, the long-time part of the $D(t)$ calculated at higher density ($\rho^* = 0.7932$ and at

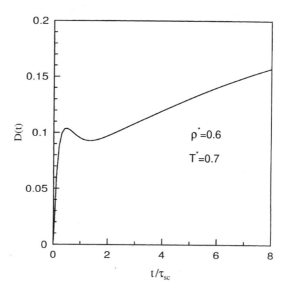

Figure 18. Time-dependent diffusion, $D(t)$, of a two dimensional neat liquid plotted against reduced time. The plot shows a faster rise in $D(t)$ with time. The plot is at $\rho^* = 0.6$ and $T^* = 0.7$. The time is scaled by $\tau_{sc} = \sqrt{m\sigma^2/\epsilon}$. $D(t)$ is scaled by σ^2/τ_{sc}. This figure has been taken from Ref. 188.

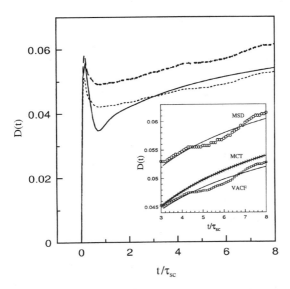

Figure 19. Time-dependent diffusion $D(t)$ of a two-dimensional system plotted against reduced time. The solid line represents the $D(t)$ obtained from the mode coupling theory (MCT) calculation, and the short-dashed line and the long-dashed line represent the $D(t)$ obtained from simulated VACF and MSD, respectively. In the inset, fits to long-time $D(t)$ to Eq. (351) are also shown. The plots are at $\rho^* = 0.7932$ and $T^* = 0.7$. The time is scaled by $\tau_{sc} = \sqrt{m\sigma^2/\epsilon}$. $D(t)$ is scaled by σ^2/τ_{sc}. This figure has been taken from Ref. 175.

$T^* = 0.7$) has been fitted (shown in the inset of Fig. 19) to the following form:

$$D(t) = A_0 + A_1 \ln t. \tag{351}$$

It is found that the $D(t)$ does show a logarithmic divergence as expected due to the presence of the t^{-1} tail in the VACF. The value of the slope (A_1) is 0.0091, which is in good agreement with those obtained from the simulations (0.0079 from VACF and 0.0086 from MSD).

3. System Size Dependence

In a recent MD simulation [175] it has been found that the time-dependent diffusion shows a system size dependence. There is a variation in the $D(t)$s calculated for different system sizes. From this simulation it is found that for smaller systems the rise in $D(t)$ is slower and $D(t)$ saturates at a finite value for very small systems.

This saturation in $D(t)$ can also be achieved from the present MCT calculation by using a lower wavenumber cutoff. Cutting off the contribution from

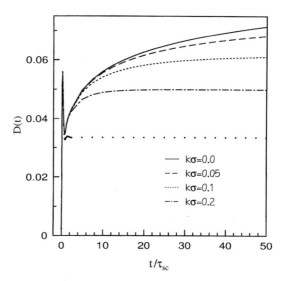

Figure 20. Time-dependent diffusion $D(t)$ of a two-dimensional system calculated using different low-wavenumber cutoff plotted against reduced time. The different low-wavenumber cutoff ($k\sigma$) values are indicated in the figure. The dotted line represents the $D(t)$ obtained neglecting the contribution from the current term. The plots are at $\rho^* = 0.7932$ and $T^* = 0.7$. The time is scaled by $\tau_{sc} = \sqrt{m\sigma^2/\epsilon}$. $D(t)$ is scaled by σ^2/τ_{sc}. This figure has been taken from Ref. 175.

the lower wavenumbers is equivalent to taking smaller systems in the simulation studies. Both of them lead to the suppression of the long-wavelength fluctuations which are otherwise present in the system. In Fig. 20 the time-dependent diffusion obtained from MCT using different lower wavenumber cutoff is plotted against reduced time. This plot shows that if we cut off or suppress long-wavelength fluctuations, then $D(t)$ does not diverge at long time but saturates to constant values depending on the lower cutoff in the $q\sigma$ integration. In the same figure the $D(t)$ obtained by neglecting the contribution from the current term is also plotted. This shows that it is only the long-wavelength *hydrodynamic* fluctuations which lead to the divergence of the diffusion coefficient. Thus this analysis seems to suggest that the saturation in the time-dependent diffusion obtained in the theoretical study is the same as system size effect obtained in the simulation studies [175]. Both can be attributed to the suppression of the contribution from the long-wavelength hydrodynamic fluctuations.

When the $D(t)$ values calculated at higher and lower densities are compared (see Figs. 18 and 19), it is found that $D(t)$ increases faster with time at lower density. This seems to suggest that at high density the $1/t$ behavior of the VACF is latent due to the slower emergence of the current mode [which is shown to be responsible for the divergence of $D(t)$] and thus the dominance of the microscopic term (the binary and the density relaxation) at short and intermediate times. This could possibly explain why the persistent hydrodynamic fluctuations that are found to be present at low-density computer simulation studies [173] are absent in studies performed at higher density [174].

The above discussion suggests the following scenario. There could be two possible reason behind obtaining a finite diffusion coefficient in the simulation studies performed at higher density [174]. For simulations with the same number of particles since the system size at higher density is smaller than that at lower density, the long-wavelength fluctuations that are present at low-density studies might be absent in studies performed at higher density. Also due to the slower emergence of the current mode at higher density, the divergence of $D(t)$ or the existence of the persistent hydrodynamic flows might show up at longer times.

Note that the above study is performed for a simple system. There exists a large body of literature on the study of diffusion in complex quasi-two-dimensional systems—for example, a collodial suspension. In these systems the diffusion can have a finite value even at long time. Schofield, Marcus, and Rice [17] have recently carried out a mode coupling theory analysis of a quasi-two-dimensional colloids. In this work, equations for the dynamics of the memory functions were derived and solved self-consistently. An important aspect of this work is a detailed calculation of wavenumber- and frequency-dependent viscosity. It was found that the functional form of the dynamics of the suspension is determined principally by the binary collisions, although the mode coupling part has significant effect on the long-time diffusion.

The results discussed in this section can be useful to understand diffusion controlled reactions on two dimensional surfaces [178–180].

XX. MODE COUPLING THEORY OF DIFFUSION IN ONE-DIMENSIONAL LENNARD-JONES RODS

The diffusion coefficient of a tagged molecule shows strong dependence on the dimensionality of the system. While it exists in three and one dimensions, it diverges in the intermediate two dimensions [57, 173, 175]. There is, however, no difficulty in defining a diffusion coefficient in arbitrary dimensions in terms of either (a) the time integral over the velocity autocorrelation function (VACF) or (b) the slope of the mean square displacement (MSD) at long times. The reason for the divergence of diffusion coefficient in two dimensions can be understood in terms of a long-time tail in the decay of the velocity time correlation function, of the form $1/t^{d/2}$ (for two and three dimensions). This long-time tail arises from the coupling of the tagged particle motion with the transverse current flow of the liquid. This gives rise to a logarithmic divergence of the MSD in two dimensions [57]. No such divergence exists for three-dimensional systems.

The behavior of VACF and of D in one-dimensional systems are, therefore, of special interest. The transverse current mode of course does not exist here, and the decay of the longitudinal current mode (related to the dynamic structure factor by a trivial time differentiation) is sufficiently fast to preclude the existence of any "dangerous" long-time tail. Actually, Jepsen [181] was the first to derive the closed-form expression for the VACF and the diffusion coefficient for *hard rods*. His study showed that in the *long time* VACF decays as $1/t^3$, in contrast to the $t^{-d/2}$ dependence reported for the two and three dimensions. Lebowitz and Percus [182] studied the short-time behavior of VACF and made an exponential approximation for VACF [i.e, $C_v(t) = e^{-2t}$], for the *short times*. Haus and Raveche [183] carried out the extensive molecular dynamic simulations to study relaxation of an initially ordered array in one dimension. This study also investigated the $1/t^3$ behavior of VACF. However, none of the above studies provides a *physical explanation* of the $1/t^3$ dependence of VACF at long times, of the type that exists for two and three dimensions.

Unlike for hard rods, no analytical solution exists for one-dimensional Lennard-Jones (LJ) rods. Molecular dynamics simulations have revealed a $1/t^3$ behavior in this system also [184, 185].

Recently a mode coupling theory study of diffusion and velocity correlation function of a one-dimensional LJ system was carried out [186]. This study reveals that the $1/t^3$ decay of the velocity correlation function could arise from the coupling of the tagged particle motion to the longitudinal current mode of the surrounding fluid. In this section a brief account of this study is presented.

A.   Mode Coupling Theory Analysis

The frequency (z)-dependent velocity correlation function $C_v(z)$ is related to the frequency dependent friction by the well-known generalized Einstein relation,

$$C_v(z) = \frac{k_B T}{m(z + \zeta(z))}, \tag{352}$$

where $\zeta(z)$ is the frequency-dependent friction. As in the mode coupling theory, the full friction is now decomposed into a short- and a long-time part. Short-time part arises from the binary collisions of tagged particle with the surrounding solvents, and the long-time part originates from the correlated recollisions. Final expression for the frequency-dependent friction used to calculate both VACF and time-dependent diffusion is given by [9]

$$\zeta(z) = \zeta^B(z) + \zeta^R(z), \tag{353}$$

where $\zeta^B(z)$ is the binary part of the zero-frequency friction and $\zeta^R(z)$ is the ring collision term, which contains the contributions from the repeated collisions to the total friction. In a one-dimensional system we can replace the ring collision term in the above expression by

$$\zeta^R(z) = R_{\rho\rho}(z) + 2\zeta^B(z)R_{\rho l}(z) + \zeta^B(z)R_{ll}(z)\zeta^B(z). \tag{354}$$

The above expression is similar to that of a three-dimensional system [9, 10] but with the absence of the term that contains the contribution from transverse current to the total friction. In the above expression, $R_{\rho\rho}(z)$ contains the coupling to the density and is given by

$$R_{\rho\rho}(t) = \frac{\rho k_B T}{m} \int [dq'/(2\pi)]\, q'^2\, [c(q')]^2\, [F^s(q',t) - F_0^s(q',t)]F(q',t). \tag{355}$$

$c(q)$ is the Fourier transform of $c(x)$. $R_{ll}(z)$ contains the longitudinal current while $R_{\rho l}$ includes the coupling of density and longitudinal current, which can be expressed by the following expressions in one dimension by following the similar procedure used in three-dimensional systems [9,187], and

$$R_{ll}(t) = -\frac{1}{\rho} \int [dq'/(2\pi)][\gamma_d^l(q') + (\rho q^2/m\beta)c(q')]^2 \omega_0^{-4}$$
$$\times [F^s(q',t) - F^0(q',t)]C_l(q',t) \tag{356}$$

and

$$R_{\rho l}(t) = -\int [dq'/(2\pi)]c(q')[\gamma_d^l(q) + (\rho q^2/m\beta)c(q')]\omega_0^{-2}$$
$$\times [F^s(q',t) - F^0(q',t)]\frac{d}{dt}F(q',t), \tag{357}$$

where $\gamma_d^l(q)$ is the distinct part of the second moment of the longitudinal current correlation function, which is given by the following equation:

$$\gamma_d^l(q) = -\frac{\rho}{m} \int dx\, \cos(qx)g(x)\frac{d^2}{dx^2}v(x) \tag{358}$$

and ω_0 is the well-known Einstein frequency in one-dimensional systems and is given by

$$\omega_0^2 = \frac{\rho}{m} \int dx \ g(x) \frac{d^2}{dx^2} v(x). \tag{359}$$

$\zeta^B(t)$ is the binary part of the friction whose expression is given by

$$\zeta^B(t) = \omega_0^2 \exp(-t^2/\tau_\zeta^2). \tag{360}$$

The relaxation time τ_ζ is determined from the second derivative of the above expression for $\zeta^B(t)$, which is given by the following equation:

$$\frac{\omega_0^2}{\tau_\zeta^2} = \frac{\rho}{3m^2} \int dx \frac{d^2}{dx^2} v(x) g(x) \frac{d^2}{dx^2} v(x) + \frac{1}{4\pi\rho} \int dq \, \gamma_d^l(q)(S(q) - 1)\gamma_d^l(q). \tag{361}$$

The static structure factor, $S(q)$, appearing in the above expression is calculated by using the one-dimensional Fourier transform of the radial distribution function. The Fourier-transformed two-particle direct correlation function $c(q)$ is obtained through the well-known Ornstein–Zernike relation [21].

Calculational procedure of all the dynamic variables appearing in the above expressions—namely, the dynamic structure factor $F(q,t)$ and its inertial part, $F_0(q,t)$, and the self-dynamic structure factor $F^s(q,t)$ and its inertial part, $F_0^s(q,t)$—is similar to that in three-dimensional systems, simply because the expressions for these quantities remains the same except for the terms that include the dimensionality. $C_v(t)$ is calculated so that it is fully consistent with the frequency-dependent friction. In order to calculate either VACF or diffusion coefficient, we need the two-particle direct correlation function, $c(x)$, and the radial distribution function, $g(x)$. Here x denotes the separation between the centers of two LJ rods. In order to make the calculations robust, we have used the $g(x)$ obtained from simulations.

The longitudinal current correlation function, $C_l(q,t)$, is related to the dynamic structure factor by the following expression:

$$C_l(q,t) = -\frac{m^2}{q^2} \frac{d^2}{dt^2} F(q,t). \tag{362}$$

It is important to note that at sufficiently long times the only significant contribution to the integral over the current mode comes from small wavenumbers.

B.   Long-Time Behavior of VACF

The long-time behavior of VACF is determined by the longitudinal current term $R_{ll}(t)$. The expression of $R_{ll}(t)$ is given by Eq. (356).

In the limit of small q (long wavelength) the following limiting condition holds:

$$\frac{[\gamma_d^l(q') + (\rho q^2/m\beta)c(q')]^2}{\omega^4} \xrightarrow{q \to 0} 1. \tag{363}$$

Thus, at the long times, the time dependence of $C_v(t)$ and hence $R_{ll}(t)$ can be expressed as

$$C_v(t) \approx R_{ll}(t) \sim \int dq' \, F^s(q',t)C_l(q',t). \tag{364}$$

The long-time behavior of the integrand in the above equation is determined by $C_l(q,t)$. Now $C_l(q,t)$ is determined by the intermediate scattering function as given by Eq. (362). When Eq. (362) is substituted in the above equation, we get

$$R_{ll}(t) \approx \int dq \frac{d^2}{dt^2} F(q,t). \tag{365}$$

The important point now to note is that in one-dimensional hard rods, $F(q,t)$ decays mostly as a Gaussian function of time. At small q decay, $F(q,t)$ can be given as

$$F(q,t) \approx \exp(-aq^2t^2), \tag{366}$$

where a is a constant.

The Gaussian decay of $F(q,t)$ at small wavenumbers at large density is a manifestation of the well-defined cage around each molecule. Thus, the relaxation of density remains nearly elastic for sufficiently long times. It is further assumed that on the time scale of decay of $F(q,t)$, the relaxation of the self term, $F_s(q,t)$, is negligible. By making this Gaussian ansatz for $F(q,t)$ and substituting $C_l(q,t)$ in Eq. (356), we get the following expression for the longitudinal current:

$$R_{ll}(t) \approx \frac{m^2}{2\sqrt{\pi}\rho t^3}, \tag{367}$$

which simply shows that in the long times longitudinal current goes as $1/t^3$.

Note that the two essential ingredients of the above derivation is the contribution of the longitudinal current term and the Gaussian decay of the intermediate scattering function. The above derivation is by no means rigorous, but numerical calculations verify both the above reasons as the origin of $1/t^3$ time dependence of VACF. The important new point is that $F(q, t)$ decays as Gaussian (in time) at small q [186].

C. Results

In order to fully analyze VACF of the 1D LJ systems, both MD simulations and MCT theory have been carried out together. In Fig. 21 the simulated long time tail of VACF is plotted after an initial decay. This figure clearly shows the t^{-3} behavior of $C_v(t)$ in the long time over a wide range of densities. This decay is more dominant, at low and intermediate densities. This can be attributed to the disappearance of local structure and the existence of positive tail in $C_v(t)$ over a longer time. On the other hand, the existence of long-time tail will be suppressed at higher densities due to the presence of the negative region and the oscillations over a longer time which masks the $1/t^3$ decay of VACF in the long time.

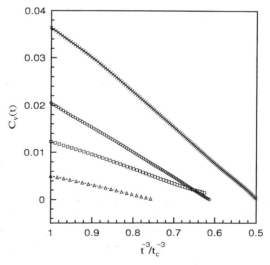

Figure 21.   The long-time tails of $C_v(t)$ obtained from simulations plotted against $(t_c/t)^3$ at various densities at $T^* = 1.0$. t is the reduced time and t_c is the time at which the long-time tail of $C_v(t)$ started approaching zero. The different symbols from top to bottom represent the $C_v(t)$ at reduced densities 0.3, 0.4, 0.6, and 0.82, respectively. The figure shows the dominance of $1/t^3$ decay in $C_v(t)$ at low and intermediate densities. This figure has been taken from Ref. 186.

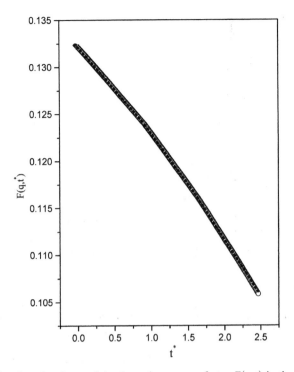

Figure 22. Gaussian decay of the dynamic structure factor $F(q,t)$ in the relevant time range. In this figure, the ratio $F(q,t)/S(q)$ has been plotted against the reduced time at a small value of the wavenumber, $ql = 0.1206$. This figure has been taken from Ref. 186.

In Fig. 22, the normalized dynamic structure factor $F(q,t)/S(q)$ is plotted at a small wavevector value, $ql = 0.1206$, in the time domain where the VACF shows pronounced t^{-3} decay. Circles show the simulated values, and the full line is the Gaussian fit. As seen from the figure, $F(q,t)$ is a Gaussian function of time in the $q \rightarrow 0$ limit. This is an important observation because it provides the key to the physical origin for the slow decay of $C_v(t)$.

Normalized VACF obtained from self-consistent mode coupling theory has been plotted in Fig. 23, against the time at a medium density $\rho^* = 0.6$ and $T^* = 1.0$. Here, we have compared the simulated $C_v(t)$ with the $C_v(t)$ obtained from MCT. As seen from the figure the theory is in good agreement with the simulation results at both short and long times. Nevertheless, the agreement is not perfect. This is because the Gaussian approximation made for the binary part is most likely at fault.

It is interesting to note that despite the existence of long-time tails of $C_v(t)$, we can still have a well-defined diffusion coefficient. We find that the decay of the longitudinal current mode is sufficiently fast to preclude

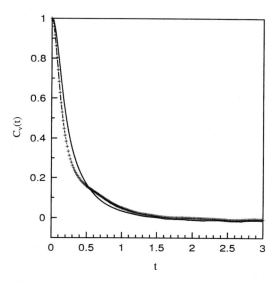

Figure 23. $C_v(t)$ obtained from the MCT and from the MD simulations plotted against the reduced time at $\rho^* = 0.6$ and $T^* = 1.0$. Symbols show the simulated $C_v(t)$, and the full line represents the $C_v(t)$ obtained by MCT. This figure has been taken from Ref. 186.

the existence of "dangerous" long-time tails of VACF, thereby avoiding the divergence of diffusion in one dimension.

The agreement of the MCT results with the simulations supports our explanation that the long-time $1/t^3$ decay of $C_v(t)$ arises from the coupling of the tagged particle motion with the longitudinal current of the surrounding fluid. To strengthen our argument, we have also calculated the coefficient multiplying the t^{-3} tail, which is equal to 0.050 from the MCT and 0.065 from simulations. Given the uncertainties in the simulations and in the $g(x)$ which is also obtained from simulations, this agreement can be regarded as satisfactory.

XXI. FUTURE PROBLEMS

In this section we briefly discuss a few of the problems that we think deserve special attention.

A. Proper Marriage Between the Short- and the Long-Time Dynamics

As discussed in Section IX, there are several approximate schemes to stitch together the short- and long-time dynamic behavior. While the initial decay is determined by the short-time expansion, the long-time decay is determined by the collective variables. Unfortunately, there is no exact way to combine these two limits. In this chapter, we have proposed an alternative scheme that is certainly more accurate than the existing one when the mass and the size of the tagged solute is larger than those of the solvent molecules. However, this scheme is still not completely satisfactory. Actually there is yet no quantitative theory to calculate (without any adjustable parameter) the diffusion coefficient even of a simple LJ fluid, except at very low density. The important point to note is that at the intermediate times, both the binary interaction and the collective modes contribute to the time-dependent friction. While one knows the binary term exactly for LJ fluids, the same is not true for the contributions of the density and the current modes. Thus, the description of the short-time behavior is problematic.

One choice could be to extend the short-time expansion to fourth or even sixth order in time and then remove the terms of such orders from the density and the current terms. Note that the transverse current contribution starts at the sixth order while the density term starts at the fourth order. Such a scheme can certainly be more successful, although it does reduce the simplicity of the present scheme.

B. Mode Coupling Theory of Dielectric and Orientational Relaxation in Dipolar Liquids

The mode coupling theory of molecular liquids could be a rich area of research because there are a large number of experimental results that are still unexplained. For example, there is still no fully self-consistent theory of orientational relaxation in dense dipolar liquids. Preliminary work in this area indicated that the long-time dynamics of the orientational time correlation functions can show highly non-exponential dynamics as a result of strong intermolecular correlations [189, 190]. The formulation of this problem, however, poses formidable difficulties. First, we need to derive an expression for the wavevector-dependent orientational correlation functions $C_{l,m}(\mathbf{k}, t)$, which are defined as

$$C_{l,m}(\mathbf{k}, t) = \langle Y_{l,m}(-\mathbf{k}, 0) Y_{l,m}(\mathbf{k}, t) \rangle \tag{368}$$

with

$$Y_{l,m}(-\mathbf{k}, t) = \Sigma_i \exp(i.\mathbf{k}.\mathbf{r}) Y_{l,m}(-\mathbf{\Omega}, t), \tag{369}$$

where $Y_{l,m}(-\mathbf{\Omega}, t)$ is the l, mth spherical harmonic with angle $\mathbf{\Omega}$. Some progress toward this have been made, but to the best of our knowledge no mode coupling theory calculation of these correlation functions has been carried out. Actually this is a very interesting problem because the decay of the orientational time correlation functions are determined partly by the translational diffusion of the molecules themselves—this effect is particularly important at the intermediate wavenumbers. Thus, the problem of translation and rotation is coupled at a microscopic level.

The Second ingredient is the expression of the rotational friction in terms of the orientational time correlation functions. We have earlier derived an expression for this which was based on Kirkwood's formula [190]. The full expression should be derived by following an approach similar to that of Sjogren and Sjolander [9]. In addition, the coupling to rotational currents (the vortices) have not been touched upon.

Third, one needs accurate static orientational correlation functions. These are now available for the first rank correlation functions for water, acetonitrile, and several other liquids [191].

C. Concentration Dependence of Ionic Mobility and Viscosity of Electrolyte Solutions

The concentration dependence of ionic mobility at high ion concentrations and also in the melt is still an unsolved problem. A mode coupling theory of ionic mobility has recently been derived which is applicable only to low concentrations [18]. In this latter theory, the solvent was replaced by a dielectric continuum and only the ions were explicitly considered. It was shown that one can describe ion atmosphere relaxation in terms of charge density relaxation and the elctrophoretic effect in terms of charge current density relaxation. This theory could explain not only the concentration dependence of ionic conductivity but also the frequency dependence of conductivity, such as the well-known Debye–Falkenhagen effect [18]. However, because the theory does not treat the solvent molecules explicitly, the detailed coupling between the ion and solvent molecules have not been taken into account. The limitation of this approach is most evident in the calculation of the viscosity. The MCT theory is found to be valid only to very low values of the concentration.

When the solvent molecules are explicitly included, one needs to treat a ternary system (two ions and the dipolar solvent molecules). The additional slow variables to be included in the mode coupling theory are the products of the ion charge and solvent densities. This will explicitly introduce terms like $F_{is}(k, t)$, which is the partial dynamic structure factor involving the ion and the solvent molecules. The calculation of the microscocpic terms of the friction, containing the density terms, does not appears to be difficult, but calculation of the current terms now appears to be formidable.

D. Mode Coupling Theory of Mixtures

This is yet another very important problem where the mode coupling theory can be used to understand a myriad of unexplained properties, the most

interesting being the pronounced nonideal behavior in the transport properties of the mixtures [192]. Among the nonideal properties, viscosity and diffusion are the two that can be addressed by using the mode coupling theory
approach. Such a theory can also be used to understand various relaxation
phenomena such as vibrational phase and energy relaxation in binary mixtures. Until now, theoretical work has been primarily directed toward understanding of the glass transition in binary mixtures which are good glass
formers [193, 194].

Mode coupling theory of binary mixtures where the constituents are of
rather different sizes is a challenging task, as we have already discussed while
addressing the mass depenence of diffusion. In addition to the problem with
proper formulation of mode coupling terms, there is an additional difficulty
of the nonavailability of the equilibrium two-particle correlation functions:
The existing integral equation theories become unstable when the size ratio
exceeds a certain (low) value, like 1.5 or so [195].

E. Mode Coupling Theory of Polyelectrolyte Solutions

Dilute polyelectrolyte solutions, such as solutions of tobacco mosaic virus
(TMV) in water and other solvents, are known to exhibit interesting dynamic
properties, such as a plateau in viscosity against concentration curve at very
low concentration [196]. It also shows a shear thinning at a shear strain rate
which is inverse of the relaxation time obtained from the Cole–Cole plot of
frequency dependence of the shear modulus, $G(\omega)$.

A simple theory of the concentration dependence of viscosity has recently
been developed by using the mode coupling theory expression of viscosity
[197]. The slow variables chosen are the center of mass density and the
charge density. The final expressions have essentially the same form as
discussed in Section X; the structure factors now involve the intermolecular
correlations among the polyelectrolyte rods. Numerical calculation shows
that the theory can explain the plateau in the concentration dependence of
the viscosity, if one takes into account the anisotropy in the motion of the
rod-like polymers. The problem, however, is far from complete. We are
also not aware of any study of the frequency-dependent properties. Work
on this problem is under progress [198].

F. Mode Coupling Theory of Polymer Dynamics

This is a large area of research, and significant progress in this direction has
already been made by Schweizer [199] and by Freed and coworkers [200].
However, these two schemes are entirely different. While the approach of
Freed pays special attention to the coupling between different Rouse modes,
the formulation of Schweizer is more in line with the mode coupling theory
of liquids with monomer density as the slow collective mode. The main

difficulty in formulating a full mode coupling theory of polymer dynamics is of course the connectivity that introduces a different set of slow variables. The correct approach would have ingredients of both the two approaches mentioned above. In addition, the theory may need to be tailored, to avoid complexity, according to the nature of a given problem. For example, for diffusion of a flexible polymer chain, the density of polymer chains becomes a slow variable. The static and dynamic correlations between the center of mass will be required. The decay of dynamic correlations between two flexible polymers can be mediated through the Rouse modes, in addition to the conventional diffusion mode.

G. Diffusion in Supercritical Fluids

As discussed before, the contribution of the current modes are not important for diffusion in dense neat liquids at low temperature. This is because at large viscosity, the decay of the transverse current mode is rapid. In the gas phase, however, the decay of the current modes can be rather slow, and this can contribute significantly to diffusion. The transverse current mode is expected to play an important role here also because it is precisely the range where the current mode plays important role in hard-sphere fluids. Preliminary work is being done to find an important role of transverse current modes in the diffusion of supercritical fluids [201].

The remnants of the critical behavior is seen even at conditions significantly away from the critical point. For example, the specific heat and compressibilities are still quite large at the reduced temperature $T^* = 1.41$ and density $\rho^* = 0.3$, whereas the critical values are $T^* = 1.32$ and $\rho^* = 0.32$. This makes study of diffusion near the critical point a challenging theoretical problem.

H. Mode Coupling Theory Calculation of Wavevector-Dependent Transport Properties

Recently, an interesting study of the molecular dynamics calculation of the wavevector dependence of the viscosity and the thermal conductivity of a Lennard-Jones fluid was reported [202]. The transport properties were found to decrease rapidly as the value of the wavevector k was increased from zero, and they were nearly zero when $k\sigma$ is larger than 5. However, we are not aware of any mode coupling theory calculation of this interesting behavior. In fact, most of the theoretical expressions exist, but the numerical calculation is formidable.

XXII. CONCLUSION

The list of problems provided above is by no means exhaustive and is certainly biased by the authors' own interests and expertise. Although the framework of mode coupling theory has been around for more than four decades, its use to understand various aspects of liquid-state dynamics, pertaining to chemistry in particular, is relatively new. In this chapter, we have tried to remove this lacuna. We have described calculations of diffusion and viscosity, with two new ingredients. First is the imposition of the full self-consistency between the frequency-dependent friction and the velocity time correlation function. The second is the use of a new decomposition to combine the short- and the long-time descriptions. While this new decomposition is still approximate, it should be more accurate when the mass of the solute becomes appreciably larger than that of the solvent. Numerical calculations verify this argument.

One of the objectives of this chapter has been to articulate the historical development of MCT in a coherent fashion to enable the nonexperts to gather the proper perspective. There are of course more advanced and specialized reviews which the reader can turn to for more details [3, 16].

We have not dwelt on the limitations of the mode coupling theory near the glass transition due to the acivated or rare events—these have been discussed at length at many places. Instead, we emphasized the applications of MCT to liquid-state dynamics.

It is fair to say that the mode coupling theory is still at its infancy, although this seems to be the only theoretical framework we currently have at our disposal to deal with the coupled single-particle and collective dynamics in a strongly correlated random system. In particular, MCT provides a unique theoretical framework to calculate both the short- and the long-time dynamics, although several difficulties remain, as mentioned above. With the availability of more accurate static and dynamic informations of complex systems, we can look forward to interesting and exciting future developments in this area.

Acknowledgment

We thank Prof. Karl Freed for suggesting the review and for several helpful discussions. It is a pleasure to thank Mr. Rajesh Murarka for much help in the preparation of the manuscript and also for collaboration in Section XVIII and for discussions. We thank Mr. G. Srinivas for help, discussions, and collaboration in sections XV, XIX, and XX. We thank Dr. N. Gayathri and Dr. R. Biswas for collaboration in Sections XVII and XVIII, respectively. We thank Dr. Kuni Miyazaki for many illuminating discussions on various aspects of mode coupling theory, and we thank Prof. A. Yethiraj for discussions. We thank Dr. Ashwini Kumar for much help in preparing the manuscript. It is a pleasure to thank Prof. Graham Fleming, Prof. Jim Skinner, Prof. Sergei Egorov, Prof. Mark Ediger, Prof. Mark Maroncelli, and Prof. Iwao Ohmine for many discussions, help, and encouragement. The work reported here was supported in part by grants from the Department of Science and Technology, India and the Council of Scientific and Industrial Research, India. SB thanks CSIR for a Senior Research Fellowship.

References

1. S.-K. Ma, *Modern Theory of Critical Phenomena*, W. A. Benjamin, New York, 1976.

2. H. E. Stanley, *Introduction to Phase Transitions and Critical Phenomena*, Oxford University Press, New York, 1971.

3. W. Gotze, *Liquids, Freezing and the Glass Transition*, edited by J. P. Hansen, D. Levesque, and J. Zinn-Justin, North-Holland, Amsterdam, 1991, p. 287.

4. W. Gotze and L. Sjogren, *Rep. Prog. Phys.* **55**, 241 (1992).

5. G. F. Mazenko, *Phys. Rev. A* **7**, 209 (1973).

6. G. F. Mazenko, *Phys. Rev. A* **7**, 222 (1973).

7. G. F. Mazenko, *Phys. Rev. A* **9**, 360 (1974).

8. J. R. Mehaffey and R. I. Cukier, *Phys. Rev. A* **17**, 1181 (1978).

9. L. Sjogren and A. Sjolander, *J. Phys. C: Solid State Phys.* **12**, 4369 (1979); L. Sjogren, *J. Phys. C: Solid State Phys.* **13**, 705 (1980).

10. T. R. Kirkpatrick, *Phys. Rev. A* **32**, 3130 (1985).

11. R. Zwanzig, *Ann. Rev. Phys. Chem.* **16**, 67 (1965).

12. R. Zwanzig, Lectures in: *Theoretical Physics*, Vol. III, edited by W. E. Britton, B. W. Downs, and J. Downs, Wiley-Interscience, New York, 1961, p. 135.

13. (a) J. L. Barrat, J. N. Roux, and J. P. Hansen, *Chem. Phys.* **149**, 197 (1990). (b) J. N. Roux, J. L. Barrat, and J. P. Hansen, *J. Phys.: Condens. Matter* **1**, 7171 (1989).

14. G. Wahnstrom, *Phys. Rev. A* **44**, 3752 (1991).

15. A. H. Marcus, J. Schofield, and S. A. Rice, *Phys. Rev. E* **60**, 5725 (1999).

16. U. Balucani and M. Zoppi, *Dynamics of the Liquid State*, Clarendon Press, Oxford, 1994; also see references therein.

17. J. Schofield, A. H. Marcus, and S. A. Rice, *J. Phys. Chem.* **100**, 18950 (1996).

18. (a) A. Chandra and B. Bagchi, *J. Chem. Phys.* **110**, 10024 (1999). (b) A. Chandra and B. Bagchi, *J. Chem. Phys.* **112**, 1876 (2000).

19. L. D. Landau and E. M. Lifshitz, *Fluid Mechanics*, Pergamon Press, London, 1963.

20. B. J. Berne and R. Pecora, *Dynamic Light Scattering*, John Wiley, New York, 1976.

21. J. P. Hansen and I. R. McDonald, *Theory of Simple Liquids*, Academic, New York, 1986; also see references therein.

22. J. P. Boon and S. Yip, *Molecular Hydrodynamics*, McGraw-Hill, New York, 1980; also see references therein.

23. R. Zwanzig and M. Bixon, *Phys. Rev. A* **2**, 2005 (1970).

24. J. R. D. Copley and S. W. Lovesey, *Rep. Prog. Phys.* **38**, 461 (1975).

25. D. Levesque, L. Verlet, and J. Kurkijarvi, *Phys. Rev. A* **7**, 1690 (1973).

26. P. G. de Gennes, *Physica* **25**, 825 (1959).

27. T. Munakata, *J. Phys. Soc. Japan* **45**, 749 (1978).

28. B. Bagchi, *J. Chem. Phys.* **82**, 5677 (1985); *Physica A* **145**, 273 (1987).

29. J. M. Gordon, *Phys. Rev.* **18**, 1272 (1973).

30. T. R. Kirkpatrick, *Phys. Rev. A* **31**, 939 (1985).

31. I. M. de Schepper and E. G. D. Cohen, *Phys. Rev. A* **22**, 287 (1980); *J. Stat. Phys.* **27**, 223 (1982).

32. W. E. Alley, B. J. Alder, and S. Yip, *Phys. Rev. A* **27**, 3174 (1983).

33. I. M. de Schepper, E. G. D. Cohen, and M. J. Zuilhof, *Phys. Lett.* **101A**, 399 (1984); M. J. Zuilhof, E. G. D. Cohen, and I. M. de Schepper, *Phys. Lett.* **103A**, 120 (1984); E. G. D. Cohen, I. M. de Schepper, and M. J. Zuilhof, *Physica B* **127**, 282 (1984).

34. E. Leutheusser, *Phys. Rev. A* **29**, 2765 (1984).

35. M. Fixman, *J. Chem. Phys.* **36**, 310 (1962).

36. K. Kawasaki, *Ann. Phys.* **61**, 1 (1970).

37. L. P. Kadanoff and J. Swift, *Phys. Rev.* **166**, 89 (1968).

38. R. Kubo, *J. Phys. Soc. Japan* **12**, 570 (1957).

39. T. Geszti, *J. Phys. C: Solid State Phys.* **16**, 5805 (1983).

40. E. P. Gross, *J. Stat. Phys.* **11**, 503 (1974); *J. Stat. Phys.* **15**, 181 (1976).

41. C. D. Boley, *Ann. Phys.* **86**, 91 (1974); *Phys. Rev.* **11**, 328 (1975).

42. P. Resibois and J. L. Lebowitz, *J. Stat. Phys.* **12**, 483 (1975).

43. L. Sjogren, *Ann. Phys.* **113**, 304 (1978).

44. M. H. Ernst and J. D. Dorfman, *Physica, (Utrecht)* **61**, 157 (1972).

45. J. R. Dorfman, H. van Beijeren and C. P. McClure, *Arch. Mech. Stosow.* **28**, 333 (1976).

46. J. A. Leegwater, *J. Chem. Phys.* **94**, 7402 (1991).

47. L. Sjogren, *Phys. Rev. A* **22**, 2883 (1980).

48. D. Levesque and L. Verlet, *Phys. Rev. A* **2**, 2514 (1970).

49. A. Rahman, *Phys. Rev. A* **136**, 405 (1964); *J. Chem. Phys.* **45**, 258 (1966).

50. A. G. Zawadzki and J. T. Hynes, *Chem. Phys. Lett.* **113**, 476 (1985).

51. A. Z. Akeazu and E. Daniels, *Phys. Rev. A* **2**, 962 (1970).

52. S. W. Lovesey, *J. Phys. C: Solid State Phys.* **4**, 3057 (1971).

53. J. R. Mehaffey, R. C. Desai, and R. Kapral, *J. Chem. Phys.* **66**, 1665 (1977).

54. B. Bagchi and A. Chandra, *Adv. Chem. Phys.* **80**, 1 (1991).

55. A. Chandra and B. Bagchi, *J. Phys. Chem.* (in press).

56. M. H. Ernst, E. H. Hauge, and J. M. J. van Leeuwen, *Phys. Rev. A* **4**, 2055 (1971).

57. Y. Pomeau and P. Resibois, *Phys. Rep.* **19**, 63 (1975).

58. J. T. Hynes, R. Kapral, and M. Weinberg, *J. Chem. Phys* **70**, 1456 (1979).

59. S. Bhattacharyya and B. Bagchi, *J. Chem. Phys.* **106**, 1757 (1997).

60. S. A. Egorov, M. D. Stephens, A. Yethiraj, and J. L. Skinner *Mol. Phys.* **88**, 477 (1996).

61. R. Zwanzig and R. D. Mountain, *J. Chem. Phys.* **43**, 4464 (1965).

62. U. Balucani, R. Vallauri, and T. Gaskell, *Phys. Rev. A* **37**, 3386 (1988).

63. P. Hutchinson, *Proc. Phys. Soc. London* **91**, 506 (1967).

64. C. G. Gray and K. E. Gubbins, *Theory of Molecular Fluids*, Vol. I, Oxford University Press, Oxford, 1984.

65. T. Gaskell and M. S. Woolfson, *J. Phys. C* **17**, 5087 (1984).

66. L. Sjogren, *Phys. Rev. A* **22**, 2866 (1980).

67. W. Hess and R. Klein, *J. Phys. A: Math. Gen.* **13**, L5 (1980).

68. U. Bengtzelius, *Phys. Rev. A* **34**, 5059 (1986).

69. U. Bengtzelius, W. Gotze, and A. Sjolander, *J. Phys. C* **17**, 5915 (1984); W. Gotze and L. Sjogren, *Chem. Phys.* **212**, 47 (1996).

70. M. Berg, *J. Phys. Chem. A* **102**, 17 (1998).

71. J. T. Hynes, in: *The Theory of Chemical Reaction Dynamics*, Vol. IV, edited by M. Baer, CRC Press, Boca Raton, FL, 1985, p. 171.

72. M. Berg, *Chem. Phys. Lett.* **228**, 317 (1994).

73. T. Joo et al., *J. Chem. Phys.* **104**, 6089 (1996); M. L. Horng et al., *J. Phys. Chem.* **99**, 17311 (1995).

74. (a) B. J. Berne, M. E. Tuckerman, J. E. Straub, and A. L. R. Bug, *J. Chem. Phys.* **93**, 5084 (1990). (b) J. E. Straub, M. Borkovec, and B. J. Berne, *J. Chem. Phys.* **89**, 4833 (1988).

75. U. Balucani, R. Vallauri, and T. Gaskell, *Ber. Bunsenges. Phys. Chem.* **94**, 261 (1990).

76. B. H. C. Chen, C. K. J. Sun, and S. H. Chen, *J. Chem. Phys.* **82**, 2052 (1985); R. Ravi and D. Ben-Amotz, *Chem. Phys.* **183**, 385 (1994).

77. U. Balucani and M. Zoppi, *Dynamics of the Liquid State*, Claredon, Oxford, 1994. Page 193, Table 5.1 for the value of the diffusion coefficient. Page 261, Table 6.1 for the viscosity value. Page 264, using the effective particle radius instead of the defined length scale *a*.

78. U. Balucani, R. Vallauri, T. Gaskell, and S. F. Duffy *J. Phys.: Condens. Matter* **2** , 5015 (1990).

79. U. Balucani, *Mol. Phys.* **71**, 123 (1990).

80. See, for example, papers in *Science* **267**, 1995).

81. M. D. Ediger, C. A. Angell, and S. R. Nagel, *J. Phys. Chem.* **100**, 13200 (1996).

82. F. Fujara, B. Geil, H. Sillescu, and G. Fleischer, *Z. Phys. B* **88**, 195 (1992).

83. G. Heuberger and H. Sillescu, *J. Phys. Chem.* **100**, 15255 (1996).

84. M. T. Cicerone and M. D. Ediger, *J. Chem. Phys.* **104**, 7210 (1996).

85. M. T. Cicerone and M. D. Ediger, *J. Chem. Phys.* **103**, 5684 (1995).

86. M. T. Cicerone, F. R. Blackburn, and M. D. Ediger, *J. Chem. Phys.* **102**, 471 (1995).

87. M. T. Cicerone, F. R. Blackburn, and M. D. Ediger, *Macromolecules* **28**, 8224 (1995).

88. D. Thirumalai and R. D. Mountain, *Phys. Rev. E* **47**, 479 (1993).

89. G. Tarjus and D. Kivelson, *J. Chem. Phys.* **103**, 3071 (1995).

90. (a) L. Sjogren and W. Gotze, in: *Dynamics of Disordered Materials: Proceedings of the ILL Workshop Grenoble, France, September 26–28*, edited by D. Richter, A. J. Dianoux, W. Petry, and J. Teixeira, Springer-Verlag, New York, 1988, p. 18. (b) F. Mezei, in: *Dynamics of Disordered Materials: Proceedings of the ILL Workshop Grenoble, France, September 26–28*, edited by D. Richter, A. J. Dianoux, W. Petry, and J. Teixeira, Springer-Verlag, New York, 1988, p. 164.

91. K. Kawasaki, *Physica A* **208**, 35 (1994).

92. S. Dattagupta and L. A. Turski, *Phys. Rev. Lett.* **54**, 2359 (1985).

93. S. Dattagupta, *J. Non-Cryst. Solids* **131**, 200 (1991).

94. B. Bagchi, *J. Chem. Phys.* **101**, 9946 (1994).

95. R. Zwanzig, *J. Chem. Phys.* **97**, 4507 (1983).

96. B. J. Alder, W. E. Alley, and J. H. Dymond, *J. Chem. Phys.* **61**, 1415 (1974).

97. P. T. Herman and B. J. Alder, *J. Chem. Phys.* **61**, 987 (1972).

98. I. Ebbsjo, P. Schofield, K. Skold, and I. Waller, *J. Phys. C* **7**, 3891 (1974).

99. R. J. Bearman and D. L. Jolly, *Mol. Phys.* **44**, 665 (1981).

100. R. J. Bearman and D. L. Jolly, *Mol. Phys.* **52**, 447 (1984).

101. K. Kerl and M. Willeke, *Mol. Phys.* **96**, 1169 (1999).

102. F. Ould-Kaddour and J. L. Barrat, *Phys. Rev. A* **45**, 2308 (1992).

103. K. Toukubo, K. Nakanishi, and W. Watanabe, *J. Chem. Phys.* **67**, 4162 (1977).

104. R. Walser, A. E. Mark, and W. F. van Gunsteren, *Chem. Phys. Lett.* **303**, 583 (1999).

105. D. Brown and J. H. R. Clarke, *J. Chem. Phys.* **86**, 6446 (1987).

106. F. P. Ricci, *Phys. Rev.* **156**, 184 (1967).

107. A. W. Castleman, Jr. and J. J. Conti, *Phys. Rev. A* **2**, 1975 (1970).

108. G. G. Allen and P. J. Dunlop, *Phys. Rev. Lett.* **30**, 316 (1973).

109. P. J. Dunlop and C. M. Bignell, *J. Chem. Phys.* **108**, 7301 (1998).

110. D. Beece et al., *Biochemistry* **19**, 5147 (1980).

111. E. W. Findsen, J. M. Friedman, and M. R. Ondrias, *Biochemistry* **27**, 8719 (1988).

112. T. G. Hiss and E. L. Cussler, *AIChE J.* **19**, 698 (1973).

113. H. J. V. Tyrrell and K. R. Harris, *Diffusion in Liquids: A Theoretical and Experimental Study*, Butterworths, London, 1984.

114. G. L. Pollack and J. J. Enyeart, *Phys. Rev. A* **31**, 980 (1985); G. L. Pollack et al., *J. Chem. Phys.* **92**, 625 (1990).

115. D. F. Evans, T. Tominaga, and H. T. Davis, *J. Chem. Phys.* **74**, 2 (1981); S.-H. Chen, D. F. Evans, and D. H. T. Davis, *AICHE J.* **29**, 640 (1983); S.-H. Chen, D. F. Evans, and D. H. T. Davis, *J. Chem. Phys.* **75**, 1422 (1981).

116. R. Zwanzig and A. K. Harrison, *J. Chem. Phys.* **83**, 5861 (1985).

117. A. J. Easteal and L. A. Woolf, *Physica A* **121**, 286 (1983); A. J. Easteal and L. A. Woolf, *Chem. Phys. Lett.* **167**, 329 (1990); A. J. Easteal and L. A. Woolf, *Chem. Phys.* **88**, 101 (1984).

118. J. D. Weeks, D. Chandler, and H. C. Anderson, *J. Chem. Phys.* **54**, 5237 (1971); D. Chandler and J. D. Weeks, *Phys. Rev. Lett.* **25**, 149 (1970).

119. J. L. Lebowitz, *Phys. Rev. A* **133**, 895 (1964).

120. P. W. Atkins, *Physical Chemistry*, ELBS with Oxford University Press, Oxford, 1994.

121. D. B. Amotz and D. R. Herschbach, *J. Phys. Chem.* **94**, 1038 (1990).

122. (a) D. W. Oxtoby, *Adv. Chem. Phys.* **40**, 1 (1979). (b) D. W. Oxtoby, *Annu. Rev. Phys. Chem.* **32**, 77 (1981); *Adv. Chem. Phys.* **47**, 487 (1981).

123. K. S. Schweizer and D. Chandler, *J. Chem. Phys.* **76**, 2296 (1982).

124. A. Myers and F. Markel, *Chem. Phys.* **149**, 21 (1990).

125. D. W. Oxtoby, *J. Chem. Phys.* **70**, 2605 (1979).

126. K. Tominaga and K. Yoshihara, *Phys. Rev. Lett.* **74**, 3061 (1995); K. Tominaga, in: *Advances in Multi-Photon Processes and Spectroscopy*, Vol. 11, in press; K. Tominaga and K. Yoshihara, *J. Chem. Phys.* **104**, 1159 (1996), K. Tominaga and K. Yoshihara, *J. Chem. Phys.* **104**, 4419 (1996).

127. W. Kiefer and H. J. Bernstein, *J. Raman Spectros.* **1**, 417 (1973).

128. M. R. Battaglia and P. A. Madden, *Mol. Phys.* **36**, 1601 (1978).

129. P. A. Madden and R. M. Lynden-Bell, *Chem. Phys. Lett.* **38**, 163 (1976).

130. S. F. Fischer and A. Laubereau, *Chem. Phys. Lett.* **35**, 6 (1975).

131. R. Kubo, *J. Math. Phys.* **4**, 174 (1963).

132. S. Bhattacharyya and B. Bagchi, *J. Chem. Phys.* **106**, 7262 (1997).

133. N. Gayathri, S. Bhattacharyya, and B. Bagchi, *J. Chem. Phys.* **106**, 1757 (1997); N. Gayathri and B. Bagchi, *J. Chem. Phys.* **110**, 539 (1999).

134. R. Kubo, *J. Phys. Soc. Japan* **17**, 1100 (1962).

135. J. C. Decius and R. M. Hexter, *Molecular Vibrations in Crystals*, McGraw-Hill, New York 1977.

136. W. L. Jorgensen, J. D. Madura, and C. J. Swenson, *J. Am. Chem. Soc.* **106**, 6638 (1984).

137. A. Michels and M. Goudeket, *Physica* **8**, 347 (1941).

220   BIMAN BAGCHI AND SARIKA BHATTACHARYYA

138. G. Srinivas, S. Bhattacharyya, and B. Bagchi, *J. Chem. Phys.* **110**, 4477 (1999).
139. R. B. Wright, M. Schwartz, and C. H. Wang, *J. Chem. Phys.* **58**, 5125 (1973).
140. C. Brodbeck, I. Rossi, Nguyen-Van-Thank, and A. Ruoff, *Mol. Phys.* **32**, 71 (1976).
141. J. Schroeder, V. Schiemann, and J. Jonas, *Mol. Phys.* **34**, 1501 (1977).
142. J. T. Hynes, *Annu. Rev. Phys. Chem.* **36**, 573 (1985); *J. Stat. Phys.* **42**, 149 (1986).
143. R. F. Grote, G. van der Zwan, and J. T. Hynes, *J. Phys. Chem.* **88**, 4676 (1984); **88**, 4767 (1984).
144. B. Bagchi and D. W. Oxtoby, *J. Chem. Phys.* **78**, 2735 (1983).
145. B. Bagchi, *Int. Rev. Phys. Chem.* **6**, 1 (1987).
146. B.Carmeli and A. Nitzan, *J. Chem. Phys.* **79**, 393 (1983); **80**, 3596 (1984).
147. A. Nitzan, *Adv. Chem. Phys.* **70**, 489 (1988).
148. H. A. Kramers, *Physica* **7**, 284 (1940).
149. R. F. Grote, and J. T. Hynes, *J. Chem. Phys.* **73**, 2715 (1980); **74**, 4465 (1981).
150. E. Pollak, *J. Chem. Phys.* **85**, 865 (1986).
151. E. Pollak, H. Grabert, and P. Hänggi, *J. Chem. Phys.* **91**, 4073 (1989); G. R. Haynes, G. A. Voth, and E. Pollak, *J. Chem. Phys.* **101**, 7811 (1994).
152. P. Hänggi, P. Talkner, and M. Borkovec, *Rev. Mod. Phys.* **62**, 251 (1990).
153. B. Berne, M. Berkovec, and J. E. Straub, *J. Phys. Chem.* **92**, 3711 (1988); J. E. Straub and B. J. Berne, *J. Chem. Phys.* **85**, 2999 (1986); J. E. Straub, M. Berkovec, and B. J. Berne, *J. Chem. Phys.* **84**, 1788 (1986).
154. D. H. Waldeck and G. R. Fleming, *J. Phys. Chem.* **85**, 2614 (1981).
155. S. P. Velsko and G. R. Fleming, *J. Chem. Phys.* **65**, 59 (1982); **76**, 3553 (1982); S. P. Velsko, D. H. Waldeck, and G. R. Fleming, *J. Chem. Phys.* **78**, 249 (1983).
156. G. R. Rothenberger, D. K. Negus, and R. M. Hochstrasser, *J. Chem. Phys.* **79**, 5360 (1983).
157. S. H. Courtney and G. R. Fleming, *J. Chem. Phys.* **83**, 215 (1985).
158. N. S. Park and D. H. Waldeck, *J. Chem. Phys.* **91**, 943 (1989).
159. D. H. Waldeck, *Chem. Rev.* **91**, 415 (1991).
160. G. R. Fleming and P. G. Wolynes, *Physics Today* **43(5)**, 36 (1990).
161. (a) J. Schroeder, D. Schwarzer, J. Troe, and F. Voβ, *J. Chem. Phys.* **93**, 2393 (1990); (b) J. Schroeder, J. Troe and P. Vöhringer, *Chem. Phys. Lett.* **203**, 255 (1993).
162. T. Asano, H. Furuta, and H. Sumi, *J. Am. Chem. Soc.* **116**, 5545 (1994); H. Sumi and T. Asano, *J. Chem. Phys.* **102**, 9565 (1995); H. Sumi, *J. Mol. Liq.* **65**, 65 (1995).
163. S. R. Folm, V. Nagarajan, and P. F. Barbara, *J. Phys. Chem.* **90**, 2085 (1986); A. M. Brearley, S. R. Folm, V. Nagarajan, and P. F. Barbara, *ibid.* **90**, 2092 (1986).
164. R. Biswas and B. Bagchi, *J. Chem. Phys.* **105**, 7543 (1996).
165. H. Metiu, D. W. Oxtoby, and K. F. Freed, *Phys. Rev. A* **15**, 361 (1977).
166. A. B. Bhatia, *Ultrasonic Absorption*, Oxford, London, 1967; T. A. Litovitz and C. M. Davis, in: *Physical Acoustics*, edited by W. P. Mason, Academic, New York, 1965, page 281.
167. R. Zwanzig, *J. Stat. Phys.* **9**, 215 (1973).
168. E. Cortes, B. J. West, and K. Lindenberg, *J. Chem. Phys.* **82**, 2708 (1985).
169. B. J. Gertner, K. R. Wilson, and J. T. Hynes, *J. Chem. Phys.* **90**, 3537 (1989).
170. R. K. Murarka, S. Bhattacharyya, R. Biswas, and B. Bagchi, *J. Chem. Phys.* **110**, 7365 (1999).

171. G. Gershinsky and E. Pollak, *J. Chem. Phys.* **105**, 4388 (1996); G. Gershinsky and E. Pollak, *J. Chem. Phys.* **107**, 812 (1997).

172. (a) J. T. Hynes, *J. Phys. Chem.* **90**, 3701 (1986). (b) *Chem. Phys. Lett.* **90**, 21 (1984).

173. B. J. Alder and T. E. Wainwright, *Phys. Rev. A* **1**, 18 (1970).

174. D. N. Perera and P. Harrowell, *Phys. Rev. Lett.* **80**, 4446 (1998); *Phys. Rev. Lett.* **81**, 120 (1998); M. M. Hurley and P. Harrowell, *Phys. Rev. E* **52**, 1694 (1995).

175. S. Bhattacharyya, G. Srinivas, and B. Bagchi, *Phys. Lett. A* **226**, 394 (2000).

176. T. Gaskell and S. Miller, *J. Phys. C: Solid State Phys.* **11**. 3749 (1978).

177. M. Baus and J. L. Colot, *J. Phys. C : Solid State Phys.* **19**, L643 (1986).

178. S. Okuyama and D. W. Oxtoby, *J. Chem. Phys.* **84**, 5824 (1986); W. Dong and J. C. Andre, *J. Chem. Phys.* **101**, 299 (1994).

179. C. A. Emis and P. L. Fehder, *J. Am. Chem. Soc.* **92**, 2246 (1970); J. Keizer, *J. Phys. Chem.* **85**, 940 (1981); D. A. McQuarrie and J. A. Keizer, in: *Physical Chemistry: An Advanced Treatise*, Vol. 6, Part A, edited by D. Henderson, Academic Press, New York, 1981, p. 165; K. R. Naqvi, *Chem. Phys. Lett.* **28**, 280 (1974); K. R. Naqvi, K. J. Mork, and S. Walderstrom, *J. Phys. Chem.* **84**, 1315 (1980).

180. D. M. Gass, *J. Chem. Phys.* **54**, 1898 (1971).

181. D. W. Jepsen, *J. Math. Phys.* **6**, 405 (1965).

182. J. L. Lebowitz and J. K. Percus, *Phys. Rev.* **155**, 122 (1967); J. L. Lebowitz, J. K. Percus, and J. Sykes, *Phys. Rev.* **171**, 224 (1968).

183. J. W. Haus and H. J. Raveche, *J. Chem. Phys.* **68**, 4969 (1978).

184. M. Bishop, *J. Chem. Phys.* **75**, 4741 (1981).

185. M. Bishop and B. J. Berne, *J. Chem. Phys.* **59**, 5337 (1973).

186. G. Srinivas and B. Bagchi, *J. Chem. Phys.* **112**, 7557 (2000).

187. S. Bhattacharyya and B. Bagchi, *Phys. Rev. E* **61**, 3850 (2000).

188. S. Bhattacharrya, Ph.D. thesis, Indian Institute of Science, Bangalore, 1999.

189. B. Bagchi and A. Chandra, *Phys. Rev. Lett.* **64**, 455 (1990).

190. B. Bagchi, *J. Mol. Liq.* **77**, 177 (1998).

191. See, for example, B. Bagchi and R. Biswas, *Adv. Chem. Phys.* **109**, 207 (1999).

192. C. G. E. Bottcher and P. Bordewijk, *Theory of Electric Polarization*, Vol. II, Elsevier, Amsterdam, 1979.

193. W. Kob and H. C. Andersen, *Phys. Rev. E* **51**, 4626 (1995).

194. J. Bosse and J. S. Thakur, *Phys. Rev. Lett.* **59**, 998 (1987).

195. G. Zerah and J. P. Hansen, *J. Chem. Phys.* **84**, 2336 (1986).

196. S. Forster and M. Schmidt, *Adv. Polym. Sci.* **120**, 51 (1995).

197. A. Yethiraj and B. Bagchi (unpublished work).

198. K. Miyazaki, A. Yethiraj, and B. Bagchi (unpublished work).

199. K. S. Schweizer, *J. Chem. Phys.* **91**, 5802 (1989); *ibid.* **91**, 5822 (1989); *J. Non-Cryst. Solids* **131**, 643 (1991); *Phys. Scr.* **T49**, 99 (1993).

200. (a) W. H. Tang, X. Y. Chang, and K. F. Freed, *J. Chem. Phys.* **103**, 9492 (1995); (b) K. S. Kostov and K. F. Freed, *J. Chem. Phys.* **106**, 771 (1997); **108**, 8277 (1998).

201. S. Egorov and B. Bagchi (unpublished work).

202. I. M. Mryglod, I. P. Omelyan, and M. V. Tokarchuk, *Mol. Phys.* **84**, 235 (1995); I. M. Mryglod and I. P. Omelyan, *Mol. Phys.* **92**, 913 (1997).

ANOMALOUS STOCHASTIC PROCESSES IN THE FRACTIONAL DYNAMICS FRAMEWORK: FOKKER–PLANCK EQUATION, DISPERSIVE TRANSPORT, AND NON-EXPONENTIAL RELAXATION

RALF METZLER

*School of Chemistry, Tel Aviv University, Tel Aviv, Israel
and Department of Physics, Massachusetts Institute of Technology,
Cambridge, MA*

JOSEPH KLAFTER

School of Chemistry, Tel Aviv University, Tel Aviv, Israel

CONTENTS

Advances in Chemical Physics, Volume 116, edited by I. Prigogine and Stuart A. Rice.
ISBN 0-471-40541-8   © 2001 John Wiley & Sons, Inc.

I. INTRODUCTION

Brownian motion [1] denotes the erroneous motion of a massive particle in a bath of molecules whose ongoing bombardments are the cause for the particle's random walk, which is displayed in Fig. 1. The compelling story of the experimental investigation of Brownian motion [1, 2], with its theoretical description [3–9] being worked out hand in glove with each other, added much to the halcyon development of physics at the end of the nineteenth and in the first half of the twentieth centuries. The importance of the understanding of Brownian motion was honored by the Nobel Prize for Jean Perrin in 1926 for his investigations leading to the determination of Avogadro's number in terms of microscopic quantities, according to Einstein's theory. Today, Brownian motion is well understood, with its continuum description drawing on the central limit theorem according to which the probability density function (pdf) to find the particle at a certain position x at a given time t is a universal Gaussian whose second moment, the mean squared displacement $\langle x^2(t) \rangle = 2dKt$, grows linearly in time, in any dimension [7–16].

Figure 1. Recorded random walk trajectories by Jean Baptiste Perrin [2]. **Upper panel:** Three designs obtained by tracing a small grain of putty (*mastic*, used for varnish) at intervals of 30 s. One of the patterns contains 50 single points. **Lower panel:** The starting point of each motion event is shifted to the origin. The figure illustrates the pdf of the traveled distance r to be in the interval $(r, r + dr)$, according to $(2\pi\xi^2)^{-1}\exp(-r^2/[2\xi^2])2\pi r\,dr$, in two dimensions, with the length variance ξ^2. These figures constitute part of the measurement of Perrin, Dabrowski, and Chaudesaigues, leading to the determination of the Avogadro number. The result given by Perrin is $70.5 \cdot 10^{22}$. The remarkable œuvre of Perrin discusses all possibilities of obtaining Avogadro's number known at that time. Concerning the trajectories displayed in the upper part of this figure, Perrin makes an interesting statement: "Si, en effet, on faisait des pointés de seconde en seconde, chacun de ces segments rectilignes se trouverait remplacé par un contour polygonal de 30 côtés relativement aussi compliqué que le dessin ici reproduit, et ainsi de suite." [If, veritably, one took the position from second to second, each of these rectilinear segments would be replaced by a polygonal contour of 30 edges, each itself being as complicated as the reproduced design, and so forth.] This already anticipates Lévy's cognizance of the self-similar nature [11], as well as of the nondifferentiability recognized by N. Wiener [7].

Brownian transport processes and the related relaxation dynamics in the presence and absence of an external potential are most conveniently described in terms of partial differential equations of the Fokker–Planck (Smoluchowski) [13, 14, 17–19], Rayleigh [13, 20], and Klein–Kramers [13, 14,

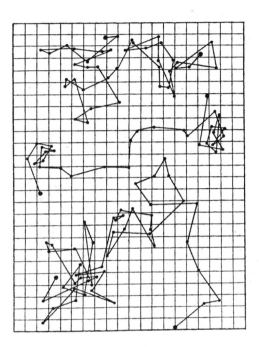

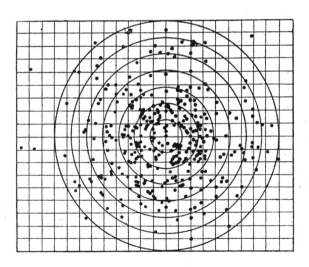

21, 22] types. These equations are closely connected to the exponential equilibration of the system and the exponential decay of the survival probability in the Kramers escape problem that is in turn connected to chemical reaction kinetics. Due to its universal character, the Brownian transport theory was believed to prevail in any nonpathological system.

Toward the end of the 1960s, however, surprising data were obtained for the charge carrier transport in amorphous semiconductors, then an important issue in the development of photocopiers and solar cells. Apparently, these data could not be satisfactorily accounted for by the traditional Brownian description of random walks that the charge carriers were a priori supposed to perform. It was Scher, then staff researcher at the Xerox company, who saw himself confronted with this conundrum. Eventually, in collaboration with Montroll, Lax, and Shlesinger, it had been realized that these puzzling data could be understood by the invocation of a random walk description in which each step of the walker occurs at a random time which is chosen from a random distribution $w(t)$ so broad that it is actually scale-free; that is, it does not possess a characteristic time scale. Coming up with the assumption of a power-law form for this distribution, $w(t) \sim A_\alpha \tau^\alpha / t^{1+\alpha}$ [23], the breakthrough was achieved. The resulting continuous-time random walk model with this "fractal time" waiting time distribution has been put on a solid mathematical and physical foundation and has been a successfully applied theory [24].

Amorphous semiconductors have been *the* testing ground for the new theoretical concepts for which the detailed physical mechanisms developed could be impressively corroborated by experiments [24–26]. In the course of time, more and more systems appeared to exhibit "strange kinetics," with a sublinearly growing mean squared displacement and with nonexponential relaxation patterns. Slow diffusion was observed for the tracer dispersion in Rayleigh–Bénard convection systems [27], for polymer dynamics and for a bead immersed in a polymeric fluid [28, 29], and for diffusion in porous media [30, 31]. Strange kinetics is related to the growth of a submonolayer film on a solid surface in the presence of repulsive impurities: The latter are supposed to give rise to anomalous diffusion of the atoms which are deposited on the surface, a process that finally leads to a typical scaling of the island density of the emerging growth pattern [32]. Recently, there has been growing interest in such slow transport in the investigation of tracer dispersion in groundwater systems which might render important new insight into the ecological impact of deposited chemicals or radioactive waste [33], and the cognisance of strange kinetics has been taken in protein dynamics [34, 35]. Another boost for the continuous-time random walk theory came from its applications to chaotic systems where broadly distributed waiting times arise for the sticking of a trajectory in deterministic maps, or close to stable islands [36]. Strange kinetics even stretches far into the nanoscale reign, being related to the power-law blinking kinetics of quantum dots [37] or to the broad distribution of waiting times in subrecoil laser cooling [38]. It is common to these complex systems that some kind of disorder—that is, the presence of spatial or temporal constraints—reduces the spatiodynamical degrees of freedom of the random motion of the particle under consideration. This presence of constraints leads to the temporally non-local behavior, to the slowly decaying memory expressed in the transport and relaxation dynamics of such systems, and consequently to the observation of strange kinetics.

Here, we present an approach for the description of such anomalous trans-port processes that is based on the continuous-time random walk theory for a power-law waiting time distribution $w(t)$ but which can be used to find the probability density function of the random walker *in the presence of an ex-ternal force field*, or in phase space. This framework is *fractional dynamics*, and we show how the traditional kinetic equations can be generalized and solved within this approach.

The anomalous transport processes on which we focus and which corre-spond to the above examples are, in the force-free limit, characterized by the power-law form [15, 16, 39–42]

$$\langle x^2(t) \rangle \sim K_\alpha^* t^\alpha, \qquad \alpha \neq 1 \tag{1}$$

of the mean squared displacement which leads to a spectrum of diffusion pro-cesses, depending on the anomalous exponent α. In Eq. (1), K_α^* is a general-ized diffusion constant of dimension cm^2/s^α which will be specified below. For the power-law form of the waiting time distribution, $w(t) \sim A_\alpha \tau^\alpha / t^{1+\alpha}$ with α ranging in the interval $\alpha \in (0, 1)$, one observes *slow diffusion*, or *sub-diffusion*. Thus, α becomes an essential characteristic quantity of the under-lying kinetic process. In that sense, anomalous transport   processes are nonuniversal. However, these processes are subject to a superordinate limit theorem that is connected with Lévy (stable) distributions [41, 43] and that guarantees the existence of a limit distribution for the anomalous process in the same way as the central limit theorem enforces the Gaussian limit distri-bution of the Brownian process. In fact, the central limit theorem is a special case located on the verge of the basin of attraction of this more general theorem. The anomalous exponent α characterizes the special system under consideration, and therefore it has to be determined independently. It may depend on thermodynamic parameters like temperature, pressure, and so on.

Among the most striking changes brought about by fractional dynamics is the substitution of the traditionally obtained exponential system equilibration of time-dependent system quantities by the Mittag–Leffler pattern [44–46] that interpolates between an initial stretched exponential and a final inverse power-law pattern,

$$\exp\left(-\frac{t}{\tau}\right) \longrightarrow E_\alpha(-(t/\tau)^\alpha) \sim \begin{cases} \exp\left(-\dfrac{t^\alpha}{\tau^\alpha\Gamma(1+\alpha)}\right), & t \ll \tau \\ (\Gamma(1-\alpha)(t/\tau)^\alpha)^{-1}, & t \gg \tau. \end{cases} \quad (2)$$

In what follows, we present a generalized stochastics framework for the description of slow $(0 < \alpha < 1)$ transport in position and phase space, on the basis of fractional dynamics. Accordingly, complex systems close to thermodynamic equilibrium which exhibit a self-similar memory are governed by the fractional Fokker–Planck, Rayleigh, and Klein–Kramers equations. We demonstrate that fractional dynamics may arise for multiple trapping systems with broadly distributed trapping times. More generally, it is equivalent to a generalized master equation with a power-law memory kernel that can be connected to a continuous-time random walk approach.

The advantage of the fractional formulation in comparison to other approaches lies in its proximity to the classical partial differential equations and their methods of solution.

Note that throughout the presentation, we concentrate on the one-dimensional case.

II. THE RISE OF FRACTIONAL DYNAMICS

A. The Master Equation

The discrete Markovian master equation [13–16]

$$W_j(t + \Delta t) = \sum_{j' \neq j} A_{j,j'} W_{j'}(t) \quad (3)$$

describes the evolution of the pdf $W_j(t)$ during the time step Δt as determined by the transfer matrix $A_{j,j'}$. $W_j(t)$ denotes the probability to find the random walker at site j at the given time t. The continuum limit with respect to the position coordinate j relies on a Taylor expansion in the step length Δx of the corresponding transfer function $A(x)$. If this expansion converges and Δx can be regarded a small parameter, one recovers the Fokker–Planck (Smoluchowski) equation [13–19, 47]

$$\frac{\partial W}{\partial t} = \left(-\frac{\partial}{\partial x}\frac{F(x)}{m\eta} + K\frac{\partial^2}{\partial x^2}\right) W(x,t) \quad (4)$$

for the pdf $W(x,t)$ in the time-independent external force field $F(x) = -\frac{d}{dx}\Phi(x)$. Here, m is the mass of the particle, η the friction constant quantifying the effective interaction with the environment, and K is the diffusion constant. The latter two are connected via the Einstein–Stokes relation $K = k_B T / (m\eta)$, where $k_B T$ is the Boltzmann temperature [5]. The coefficients in Eq. (4) are given by

$$\frac{F(x)}{m\eta} \equiv \lim_{\Delta x \to 0, \Delta t \to 0} \frac{\Delta x}{\Delta t}(A_+(x) - A_-(x)), \qquad K \equiv \lim_{\Delta x \to 0, \Delta t \to 0} \frac{(\Delta x)^2}{2\Delta t}, \qquad (5)$$

where the continuum version $A_\pm(x)$ of the transfer matrix $A_{j,j'}$ denotes the probability of coming from the left or right of the position x. For taking these limits, the normalization condition $A_+(x) + A_-(x) = 1$ was imposed, and we assumed that the system is close to thermal Gibbs–Boltzmann equilibrium; that is, $A_+(x - \Delta x) \sim (1 - 2\beta \Delta x F(x))A_-(x)$, where $\beta \equiv (k_B T)^{-1}$ is the Boltzmann factor.

B. Long-Tailed Waiting Times Processes and the Generalized Master Equation

The emergence of slow kinetics with its typical slowly decaying memory effects is tightly connected to a scale-free waiting time pdf; that is, the temporal occurrence of the motion events performed by the random walking particle is broadly distributed such that no characteristic waiting time exists. It has been demonstrated that it is the assumption of the power-law form for the waiting time pdf which leads to the explanation of the kinetics of a broad diversity of systems such as the examples quoted above.

Systems that display strange kinetics no longer fall into the basin of attraction of the central limit theorem, as can be anticipated from the anomalous form (1) of the mean squared displacement. Instead, they are connected with the Lévy–Gnedenko generalized central limit theorem, and consequently with Lévy distributions [43]. The latter feature asymptotic power-law behaviors, and thus the asymptotic power-law form of the waiting time pdf, $w(t) \sim A_\alpha \tau^\alpha / t^{1+\alpha}$, may belong to the family of completely asymmetric or one-sided Lévy distributions $\mathbf{L}_\alpha^+$, that is,

$$w(t) = \mathbf{L}_\alpha^+(t/\tau) \sim A_\alpha \frac{\tau^\alpha}{t^{1+\alpha}}, \qquad 0 < \alpha < 1, \qquad (6)$$

see the compilation in Appendix A.[1] Due to its long-tailed nature, the waiting time pdf (6) fulfills the criterion that it possesses no characteristic time scale:

---

[1] We choose the representation in terms of a Lévy distribution for convenience because it includes the Brownian limit. Indeed, *any* waiting time pdf $w(t)$ with the asymptotic power-law trend following Eq. (6) leads to the same results as obtained in the following for $0 < \alpha < 1$.

$T \equiv \int_0^\infty w(t)t\,dt \to \infty$, manifesting the self-similar nature of this waiting process that has also prompted the coinage of "fractal time" processes [48]. Note that in the limit $\alpha \to 1$, this waiting time pdf reduces to the singular form $\mathbf{L}_1^+(t/\tau) = \delta(t - \tau)$ with finite $T = \tau$ that leads back to the temporally local Markovian formulation of classical Brownian transport. In fact, for *any* waiting time pdf with a finite characteristic time T, one recovers the Brownian picture, such as for the Poissonian form $w(t) = \tau^{-1}e^{-t/\tau}$.

In the continuous-time random walk model, a random walker is pictured to execute jumps at time steps chosen from the waiting time pdf $w(t)$. In the isotropic and homogeneous (that is, force-free) case, the distance covered in a single jump event can be drawn from the jump length pdf $\lambda(x)$. Then, the probability $\eta(x, t)$ of just having arrived at position x is given through [49]

$$\eta(x, t) = \int_{-\infty}^{\infty} dx' \int_0^{\infty} dt' \eta(x', t')\lambda(x - x')w(t - t') + \delta(x)\delta(t), \qquad (7)$$

where the initial condition is $\delta(x)$. Consequently, the pdf $W(x, t)$ of being in x at time t is given by

$$W(x, t) = \int_0^t dt\eta(x, t')\Psi(t - t'), \qquad (8)$$

in terms of the convoluted pdf $\eta(x, t)$ of just having arrived in x at time t', and not having moved since. The latter is defined by the cumulative probability

$$\Psi(t) = 1 - \int_0^t dt'w(t') \qquad (9)$$

assigned to the probability of no jump event during the time interval $(0, t)$. Converting Eq. (9) to Fourier–Laplace space which is defined through

$$f(u) \equiv \mathscr{L}\{f(t)\} = \int_0^\infty f(t)e^{-ut}dt \qquad (10a)$$

$$g(k) \equiv \mathscr{F}\{g(x)\} = \int_{-\infty}^\infty g(x)e^{ikx}dx, \qquad (10b)$$

the transformed pdf $W(x, t)$ obeys the algebraic relation [49]

$$W(k, u) = \frac{1 - w(u)}{u}\frac{W_0(k)}{1 - \lambda(k)w(u)}, \qquad (11)$$

where $W_0(k)$ denotes the Fourier transform of the initial condition $W_0(x) \equiv \lim_{t \to 0^+} W(x, t)$. If now the jump length pdf is such that it possesses a finite variance, its Fourier transform is accordingly given through $\lambda(k) \sim 1 - Ck^2$ for small wavenumber k. Conversely, the small u expansion of our waiting time pdf $w(t) = L_\alpha^+(t/\tau)$ follows:

$$w(u) \sim 1 - (u\tau)^\alpha. \tag{12}$$

If the two pdfs λ and w are plugged into Eq. (11), the short wavenumber and short Laplace frequency limit $(k, u) \to (0, 0)$ reveals

$$W(k, u) - \frac{1}{u} = -u^{-\alpha} \frac{C}{\tau^\alpha} k^2 W(k, u), \tag{13}$$

for $W_0(x) = \delta(x)$. The Laplace inversion of Eq. (13) involves the term $u^{-\alpha} W(k, u)$. With the definition of the Riemann–Liouville fractional operator [Eq. (21) below], this expression corresponds to the Riemann–Liouville fractional integral $_0D_t^{-\alpha} W(x, t)$ in (x, t) space. Consequently, one recovers by Fourier–Laplace inversion of Eq. (13) the fractional diffusion equation [50–54]

$$W(x, t) - W_0(x) = {}_0D_t^{-\alpha} K_\alpha \frac{\partial^2}{\partial x^2} W(x, t) \tag{14}$$

in the integral formulation. Equation (14) was first discussed and solved by Schneider and Wyss [50]. Note that we identified $K_\alpha \equiv C/\tau^\alpha$ [49, 53, 54]. Alternatively, Eq. (14) can be rephrased in the differential form

$$\frac{\partial W}{\partial t} = {}_0D_t^{1-\alpha} K_\alpha \frac{\partial^2}{\partial x^2} W(x, t) \tag{15}$$

by operation of the ordinary differential $\partial/\partial t$. The fractional diffusion equation (15) for the initial condition $W_0(x) = \delta(x)$ can be solved in closed form, invoking Fox's H-functions $H_{p,q}^{m,n}$ [55, 56], to obtain

$$W(x, t) = \frac{1}{\sqrt{4K_\alpha t^\alpha}} H_{1,1}^{1,0} \left[\frac{|x|}{\sqrt{K_\alpha t^\alpha}} \,\middle|\, \begin{matrix} (1 - \frac{\alpha}{2}, \frac{\alpha}{2}) \\ (0, 1) \end{matrix} \right], \tag{16}$$

which is equivalent to the result found by Schneider and Wyss in terms of $H_{1,2}^{2,0}$ [50]. Through the Fox function formulation one finds the series representation

$$W(x, t) = \frac{1}{\sqrt{4K_\alpha t^\alpha}} \sum_{n=0}^{\infty} \frac{(-1)^n}{n! \Gamma(1 - \alpha[n+1]/2)} \left(\frac{x^2}{K_\alpha t^\alpha} \right)^{n/2} \tag{17}$$

and the asymptotic expansion

$$
\begin{aligned}
W(x,t) \sim{} & \frac{1}{\sqrt{4\pi K_\alpha t^\alpha}} \sqrt{\frac{1}{2-\alpha}} \left(\frac{2}{\alpha}\right)^{(1-\alpha)/(2-\alpha)} \left(\frac{|x|}{\sqrt{K_\alpha t^\alpha}}\right)^{-(1-\alpha)/(2-\alpha)} \\
& \times \exp\left(-\frac{2-\alpha}{2}\left(\frac{\alpha}{2}\right)^{\alpha/(2-\alpha)} \left[\frac{|x|}{\sqrt{K_\alpha t^\alpha}}\right]^{1/(1-\alpha/2)}\right)
\end{aligned}
\tag{18}
$$

valid for $|x| \gg \sqrt{K_\alpha t^\alpha}$. The functional form of the result (18) has prompted the coinage of *stretched Gaussian* form, and it is equivalent to the continuous-time random walk findings reported by Zumofen and Klafter [57].

If an external force field acts on the random walker, it has been shown [58, 59] that in the diffusion limit, this broad waiting time process is governed by the fractional Fokker–Planck equation (FFPE) [60]

$$
\frac{\partial W}{\partial t} = {}_0D_t^{1-\alpha}\left(-\frac{\partial}{\partial x}\frac{F(x)}{m\eta_\alpha} + K_\alpha \frac{\partial^2}{\partial x^2}\right)W(x,t),
\tag{19}
$$

which is discussed in detail in the next section. Equations (15) and (19) feature the Riemann–Liouville operator

$$
{}_0D_t^{1-\alpha} = \frac{\partial}{\partial t}{}_0D_t^{-\alpha}
\tag{20}
$$

whose definition is given in terms of the convolution [61]

$$
{}_0D_t^{-\alpha}W(x,t) = \frac{1}{\Gamma(\alpha)}\int_0^t dt' \frac{W(x,t')}{(t-t')^{1-\alpha}}.
\tag{21}
$$

It is interesting to note that the notion of noninteger order differentials goes back to one of the founders of classical calculus, Leibniz, who mentions the problem as an interesting topic in a letter to de l'Hôspital in 1695. Numerous famous mathematicians have worked on the field which eventually was to become fractional calculus. Of the several different definitions in use, the Riemann–Liouville version defined through Eq. (21) corresponds to physical problems with a defined initial condition at $t = 0$. The (Riemann–Liouville) fractional integral (21) generalizes the Cauchy multiple integral to a "real-value folded" integration. Fractional differentiation is defined as a fractional integration, followed by an ordinary differentiation:

$$
{}_0D_t^q f(t) \equiv \frac{d^n}{dt^n}\left({}_0D_t^{-(n-q)}f(t)\right), \qquad n-1 \geq q < n.
\tag{22}
$$

The fractional differentiation of a power

$$_0D_t^q t^p = \frac{\Gamma(1+p)}{\Gamma(1+p-q)} t^{p-q} \tag{23}$$

is analogous to the standard case; however, note that the fractional differentiation of a constant does not vanish:

$$_0D_t^q 1 = \frac{1}{\Gamma(1-q)} t^{-q}. \tag{24}$$

Expression (24) reduces to the standard $\frac{d^n}{dt^n} 1 = 0$ for $q \to n$, due to the divergence of the gamma function $\Gamma(z)$ for nonpositive integers. The fractional Riemann–Liouville integral operator $_0D_t^{-q}$ fulfills the generalized integration theorem of the Laplace transformation:

$$\mathscr{L}\{_0D_t^{-q}f(t)\} = u^{-q}f(u), \qquad q \in \mathbb{R}_0^+. \tag{25}$$

In the limit $\alpha \to 1$, the Riemann–Liouville fractional integral $_0D_t^{-\alpha}$ reduces to an ordinary integration so that $\lim_{\alpha \to 1} {}_0D_t^{1-\alpha} \equiv \frac{\partial}{\partial t} \int_0^t dt'$ becomes the identity operator; that is, Eqs. (15) and (19) simplify to the standard diffusion and Fokker–Planck equations, respectively.

According to Eq. (21), the FFPE (19) involves a slowly decaying, self–similar memory so that the present state $W(x,t)$ of the system depends strongly on its history $W(x,t')$, $t' < t$, in contrast to its Brownian counterpart which is local in time. In the force-free case, $F(x) = 0$, the FFPE (19) reduces to the fractional diffusion equation (15).

It has been shown that the FFPE (19) is equivalent to the generalized master equation [58]

$$\frac{\partial W}{\partial t} = \int_{-\infty}^{\infty} dx' \int_0^t dt' K(x,x';t-t')W(x',t') \tag{26}$$

with the kernel

$$K(x,x';u) \equiv uw(u)\frac{\Lambda(x,x') - \delta(x)}{1 - w(u)} \tag{27}$$

given in Laplace space. Here, the transfer function $\Lambda(x,x')$ generalizes the homogeneous jump length pdf $\lambda(x-x')$ of the standard continuous-time random walk model, and thus it quantifies the local anisotropy and gives rise to the drift and diffusion terms [58]. Λ may be defined through $\Lambda(x,x') \equiv$

$\lambda(x - x')(A(x')\Theta(x - x') + B(x')\Theta(x' - x))$ in terms of the probabilities to jump left or right, as introduced through the Heaviside functions $\Theta(x)$ [58].

For systems that exhibit slow anomalous transport, the incorporation of external fields is in complete analogy to the existing Brownian framework which itself is included in the fractional formulation for the limit $\alpha \to 1$: The FFPE (19) combines the linear competition of drift and diffusion of the classical Fokker–Planck equation with the prevalence of a new relaxation pattern. As we are going to show, also the solution methods for fractional equations are similar to the known methods from standard partial differential equations. However, the temporal behavior of systems ruled by fractional dynamics mirrors the self–similar nature of its nonlocal formulation, manifested in the Mittag-Leffler pattern dominating the system equilibration.

C. Boundary Value Problems for the Fractional Diffusion Equation

Exemplifying the convenience of the fractional approach, we address the imposition of boundary value problems on the fractional diffusion equation which was demonstrated in Ref. 62. In this force-free case for which the kernel, Eq. (27), takes on the homogeneous form $K(x, x'; u) = uw(u)$ $(\lambda(x - x') - \delta(x))/(1 - w(u))$, one can apply the method of images in order to construct the solution [12].

Let us address the example of subdiffusion modelled through Eq. (15) in a box with absorbing boundary conditions, located at $x = \pm a$, and the symmetric initial condition $W_0(x) = \delta(x)$; that is, the corresponding pdf $Q(x, t)$ has to fulfill $Q(\pm a, t) = 0$ and $Q_0(x) = \delta(x)$. The solution to this boundary value problem is obtained in the same way as for the classical case [12], resulting in [62]

$$Q(x, t) = \sum_{m=-\infty}^{\infty} (W(x + 4ma, t) - W(4ma - x + 2a, t)), \qquad (28)$$

where the image function $Q(x, t)$ fulfills the boundary condition. The sum (28) can be transformed to the more convenient representation [62]

$$Q(x, t) = \frac{1}{a} \sum_{m=0}^{\infty} e^{\pi i (2m+1)x/(2a)} E_\alpha \left(-K_\alpha \frac{(2m + 1)^2 \pi^2}{4a^2} t^\alpha \right), \qquad (29)$$

which can be evaluated numerically. Equation (29) involves the Mittag-Leffler function $E_\alpha(z)$ instead of the classical exponential functions. We will comment on the Mittag-Leffler function in Appendix B. The result for $\alpha = 1/2$ is displayed in Fig. 2, in comparison to the Brownian result. The subdiffusive solution features distinct humps in the center, close to the

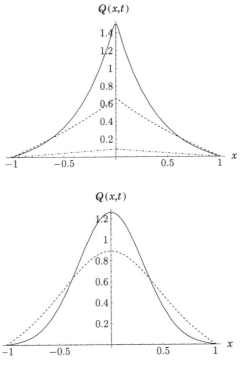

Figure 2. Probability density function $Q(x,t)$ for absorbing boundaries in $x = \pm 1$. Top: The subdiffusive case, $\alpha = 1/2$. Bottom: The Brownian case, $\alpha = 1$. The curves are drawn for the times $t = 0.005, 0.1, 10$ on the top and for $t = 0.05, 0.1, 10$ on the bottom. Note the distinct cusp-like shape of the subdiffusive solution in comparison to the smooth Brownian counterpart. For the longest time, the Brownian solution has almost completely decayed.

initial condition that persists on that point, in contrast to the fast smoothening in the Brownian case. The temporal decay of the survival probability (i.e., the overall probability of not having been absorbed), is given through the integral

$$p_a(t) = \int_{-a}^{a} dx Q(x, t), \tag{30}$$

which for Eq. (29) becomes

$$p_a(t) = \frac{4}{\pi} \sum_{m=0}^{\infty} \frac{(-1)^m}{2m+1} E_\alpha \left(-K_\alpha \frac{(2m+1)^2 \pi^2}{4a^2} t^\alpha \right). \tag{31}$$

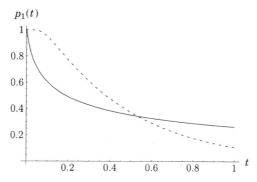

Figure 3. Survival probability for absorbing boundary conditions positioned at $x = \pm 1$, plotted for the subdiffusive case $\alpha = 1/2$ and the Brownian case $\alpha = 1$ (dashed curve). For longer times, the faster (exponential) decay of the Brownian solution, in comparison to the power-law asymptotic of the Mittag-Leffler behavior, is obvious.

This function has the long-time behavior $p_a(t) \sim C_\alpha t^{-\alpha}$, where C_α is a constant. The survival probability for the subdiffusive case is plotted in Fig. 3 and compared with the Brownian survival. Clearly, for long times, the survival probability in the subdiffusive system decays in a much slower fashion.

A recent work has demonstrated that the formulation of reaction–diffusion problems in systems that display slow diffusion within a continuous-time random walk model with a broad waiting time pdf of the form (6) leads to a fractional reaction–diffusion equation that includes a source or sink term in the same additive way as in the Brownian limit [63]. With the fractional formulation for single-species slow reaction–diffusion obtained by the authors still being linear, no pattern formation due to Turing instabilities can arise. This is due to the fact that fractional systems of the type (15) are close to Gibbs–Boltzmann thermodynamic equilibrium as shown in the next section.

III. THE FRACTIONAL FOKKER–PLANCK EQUATION

A. The Classical Fokker–Planck Equation

The formulation of the Fokker–Planck equation is due to Fokker's and Planck's independent works on the description of the Brownian motion of particles [17, 18]. Commonly, an N variables equation of the type

$$\frac{\partial W}{\partial t} = \left(-\sum_{i=1}^{N} \frac{\partial}{\partial x_i} D_i^{(1)}(\{x\}) + \sum_{i,j=1}^{N} \frac{\partial^2}{\partial x_i \partial x_j} D_{ij}^{(2)}(\{x\}) \right) W(\{x\}, t), \quad (32)$$

where the drift vector $D_i^{(1)}$ and the diffusion tensor $D_{ij}^{(2)}$ may depend on the position $x \equiv \{x_1, x_2, \ldots, x_N\}$, is called Fokker–Planck equation [14]. In the following, we deal with the monovariate $(N = 1)$ and bivariate $(N = 2)$ cases, and we assume a constant, purely diagonal diffusion tensor.

The monovariate Fokker–Planck equation with a position dependent diffusion coefficient $D^{(2)}(x)$,

$$\frac{\partial W}{\partial t} = \left(-\frac{\partial}{\partial x} D^{(1)}(x) + \frac{\partial^2}{\partial x^2} D^{(2)}(x) \right) W(x, t), \tag{33}$$

can be transformed in general onto the so-called normalized Fokker–Planck equation (4) with the diffusion *constant* K [14]. In the latter formulation, we have chosen the coefficients according to the Smoluchowski model in which the drift caused by the external force field $F(x)$ is moderated through the friction constant η. In Eq. (4), the pdf W approaches the Gibbs–Boltzmann equilibrium

$$W_{\text{st}}(x) \equiv \lim_{t \to \infty} W(x, t) = N \exp(-\beta \Phi(x)), \tag{34}$$

where $\beta \equiv (k_B T)^{-1}$ is the Boltzmann factor, and the normalization constant N explicitly depends on the potential $\Phi(x)$ that is defined through $\Phi(x) = -\int^x dx' F(x')$. The diffusion and friction constants are related through the Einstein–Stokes relation $K = k_B T / m\eta$, fulfilling the fluctuation–dissipation condition.

A one-dimensional Fokker–Planck equation was used by Smoluchowski [19], and the bivariate Fokker–Planck equation in phase space was investigated by Klein [21] and Kramers [22]. Note that, in essence, the Rayleigh equation [23] is a monovariate Fokker–Planck equation in velocity space. Physically, the Fokker–Planck equation describes the temporal change of the pdf of a particle subjected to diffusive motion and an external drift, manifest in the second- and first-order spatial derivatives, respectively. Mathematically, it is a linear second-order parabolic partial differential equation, and it is also referred to as a forward Kolmogorov equation. The most comprehensive reference for Fokker–Planck equations is probably Risken's monograph [14].

B. Basic Properties

Before discussing the FFPE (19) in detail, we note that the fractional approach meets the following requirements: **(i)** In the absence of an external force field, Eq. (1) is satisfied; **(ii)** in the presence of an external nonlinear and time-independent field the stationary solution is the Boltzmann distribution; **(iii)** generalized Einstein relations are satisfied; and **(iv)** in the limit $\alpha \to 1$, the standard FPE is recovered. While other approaches like fractional Brownian motion [64], the continuous time random walk [23, 25, 65], modified diffusion equations [66], generalized Langevin equations [67], fractional equations for Lévy flights in external fields [68–71], local fractional equations [72], or generalized thermostatistics [73] fulfill part of these requirements, we know of no other simple approach that meets all of these physical demands. Together with the straightforward mathematical tractability of fractional equations, these are the major criteria why we regard the fractional approach as especially suited.

The FFPE (19) contains the generalized friction constant η_α and the generalized diffusion constant K_α, of dimensions $[\eta_\alpha] = \mathrm{s}^{\alpha-2}$ and $[K_\alpha] = \mathrm{cm}^2\,\mathrm{s}^{-\alpha}$. The physical origin of these fractional dimensions will be explained in the next section. In what follows, we assume natural boundary conditions, that is, $\lim_{|x| \to \infty} W(x, t) = 0$. The FFPE (19) describes a physical problem, where the system is prepared at $t_0 = 0$ in the state $W(x, 0)$.

The right-hand side of the FFPE (19) is equivalent to the fractional expression

$$-{}_0D_t^{1-\alpha}\frac{\partial S(x,t)}{\partial x},\qquad(35)$$

where

$$S(x,t) = \left(\frac{F(x)}{m\eta_\alpha} - K_\alpha\frac{\partial}{\partial x}\right)W(x,t)\qquad(36)$$

is the probability current. If a stationary state is reached, S must be constant. Thus, if $S = 0$ for any x, it vanishes for all x [14], and the stationary solution is defined by $-F(x)W_{st}/[m\eta_\alpha] + K_\alpha W'_{st} = 0$. Comparing the resulting expression $W_{st}(x) \propto \exp(-\Phi(x)/[K_\alpha m\eta_\alpha])$ to the required Boltzmann distribution $W_{st} \propto \exp(-\Phi(x)/[k_BT])$, we find a generalization of the Einstein–Stokes relation, also referred to as the Stokes–Einstein–Smoluchowski relation [16],

$$K_\alpha = \frac{k_BT}{m\eta_\alpha},\qquad(37)$$

for the generalized coefficients K_α and η_α [60]. Thus, processes described by Eq. (19) fulfill the linear relation between generalized friction and diffusion coefficients, reflecting the fluctuation–dissipation theorem. In the presence of a uniform force field, given by $\Phi(x) = -Fx$, a net drift occurs. Calculating the first moment $\frac{d}{dt}\langle x(t)\rangle_F = \int dx\, x\frac{\partial}{\partial t}W$ via the FFPE (19), we obtain

$$\langle x(t)\rangle_F = \frac{F}{m\eta_\alpha}\frac{t^\alpha}{\Gamma(1+\alpha)}. \qquad (38)$$

The mean squared displacement for the FFPE (19) in the absence of a force can be calculated similarly:

$$\langle x^2(t)\rangle_0 = \frac{2K_\alpha t^\alpha}{\Gamma(1+\alpha)}. \qquad (39)$$

Note the subscripts F and 0 to indicate presence and absence of the force field. Using Eq. (37), we recover the relation [60]

$$\langle x(t)\rangle_F = \frac{1}{2}\frac{F\langle x^2(t)\rangle_0}{k_B T}, \qquad (40)$$

connecting the first moment in the presence of the uniform force field with the second moment in absence of the force. Relation (40) is the second Einstein relation discussed in Refs. 41 and 74. It can be derived from first principles, using a Hamiltonian description of the system, within the linear response régime. Recent experimental results corroborate the validity of Eq. (37) in polymeric systems in the subdiffusive domain (see Ref. 29). The investigation of charge carrier transport in semiconductors in Ref. 26 showed that, up to a prefactor 2 that could not be determined exactly, Eq. (40) is valid.

The temporal evolution of the pdf $W(x,t)$ in Eq. (19), in the presence of the arbitrary external force field $F(x)$, is formally given in terms of the operator expression [60]

$$W(x,t) = E_\alpha(L(x)t^\alpha)W(x,0), \qquad L(x) \equiv \left(-\frac{\partial}{\partial x}\frac{F(x)}{m\eta_\alpha} + K_\alpha\frac{\partial^2}{\partial x^2}\right), \qquad (41)$$

which is the Mittag-Leffler generalization of the traditional exponential relation $W(x,t) = \exp(L(x)t)$. The Mittag-Leffler function E_α [44–46] that appears in Eq. (41) is defined in Appendix B, and it includes the exponential in the Brownian limit $\alpha \to 1$. We note that the FFPE (19) was derived from a generalized master equation that was based upon a nonhomogeneous contin-uous-time random walk model in Ref. 58, as well as from a continuous-time master equation in Ref. 59. It was obtained as the diffusion limit of a frac-tional Klein–Kramers equation from a multiple trapping model in Ref. 75, as reviewed below.

C. Methods of Solution and the Nonexponential Mode Relaxation

For a specified form of the Fokker–Planck operator $L(x)$, one can find an explicit solution $W(x, t)$ of the FFPE (19) through separation of variables. Indeed, inserting the separation ansatz

$$W_n(x, t) = \varphi_n(x) T_n(t), \tag{42}$$

where n labels a certain eigenvalue of the Fokker–Planck operator $L(x)$, into the FFPE (19) yields the factorized equation

$$\frac{dT_n}{dt} \varphi_n = \left({}_0D_t^{1-\alpha} T_n \right) L(x) \varphi_n. \tag{43}$$

The complete solution $W(x, t)$ is then expressed through the sum over the particular solutions $W_n(x, t)$, over the set of eigenvalues $\{n\}$. After the separation of Eq. (43) through division by $\left({}_0D_t^{1-\alpha} T_n \right) \varphi_n$, one arrives at the two eigenequations

$$\frac{dT_n}{dt} = -\lambda_{n,\alpha} \, {}_0D_t^{1-\alpha} T_n, \tag{44a}$$

$$L(x) \varphi_n = -\lambda_{n,\alpha} \varphi_n \tag{44b}$$

for the eigenvalue $\lambda_{n,\alpha}$. The latter are related to their Brownian counterparts $\lambda_{n,1}$ for the same external potential field $\Phi(x)$ by the dimensional prefactor $\lambda_{n,\alpha} = (\eta/\eta_\alpha)\lambda_{n,1}$. The temporal eigensolution to Eq. (44a) is given in terms of the Mittag-Leffler function

$$T_n(t) = E_\alpha(-\lambda_{n,\alpha} t^\alpha). \tag{45}$$

The complete solution of the FFPE (19) is thus composed by the sum

$$W(x, t|x', 0) = \exp(\tilde{\Phi}(x')/2 - \tilde{\Phi}(x)/2) \sum_{\{n\}} \psi_n(x) \psi_n(x') E_\alpha(-\lambda_{n,\alpha} t^\alpha) \tag{46}$$

over the set of eigenvalues, $\{n\}$, for an initial distribution concentrated in x'. In Eq. (46), the functions $\psi_n(x) \equiv e^{\tilde{\Phi}(x)/2} \varphi_n(x)$ are related to the eigenfunctions of the Fokker–Planck operator $L(x)$, $\varphi_n(x)$, via the scaled potential $\tilde{\Phi}(x) \equiv \Phi(x)/(k_B T)$. Note that the ψ_n are eigenfunctions to the Hermitian operator $\tilde{L} = e^{-\tilde{\Phi}} L(x) e^{\tilde{\Phi}}$, where $\tilde{L}(x)$ and $L(x)$ have the same eigenvalues $\lambda_{n,\alpha}$ [14]. For a nonpathological case, the set of eigenvalues, $\{n\}$, is discrete and the eigenvalues are nonnegative. Thus, on arranging the eigenvalues in increasing order, (i.e., $0 \leq \lambda_{0,\alpha} < \lambda_{1,\alpha} < \lambda_{2,\alpha} < \ldots$), the first eigenvalue is zero iff there exists a stationary solution fulfilling the condition

$$\frac{\partial W(x,t)}{\partial t} = 0. \tag{47}$$

This stationary solution is then necessarily given through

$$W_{\mathrm{st}}(x) = \lim_{t \to \infty} W(x,t) = e^{\tilde{\Phi}(x')/2 - \tilde{\Phi}(x)/2} \psi_0(x)\psi_0(x'), \tag{48}$$

in full accordance with the Brownian case $\alpha = 1$: It is the Gibbs–Boltzmann distribution. In contrast to this equivalence, it should be emphasized, the relaxation of a single mode n is subexponential, decaying slowly in the Mittag-Leffler fashion. This new relaxation pattern is distinguished by its interpolation between an initial stretched exponential (Kohlrausch) function and a final inverse power-law behavior (compare Appendix B).

An important property of the FFPE (19) is the functional scaling relation [60, 76]

$$W_\alpha(x,u) = \frac{\eta_\alpha}{\eta} u^{\alpha-1} W_1\left(x, \frac{\eta_\alpha}{\eta} u^\alpha\right) \tag{49}$$

in Laplace space, connecting the fractional solution $W_\alpha(x,t)$ determined by the FFPE (19), with its Brownian analogue $W_1(x,t)$. In order for Eq. (49) to hold, the initial conditions $W_\alpha(x,0)$ and $W_1(x,0)$ must, of course, be identical. That means that the fractional solution $W_\alpha(x,t)$ exists iff the Brownian solution $W_1(x)$ exists. In Laplace space, $W_\alpha(x,u)$ is the same distribution on x as $W_1(x, (\eta_\alpha/\eta)u^\alpha)$ for the scaled Laplace variable $(\eta_\alpha/\eta)u^\alpha$, only rescaled by the factor $(\eta_\alpha/\eta)u^{\alpha-1}$. As suggested by Barkai and Silbey [77], relation (49) can be reformulated in terms of the transformation

$$W_\alpha(v,t) = \int_0^\infty E(s,t)W_1(v,s)\,ds. \tag{50}$$

The kernel $E(s,t)$ is given through [78]

$$E(s,t) = \frac{t}{\alpha s}\mathbf{L}_\alpha^+\left(\frac{t}{(\eta_\alpha s/\eta)^{1/\alpha}}\right) \tag{51}$$

involving the one-sided Lévy distribution $\mathbf{L}_\alpha^+$; that is, $E(s, t)$ has the characteristic function

$$E(s, u) = \frac{\eta_\alpha}{\eta} u^{\alpha-1} \exp\left(-\frac{\eta_\alpha}{\eta} u^\alpha s\right) \tag{52}$$

whose Laplace inversion leads to the representation [78]

$$E(s, t) = \frac{1}{\alpha s} H_{1,1}^{1,0}\left[\frac{(\eta_\alpha s/\eta)^{1/\alpha}}{t}\middle|\begin{array}{c}(1,1)\\\left(\frac{1}{\alpha}, \frac{1}{\alpha}\right)\end{array}\right] \tag{53}$$

in terms of the Fox function $H_{1,1}^{1,0}$ [55, 56], with the series expansion

$$E(s, t) = \frac{1}{s} \sum_{n=0}^{\infty} \frac{(-1)^n}{\Gamma(1 - \alpha - \alpha n)\Gamma(1 + n)} \left(\frac{\eta_\alpha s/\eta}{t^\alpha}\right)^{1+n}. \tag{54}$$

The transformation defined through Eq. (50) is a convenient tool for the numerical evaluation of the solution of the FFPE (19), once the Brownian solution is known. The scaling connection (49) and the transformation (50) guarantee the positivity of the fractional solution if only the corresponding Brownian solution is a proper pdf.

D.   A Word on the Mittag-Leffler Function

Let us briefly examine the importance of the Mittag-Leffler function in relaxation modelling. The mathematical properties of the Mittag-Leffler function are compiled in Appendix B. Besides via the series representation, the Mittag-Leffler function is defined through its Laplace transform

$$\mathscr{L}\{E_\alpha(-(t/\tau)^\alpha)\} = \left(u + \tau^{-\alpha} u^{1-\alpha}\right)^{-1}, \tag{55}$$

or through the fractional relaxation equation [79]

$$f(t) - 1 = -\tau^{-\alpha} {}_0 D_t^{-\alpha} f(t). \tag{56}$$

In Refs. 80 and 81 it is shown that the Mittag-Leffler function is the exact relaxation function for an underlying fractal time random walk process, and that this function directly leads to the Cole–Cole behavior [82] for the complex susceptibility, which is broadly used to describe experimental results. Furthermore, the Mittag–Leffler function can be decomposed into single Debye processes, the relaxation time distribution of which is given by a modified, completely asymmetric Lévy distribution [43, 81]. This last observation is related to the formulation of Mittag-Leffler relaxation described in Ref. 80. In Refs. 83 and 84, the significance of the Mittag-Leffler function was shown, where its Laplace transform was obtained as a general result for a collision model in the Rayleigh limit.

The Mittag-Leffler function, or combinations thereof, has been obtained from fractional rheological models, and it convincingly describes the behavior of a number of rubbery and nonrubbery polymeric substances [79, 85]. The numerical behavior of the Mittag-Leffler function is equivalent to asymptotic power-law patterns that are often used to fit experimental data, see the comparative discussion of data from early events in peptide folding in Ref. 86, where the asymptotic power-law was confronted with the stretched exponential fit function.

E. The Fractional Ornstein–Uhlenbeck Process

As an application of the method of the separation of variables, we consider the nonstationary behavior in the generalized, fractional version of the Brownian harmonic oscillator with the parabolic potential

$$\Phi(x) = \frac{1}{2} m\omega^2 x^2, \tag{57}$$

which leads to the FFPE

$$\frac{\partial W}{\partial t} = {}_0 D_t^{1-\alpha} \left(\frac{\partial}{\partial x} \frac{\omega^2}{\eta_\alpha} x + \frac{k_B T}{m \eta_\alpha} \frac{\partial^2}{\partial x^2} \right) W(x,t). \tag{58}$$

Equation (58) is equivalent to the fractional Rayleigh equation [75, 77], and therefore we refer to Eq. (58) as the fractional Ornstein–Uhlenbeck process. For the sharp initial condition $W_0(x) = \delta(x - x_0)$, the solution to this process is, according to Eq. (46), given by

$$W = \sqrt{\frac{m\omega^2}{2\pi k_B T}} \sum_0^\infty \frac{1}{2^n n!} E_\alpha(-n\tilde{t}^\alpha) H_n\left(\frac{\tilde{x}_0}{\sqrt{2}}\right) H_n\left(\frac{\tilde{x}}{\sqrt{2}}\right) e^{-\tilde{x}^2/2}, \tag{59}$$

employing the reduced variables $\tilde{t} = t/\tau$ and $\tilde{x} = x\sqrt{m\omega^2/(k_B T)}$, as well as $\tau^{-\alpha} \equiv \omega^2/\eta_\alpha$. H_n denotes the Hermite polynomials, and the eigenvalues are $\lambda_{n,\alpha} = n\omega^2/\eta_\alpha$. The stationary solution corresponding to the lowest eigenvalue $n = 0$, is constant and independent of α. The remaining terms decay in the course of time. Thus, for all α, the stationary solution is the Gibbs–Boltzmann distribution, as anticipated by the stationarity condition. The fractional solution (59) is displayed in Fig. 4, where we compare between the convergence of the series representation (59), and the numerical evaluation using the transformation (50).

The first moment of the fractional Ornstein–Uhlenbeck process can be calculated from Eq. (58). It evolves in time like

$$\langle x(t) \rangle = x_0 E_\alpha(-(t/\tau)^\alpha), \tag{60}$$

reducing to the usual exponential relaxation behavior for $\alpha \to 1$. The second moment is given through

$$\langle x^2(t) \rangle = \langle x^2 \rangle_{th} + (x_0^2 - \langle x^2 \rangle_{th}) E_\alpha(-2(t/\tau)^\alpha). \tag{61}$$

Accordingly, the thermal equilibrium value $\langle x^2 \rangle_{th} = k_B T/[m\omega^2]$ is reached for $t \to \infty$. The Mittag-Leffler function $E_\alpha(-2(t/\tau)^\alpha)$ behaves like $1 - 2(t/\tau)^\alpha/\Gamma(1+\alpha)$ for short times and like $(t/\tau)^{-\alpha}/[2\Gamma(1-\alpha)] - (t/\tau)^{-2\alpha}/[4\Gamma(1-2\alpha)]$ for long times. Thus, for $x_0 = 0$ the short-time behavior of Eq. (61) follows Eq. (1) exactly and is independent of ω. For long times, the thermal equilibrium value $\langle x^2 \rangle_{th}$ is approached slowly, in power-law form. This is illustrated in Fig. 5.

IV.   THE FRACTIONAL KRAMERS ESCAPE PROBLEM

A.   The Classical Kramers Escape Problem

An interesting application of fractional dynamics is the modeling of the Kramers escape rate in complex systems where fractional dynamics prevails. Traditionally, such reaction rate problems [87] are formulated through the Smoluchowski [19, 88] and Onsager [89] models in terms of diffusion in the presence of absorbing bodies, or in terms of the famed Kramers model dating back to his seminal paper of 1940 [22]. Kramers considered a point particle in phase space diffusing in the potential $V(x)$. Being initially caught in a potential well, the particle can only escape over a potential barrier. Kramers promoted this model for the study of the dependence of the escape rate on temperature and viscosity. Alternative approaches for calculating rate reactions include the consideration of Markovian first passage time problems by Pontryagin, Andronow, and Witt [90], as well as first passage time problems for the master equation considered by Landau and Teller [91], Montroll and Shuler [92], Weiss [93], and, more recently, Bar-Haim and Klafter [94], Benichou et al. [95], and Abe et al. [96].

Extensions of the Kramers model are considered necessary [92–94, 97–99] although there are refined versions of the original formulation [100, 101]. Such non-Markovian dynamics has been taken into consideration

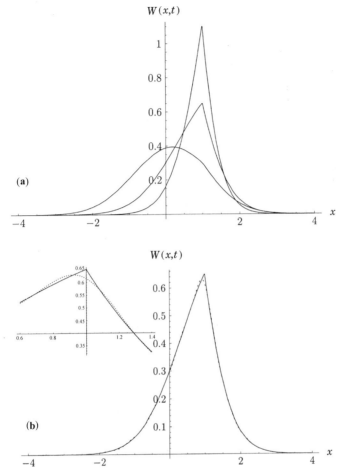

Figure 4.   (a) Pdf $W(x,t)$, Eq. (59), of the fractional Ornstein–Uhlenbeck process, for the anomalous diffusion exponent $\alpha = 1/2$. The initial value is chosen to be $W_0(x) = \delta(x-1)$. The maximum clearly slides toward the origin, acquiring an inversion symmetric shape. The curves are drawn for the times $t = 0.02$, 0.2, and 20, employing the integral relation with the Brownian solution. Note the distinct cusps around the initial position. (b) Comparison of the numerical behavior of the summation representation (dashed) with 151 summation terms and the integral representation (A transform). The latter is obtained by Mathematica employing the numerical integration command NIntegrate. The cusp which is a typical feature for subdiffusive processes is much more pronounced in the curve obtained through the integral transformation. The computation time for the latter is even shorter than for the calculation of the truncated sum so that this representation is preferrable for numerical purposes.

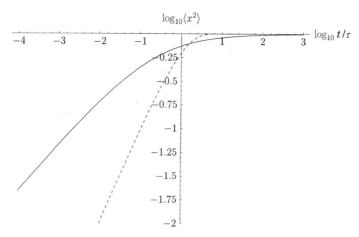

Figure 5. Mean squared displacement for the fractional ($\alpha = 1/2$, full line) and normal (dashed) Ornstein–Uhlenbeck process. The Brownian process shows the typical proportionality to t for small times; it approaches the saturation value much faster than its subdiffusive analogue, which starts off with the $t^{1/2}$ behavior and approaches the thermal equilibrium value by a power law, compare Eq. (61)

through generalized Langevin equations in the well-known models by Grote and Hynes [97] and by Hänggi et al. [98]. On the level of the Kramers equation, these generalized models lead to a formulation that is local in time and contains time-dependent coefficients. Consequently the associated Kramers survival probability still decays exponentially, but with a frequency-dependent rate [67, 97, 98]. This contrasts the dynamic descriptions related to the generalized master equation that are nonlocal in time, and exhibit memory on the macroscopic level of the pdf [102]. In systems where this memory decays in the long-tailed, self-similar power-law fashion, fractional dynamics leads to the Mittag-Leffler survival pattern, as we are going to discuss in this section.

In the standard overdamped version of the Kramers problem, the escape of a particle subject to a Gaussian white noise over a potential barrier is considered in the limit of low diffusivity—that is, where the barrier height ΔV is large in comparison to the diffusion constant K [14] (compare Fig.6). Then, the probability current over the potential barrier top near x_{max} is small, and the time change of the pdf is equally small. In this quasi-stationary situation, the probability current is approximately position independent. The temporal decay of the probability to find the particle within the potential well is then given by the exponential function [14, 22]

$$p(t) = e^{-r_K t}, \tag{62a}$$

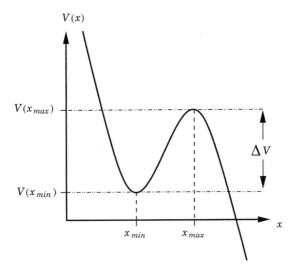

Figure 6. Potential well in the Kramers rate model. Initially the particle is assumed to be caught in the potential hole of depth $\Delta V \equiv V(x_{max}) - V(x_{min})$. The x-axis corresponds to a reaction coordinate.

where the Kramers rate is defined through [14, 22]

$$r_K = \frac{1}{2\pi m\eta} \sqrt{V''(x_{min})|V''(x_{max})|} \, \exp(-\beta\Delta V) \qquad (62b)$$

with $\Delta V = V(x_{max}) - V(x_{min})$. In Eq. (62b), the exponential function contains the Boltzmann factor $\beta \equiv (k_B T)^{-1}$ so that the inverse Kramers rate follows the Arrhenius activation [103] $r_K^{-1} \propto e^{E^*/T}$ with $E^* \equiv \Delta V/k_B$. Similarly, the Kramers rate in the low viscosity limit is given through [22]

$$r_K = \eta\beta\Delta V \exp(-\beta\Delta V). \qquad (62c)$$

According to Kramers' treatment, the proportionality of the Kramers rate to η in the low viscosity limit turns over to the inverse proportionality in the high viscosity. The interpolating behavior for arbitrary η was studied by Mel'nikov and Meshkov [104].

B. The Fractional Generalization of the Kramers Escape Problem: Mittag-Leffler Decay of the Survival Probability

Let us now derive the fractional counterpart to the exponential decay pattern (62a). To this end, we note that the solution W_α of the fractional Klein-

Kramers and Fokker–Planck equations can be expressed in terms of its Brownian analogue, W_1, according to Eq. (49). Application of relation (49) to the Laplace transform $p(u) = (r_K + u)^{-1}$ of the exponential survival probability, Eq. (62a), produces

$$p_\alpha(u) = \frac{1}{u + r_K^{(\alpha)} u^{1-\alpha}}. \tag{63}$$

By comparison with Eq. (55), the Laplace inversion of Eq. (63) leads to the Mittag-Leffler shape [105]

$$p_\alpha(t) = E_\alpha\left(-r_K^{(\alpha)} t^\alpha\right) \tag{64a}$$

of the survival probability which includes the fractional Kramers rate

$$r_K^{(\alpha)} = \frac{\eta}{\eta_\alpha} r_K. \tag{64b}$$

The fraction $\eta/\eta_\alpha = \vartheta$ is thus the rescaling of the classical Kramers rate according to the parameters classifying the multiple trapping system with broadly distributed waiting times. Similarly, in the underdamped case, one finds the fractional Kramers rate

$$r_K^{(\alpha)} = \frac{\eta^*}{\eta} r_K, \tag{64c}$$

where $\eta^* \equiv \vartheta\eta$ replaces the classical friction η. According to Eqs. (64b) and (64c), our fractional Kramers model leads to the turnover in the friction dependence from $r_K^{(\alpha)} \propto \eta^*$ to $r_K^{(\alpha)} \propto 1/\eta_\alpha$. This seemingly more complicated turnover can be reconciled with the standard picture. Indeed, on combining the elementary constant ϑ with the other constants in expressions (64b) and (64c), the traditional turnover $r_K^{(\alpha)} \propto \eta$ to $r_K^{(\alpha)} \propto 1/\eta$ is recovered for the fractional Kramers rate. This latter observation is due to the linearity of the fractional operator. As a consequence we note that the Arrhenius activation nature of the Kramers rate is preserved in systems controlled by fractional dynamics.

Often, one defines nonexponential relaxations in terms of a time-dependent rate coefficient $k(t)$ through $p(t) = \exp(-k(t)t)$. For the fractional Kramers model one therefore obtains the rate coefficient $k(t) = |\ln E_\alpha(-r_K^{(\alpha)} t^\alpha)|/t$ which leads to two limiting cases, the short-time self-similar behavior

$$k(t) \sim \frac{r_K^{(\alpha)}}{t^{1-\alpha}\Gamma(1+\alpha)}, \qquad t^\alpha \ll r_K^{(\alpha)}, \tag{65a}$$

and the long-time logarithmic pattern

$$k(t) \sim \frac{\alpha}{t}\ln(t[r_K^{(\alpha)}\Gamma(1-\alpha)]^{1/\alpha}), \qquad t^\alpha \gg r_K^{(\alpha)}. \tag{65b}$$

It is interesting to note that the latter, up to some constants, is given by $k(t) \sim \ln t/t$ which is in this sense universal; that is, the functional form is independent of the waiting time index α.

C. An Example from Protein Dynamics

Let us discuss a possible application of this fractional Kramers model to protein dynamics. It was noted before [106–109] that the dynamic process in proteins, like the rebinding process of carbon monoxide CO to myoglobin Mb,

$$Mb + CO \rightarrow MbCO, \tag{66}$$

after photodissociation is highly nonexponential and can be fitted by either a stretched exponential or an asymptotic power law, the latter being numerically equivalent to the Mittag-Leffler behavior. Glöckle and Nonnenmacher [110] have investigated the data within a fractional relaxation model; that is, they applied a Mittag-Leffler fit, as reproduced in Fig. 7. Interestingly, they observed an Arrhenius dependence of the rate of the process. In more detail, these authors assume that the fractional parameter α features a linear temperature dependence, $\alpha(T) = 0.41T/120\,\mathrm{K}$ which might take the change of the protein–solvent system into account. From the data analysis they find a remarkable agreement with the Mittag-Leffler behavior, and the Arrhenius activation is given by $\tau = \tau_m e^{E_\tau/T}$ for the characteristic time which is related to the fractional Kramers rate through $r_K^{(\alpha)} \equiv \tau^{-\alpha}$. Conversely, the associated fractional Kramers rate is temperature-independent for this $\alpha(T) \propto T$ model! The activation energy obtained from the data analysis according to the Mittag-Leffler/Arrhenius model, E_τ, complies well with the generally accepted values [110]. Thus, selecting out the temperature dependence of α, one exactly finds the Arrhenius dependence as predicted by the fractional Fokker–Planck model. The insert in Fig. 7 shows this Arrhenius activation of τ as found in Ref. 110.

It has been claimed that reactions in proteins can, as an approximation, be formulated within the Kramers reaction theory of barrier crossing [106]. The highly nonexponential relaxation pattern can now be explained by our model,

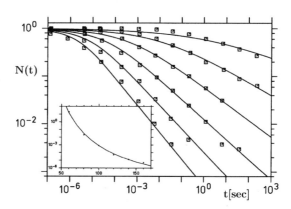

Figure 7. Mittag-Leffler model for the rebinding of CO to Mb, after a photodissociation. Data from Austin et al. [108]. The temperature dependence of τ_0 follows the Arrhenius law shown in the insert, with the parameters $\tau_m = 3.4 \times 10^{-10}$ s and $E_\tau = 1470$ K.

bearing in mind that the dynamic topology caused by the protein strand is basically equivalent to the polymer chains leading to the observation of sub-diffusion in the experiment of Amblard et al. [29]. Consequently, both the fractional relaxation (Mittag-Leffler) behavior and the Arrhenius dependence found by Glöckle and Nonnenmacher can now be understood from our fractional dynamics model.

In addition to the Kramers escape approach to ligand rebinding, a stochastic model was developed by Zwanzig [111] and by Eizenberg and Klafter [112], who assumed a fluctuating bottleneck through which the ligand passes, the process leading to an exponential survival for a white Gaussian noise. We have generalized this model in Ref. 62 within fractional dynamics, recovering the Mittag-Leffler behavior found above.

V. THE FRACTIONAL KLEIN–KRAMERS EQUATION: FRACTIONAL DYNAMICS IN PHASE SPACE

A.   The Multiple Trapping Model

We now turn toward the phase space description of a test particle performing fractional dynamics, drawing upon Langevin's treatment of the Brownian motion of a scalar test particle in a bath of smaller atoms or molecules exerting random collisions upon that particle. In that course, Langevin [6] amended Newton's law of motion with a fluctuating force $m\Gamma(t)$. From this Langevin equation, it follows that the fluctuation-averaged phase space dynamics is governed by the Klein–Kramers equation [9, 13, 14, 21], the solution of which, the pdf $W(x, v, t)$ to find the test particle at the position

$x, \ldots, x + dx$ with the velocity $v, \ldots, v + dv$, at time t, describes the macroscopic dynamics of the system. Thus, we can distinguish the two limiting cases of (a) the Rayleigh equation controlling the velocity distribution $W(v, t)$ in the force-free limit and (b) the Fokker–Planck–Smoluchowski equation.

Indeed, a similar theory can be developed for systems displaying fractional dynamics. Starting off from the classical Langevin equation with δ-correlated Gaussian noise, fractional dynamics emerges from the competition of subsequent Langevin dominated motion events of average duration τ^*, interrupted through trapping events whose duration is broadly distributed. This multiple trapping scenario is of a quite general nature, and it offers physical insight into fractional dynamics as described by the fractional Fokker–Planck equation.

During the Langevin sections of the particle motion, the dynamics is controlled by the Langevin equation in the external force field $F(x) = -\Phi'(x)$,

$$m \frac{d^2 x}{dt^2} = -m\eta v + F(x) + m\Gamma(t), \qquad v = \frac{dx}{dt}, \tag{67}$$

which describes the ongoing erroneous bombardment through small surrounding atoms or molecules via the fluctuating, δ-correlated Gaussian noise $\Gamma(t)$. The velocity-proportional damping caused by effective interactions with the environment is characterized by the friction constant η. Averaging out the fluctuations, one finds the moments of the mean velocity increments [9],

$$\langle \Delta v \rangle = \left(\eta v - \frac{F(x)}{m} \right) \Delta t, \qquad \langle (\Delta v)^2 \rangle = \frac{\eta k_B T}{m} \Delta t + O([\Delta t]^2), \tag{68}$$

from the Langevin equation (67) which are both proportional to the mean jump time Δt. Note that the noise-averaged Eq. (67), $m \langle \ddot{x} \rangle_\Gamma = -m\eta \langle v \rangle_\Gamma + F(x)$, corresponds to Newton's law of motion.

In the usual derivations of the Klein–Kramers equation, the moments of the velocity increments, Eq. (68), are taken as expansion coefficients in the Chapman–Kolmogorov equation [9]. Generalizations of this procedure start off with the assumption of a memory integral in the Langevin equation to finally produce a Fokker–Planck equation with time-dependent coefficients [67]. We are now going to describe an alternative approach based on the Langevin equation (67) which leads to a fractional Klein–Kramers equation— that is, a temporally nonlocal behavior.

Let us define the (multiple) trapping events. As mentioned before, trapping describes the occasional immobilization of the random walking test particle for a waiting time which rules the time span elapsing between the immobilization and the subsequent release of the test particle. This waiting time is drawn from the waiting time pdf $w(t)$. Such trapping has been recognized as the mechanism underlying the dispersive charge carrier transport in amorphous semiconductors [23, 25, 26] and the motion of excess electrons in liquids [113], and it occurs in the phase space dynamics of chaotic Hamiltonian systems [36].

In addition to trapping, it is supposed that the kinetic energy of a particle during a trapping–detrapping event is conserved, and that each trapping period is followed by a motion event during which the particle moves in the bath of surrounding smaller particles in which it undergoes the same collisions as underlie the standard Brownian counterpart. Each of these motion events following release from the trap is supposed to endure for the mean time τ^*. This means that while not being trapped, the test particle features a Markov behavior described by the Langevin equation (67). The immobilizing–release–walking scenario therefore combines trapping periods and Langevin dynamics in a sequential manner. The overall process is, in essence, the phase space extension of the multiple trapping model conceived in Ref. 23. In our model, the length $x^* = v_{th}\tau^*$, where $v_{th} \equiv \sqrt{\lim_{t\to\infty}\langle v^2(t)\rangle}$ is the thermal velocity, is the average distance between adjacent traps visited.

B. The Fractional Klein–Kramers Equation and the Related Equations

Fractional dynamics emerges as the macroscopic limit of the combination of the Langevin and the trapping processes. After straightforward calculations based on the continuous-time version of the Chapman–Kolmogorov equation [75, 114] which are valid in the long-time limit $t \gg \max\{\tau, \tau^*\}$, one obtains the fractional Klein–Kramers equation

$$\frac{\partial W}{\partial t} = {}_0D_t^{1-\alpha}\left(-v^*\frac{\partial}{\partial x} + \frac{\partial}{\partial v}\left(\eta^* v - \frac{F^*(x)}{m}\right) + \eta^*\frac{k_B T}{m}\frac{\partial^2}{\partial v^2}\right)W(x, v, t). \quad (69)$$

Here, the Klein–Kramers operator has the same structure as in the Brownian case, except for the occurrence of the asterisked quantities that are defined through $v^* \equiv v\vartheta$, $\eta^* \equiv \eta\vartheta$, and $F^*(x) \equiv F(x)\vartheta$ whereby the factor ϑ is the ratio

$$\vartheta \equiv \frac{\tau^*}{\tau^\alpha} \quad (70)$$

of the intertrapping time scale τ^* and the internal waiting time scale τ from Eq. (6). The stationary solution of the fractional Klein–Kramers equation

[Eq. (69)], $W_{st}(x, v) \equiv \lim_{t\to\infty} W(x, v, t)$ is given by the Gibbs–Boltzmann equilibrium distribution $W_{st}(x, v) = N \exp\{-\beta E\}$ where $E = \frac{1}{2}mv^2 + \Phi(x)$, and N is the appropriate normalization constant. In the Brownian limit $\lim_{\alpha\to 1}$, Eq. (69) reduces to the standard Klein–Kramers equation, as expected.

Integration of Eq. (69) over velocity, and of v times Eq. (69) over velocity, results in two equations whose combination leads to the fractional equation [75, 115]

$$\frac{\partial W}{\partial t} + {}_0D_t^{1+\alpha}\frac{1}{\eta^*}W = {}_0D_t^{1-\alpha}\left(-\frac{\partial}{\partial x}\frac{F(x)}{m\eta_\alpha} + K_\alpha\frac{\partial^2}{\partial x^2}\right)W(x,t). \quad (71)$$

Equation (71) reduces to the telegrapher's-type equation found in the Brownian limit $\alpha = 1$ [115]. In the usual high-friction or long-time limit, one recovers the fractional Fokker–Planck equation (19). The generalized friction and diffusion coefficients in Eq. (19) are defined by [75]

$$\eta_\alpha \equiv \frac{\eta}{\vartheta} \quad (72a)$$

and

$$K_\alpha \equiv \frac{k_B T}{m\eta_\alpha} = \vartheta K \quad (72b)$$

and are thus to be understood as a rescaled version of the well-defined physical quantities η and K. Moreover, the generalized Einstein–Stokes relation (37) has now been obtained as a direct consequence of the interplay between the Langevin diffusion with the long-tailed trapping process.

The integration of the fractional Klein–Kramers equation (69) over the position coordinate leads in, the force-free limit, to the fractional Rayleigh equation

$$\frac{\partial W}{\partial t} = {}_0D_t^{1-\alpha}\eta^*\left(\frac{\partial}{\partial v}v + \frac{k_B T}{m}\frac{\partial^2}{\partial v^2}\right)W(v,t), \quad (73)$$

which is an example of the fractional Ornstein–Uhlenbeck process discussed in the preceding section. The solution of Eq. (73), the pdf $W(v, t)$, describes the equilibration of the velocity distribution toward the Maxwell distribution $W_{st}(v) = \sqrt{\frac{\beta m}{2\pi}} \exp\left(-\beta\frac{m}{2}v^2\right)$ with the thermal velocity $v_{th}^2 = k_B T/m$.

This model for subdiffusion in the external force field $F(x) = -\Phi'(x)$ provides a basis for fractional evolution equations, starting from Langevin dynamics that is combined with long–tailed trapping events possessing a diverging characteristic waiting time T. The first stage hosts the microscopic Brownian process, characterized by the mean stepping time τ^* that is basically equivalent to a multiple of the mean jump time Δt in Eq. (68). If the characteristic waiting time is finite, $T < \infty$, the trapping mechanism also possesses its characteristic time scale. Therefore, the second stage brings two Poissonian processes together, and the macroscopic process defined as the long-time limit with respect to $\max\{\tau^*, T\}$ is Markovian, being determined by the standard Brownian Klein–Kramers equation. Conversely, if $T \to \infty$ diverges, the time scales of the microscopic Brownian motion, τ^*, separates from the combined trapping–detrapping process. The latter occasionally features very long waiting times so that individual trapping events do not have a typical time scale and cannot be distinguished from the sampling of many trapping events on the macroscopic level, a situation that is typical for self-similar processes. The overall dynamics becomes fractional.

The combined process is governed by the long-tailed form of the waiting time pdf, manifested in the fractional nature of the associated Eq. (69). Physically, this causes the rescaling of the fundamental quantity η by the scaling factor ϑ to result in the generalized friction constant $\eta_\alpha = \eta/\vartheta$. It is interesting to note that the kinematics level, force-free multiple trapping process from Ref. 113 in (x, t)-coordinates reveals the subdiffusive mean square displacement $\langle x^2(t) \rangle \propto t^\alpha$.

The Langevin picture rules the Markov motion parts in between successive trapping states. On this stage the test particle consequently obeys Newton's law, in the noise-averaged sense defined above. Conversely, averaging the fractional Klein–Kramers equation (69) over velocity and position coordinates, one recovers the memory relation $\frac{d}{dt}\langle\langle x(t) \rangle\rangle = \vartheta \, _0D_t^{1-\alpha}\langle\langle v(t) \rangle\rangle$ between the mean position $\langle\langle x(t) \rangle\rangle$ and the mean velocity $\langle\langle v(t) \rangle\rangle$. This "violation" is only due to the additional waiting time averaging that camouflages the Langevin-dominated motion events.

VI. CONCLUSIONS

Anomalous transport features have been reported for an increasing number of (complex) systems. Many of these systems underlie some sort of a generalized limit theorem that is connected with Lévy statistics and thus with self-similar evolution patterns. This fact is mirrored in the long-time prevalence of power-law time behaviors of the related physical quantities.

Fractional dynamics is a made-to-measure approach to the description of temporally nonlocal systems, the kinetics of which is governed by a self-similar memory. Fractional kinetic equations are operator equations that are mathematically close to the well-studied, analogous Brownian evolution equations of the Klein–Kramers, Rayleigh, or Fokker–Planck types. Consequently, methods such as the separation of variables can be applied. Moreover, the fractional solution exists if only the corresponding Brownian solution exists.

The characteristic changes brought about by fractional dynamics in comparison to the Brownian case include the temporal nonlocality of the approach manifest in the convolution character of the fractional Riemann–Liouville operator. Initial conditions relax slowly, and thus they influence the evolution of the system even for long times [62, 116]; furthermore, the Mittag-Leffler behavior replaces the exponential relaxation patterns of Brownian systems. Still, the associated fractional equations are linear and thus extensive, and the limit solution equilibrates toward the classical Gibbs–Boltzmann and Maxwell distributions, and thus the processes are close to equilibrium, in contrast to the Lévy flight or generalised thermostatistics models under discussion.

The physical foundation of fractional dynamics has been developed. Multiple trapping featuring a competition between Langevin-dominated motion events possessing a characteristic time scale and broadly distributed trapping events has been shown to give rise to fractional dynamics in the long-time limit. Consequently, the coming into existence of the generalized diffusion and friction constants can be understood as a rescaling resulting from this competition between Langevin and trapping regimes in which finally the generalized central limit theorem guarantees the dominance of the broad waiting time pdf.

In our presentation, we concentrated on the modeling of subdiffusive phenomena—that is, modeling of processes whose mean squared displacement in the force-free limit follows the power-law dependence $\langle x^2(t) \rangle \propto t^\kappa$ for $0 < \kappa < 1$. The extension of fractional dynamics to systems where the transport is subballistic but superdiffusive, $1 < \kappa < 2$, is presently under discussion [77, 78], (compare also Ref. 117).

We finally note that a more mathematically oriented account of fractional equations in the description of anomalous kinetic processes has recently been published [118].

Acknowledgments

We gratefully acknowledge discussions with Eli Barkai, Brian Berkowitz, Joshua Jortner, Shaul Mukamel, Theo Nonnenmacher, Ilia Rips, Harvey Scher, Michael Shlesinger, Igor Sokolov, and Gert Zumofen. Financial assistance from the German–Israeli Foundation (GIF) and the TMR programme of the European Commission (SISITOMAS) is acknowledged as well. RM was supported in part by a Feodor–Lynen fellowship from the Alexander von Humboldt Stiftung, by an Amos de Shalit fellowship from the Minerva foundation, and by an Emmy Noether fellowship from the DFG.

APPENDIX A: A PRIMER ON LÉVY DISTRIBUTIONS

Historically, the central limit theorem, which guarantees the existence of the all important Gaussian limit distribution for processes with a finite variance, had grown out of the inequality of Bienaymé, the theorems of Bernoulli and de Moivre-Laplace, and the law of large numbers. The central limit theorem has received a central role in the exact sciences and beyond, over centuries. Toward the turn of the twentieth century, mathematicians became interested in the possibility of a limit theorem for processes without a finite variance, ideas whose fundamentals actually go back to Cauchy. Such a generalized framework was conceived by Paul Lévy (after which the generalized normal distributions are named), A. Ya. Khintchine, W. Feller, A. M. Kolmogorov, and B. V. Gnedenko, among others.

According to Lévy [43], a distribution F is stable iff for the two positive constants c_1 and c_2 there exists a positive constant c such that the random variable X given by

$$c_1 X_1 + c_2 X_2 = cX \qquad (A1)$$

is a random variable following the same distribution F as the independent, identically distributed (iid) random variables X_1 and X_2. Alternatively, if

$$\varphi(z) \equiv \langle e^{iXz} \rangle = \int_{-\infty}^{\infty} e^{iXz} dF(X) \qquad (A2)$$

denotes the characteristic function of the distribution F, then F is stable iff

$$\varphi(c_1 z)\varphi(c_2 z) = \varphi(cz). \qquad (A3)$$

A more general definition is given by Feller [12]. We denote a Lévy stable pdf $dF(x)$ as $\mathbf{L}_\alpha(x)$ and call α the Lévy (stable) index. It can be shown that a stable law has a characteristic function of the form

$$\psi(z) = \log \varphi(z) = i\gamma z - c|z|^\alpha \left\{ 1 + i\beta \frac{z}{|z|} \omega(z, \alpha) \right\}, \qquad (A4)$$

where α, β, γ, c are constants (γ is any real number, $0 < \alpha \le 2$, $-1 < \beta < 1$, and $c > 0$), and

$$\omega(z, \alpha) = \begin{cases} \tan \frac{\pi\alpha}{2}, & \text{if } \alpha \ne 1 \\ \frac{2}{\pi} \log|z|, & \text{if } \alpha = 1. \end{cases} \qquad (A5)$$

From Eq. (A4) it follows that the limiting case $\alpha = 2$ corresponds to the Gaussian normal distribution governed by the central limit theorem. For $\beta = 0$, the distribution is symmetric. γ translates the distribution, and c is a scaling factor for X. Thus, γ and c are not essential parameters; if we disregard them, the characteristic function fulfills

$$|\varphi(z)| = e^{-|z|^\alpha}, \qquad \alpha \ne 1. \qquad (A6)$$

Thus, one can write

$$\psi(z) = -|z|^{\alpha} \exp\left(i\frac{\pi\beta}{2}\mathrm{sign}(z)\right) \tag{A7}$$

with the new centering constant β that is restricted in the following region:

$$|\beta| \leq \begin{cases} \alpha, & \text{if } 0 < \alpha < 1 \\ 2 - \alpha, & \text{if } 1 < \alpha < 2 \end{cases}. \tag{A8}$$

The pdf $f_{\alpha,\beta}(x)$ is the Fourier transform of $\varphi(z)$, defined by Eq. (A7):

$$f_{\alpha,\beta}(x) = \frac{1}{\pi}\mathrm{Re}\int_0^\infty \exp\left(-ixz - z^\alpha \exp\left(i\frac{\pi\beta}{2}\right)\right)dz. \tag{A9}$$

Thus,

$$f_{\alpha,\beta}(x) = f_{\alpha,-\beta}(-x) \tag{A10}$$

so that

$$f_{\alpha,0}(x) = f_{\alpha,0}(-x) \tag{A11}$$

is symmetric in x.

The asymptotic behavior of a Lévy stable distribution follows the inverse power-law

$$f_{\alpha,\beta}(x) \sim \frac{A_{\alpha,\beta}}{|x|^{1+\alpha}}, \qquad \alpha < 2; \tag{A12}$$

consequently, for all Lévy stable laws with $0 < \mu < 2$, the variance diverges:

$$\langle x^2 \rangle \to \infty. \tag{A13}$$

Special cases include the Cauchy or Lorenz distribution

$$f_{1,0}(x) = \frac{a}{\pi(a^2 + x^2)} \tag{A14}$$

for $\alpha = 1$ and $\beta = 0$, as well as the one-sided or completely asymmetric distribution $\mathbf{L}_\alpha^+(x) \equiv f_{\alpha,-\alpha}(x)$ if $0 < \alpha < 1$ and $\beta = -\alpha$. For instance, the one-sided stable density for $\alpha = 1/2$ and $\beta = -1/2$ is given by

$$\mathbf{L}_{1/2}^+(x) = \frac{1}{2\sqrt{\pi}}x^{-3/2}e^{-1/4x}. \tag{A15}$$

To obtain the characteristic function of a one-sided stable law, one calculates the Laplace transform.

Let us examine the one-sided Lévy distribution $dF(X) \equiv \mathbf{L}_\alpha^+(t/\tau)$ with the characteristic function

$$\varphi(u) \equiv \langle e^{-Xu} \rangle = \int_0^\infty e^{-Xu} dF(X) = \exp(-(u\tau)^\alpha) \qquad (A16)$$

in more detail. Identification with the corresponding Fox function [55, 56],

$$\mathbf{L}_\alpha^+(u\tau) = H_{0,1}^{1,0}\left[(t/\tau)^\alpha \,\middle|\, \overline{(0,1)} \right], \qquad (A17)$$

allows for the direct Laplace inversion, resulting in [56, 84]

$$\mathbf{L}_\alpha^+(t/\tau) = \frac{1}{\tau} H_{1,1}^{1,0}\left[\left(\frac{\tau}{t}\right)^\alpha \,\middle|\, \begin{matrix} (1,\alpha) \\ (1/\alpha, 1) \end{matrix} \right] \qquad (A18)$$

with the corresponding series expansion

$$\mathbf{L}_\alpha^+(t/\tau) = \frac{1}{\tau} \sum_{n=1}^\infty \frac{(-1)^n}{n!\Gamma(-\alpha n)} \left(\frac{\tau}{t}\right)^{1+\alpha n} \qquad (A19)$$

from which we find the asymptotic behavior

$$\mathbf{L}_\alpha^+(t/\tau) \sim A_\alpha \frac{\tau^\alpha}{t^{1+\alpha}}, \qquad (A20)$$

with $A_\alpha \equiv 1/|\Gamma(-\alpha)|$, for $0 < \alpha < 1$. For short times $t \ll \tau$, the one-sided Lévy distribution $\mathbf{L}_\alpha^+(t/\tau)$ becomes exponentially small (compare the discussion of nonexponential relaxation in Ref. 86).

APPENDIX B: THE UBIQUITOUS MITTAG-LEFFLER FUNCTION

The Mittag-Leffler function [44–46] can be viewed as a natural generalization of the exponential function. Within fractional dynamics, it replaces the traditional exponential relaxation patterns of moments, modes, or of the Kramers survival. It is an entire function that decays completely monotonically for $0 < \alpha < 1$. It is the exact relaxation function for the underlying multiscale process, and it leads to the Cole–Cole behavior for the complex susceptibility, which is broadly used to describe experimental results. It can be decomposed into single Debye processes, the relaxation time distribution of which is given by a one-sided Lévy distribution [80].

The Mittag-Leffler function is defined through the inverse Laplace transform

$$E_\alpha(-(t/\tau)^\alpha) = \mathscr{L}^{-1}\left\{\frac{1}{u + \tau^{-\alpha}u^{1-\alpha}}\right\}, \tag{B1}$$

from which the series expansion

$$E_\alpha(-(t/\tau)^\alpha) = \sum_{n=0}^{\infty}\frac{(-(t/\tau)^\alpha)^n}{\Gamma(1 + \alpha n)} \tag{B2}$$

can be deduced. The asymptotic behavior is

$$E_\alpha(-(t/\tau)^\alpha) \sim ((t/\tau)^\alpha\Gamma(1-\alpha))^{-1} \tag{B3}$$

for $t \gg \tau$, $0 < \alpha < 1$. Special cases of the Mittag-Leffler function are the exponential function

$$E_1(-t/\tau) = e^{-t/\tau} \tag{B4}$$

and the product of the exponential and the complementary error function

$$E_{1/2}\left(-(t/\tau)^{1/2}\right) = e^{t/\tau}\mathrm{erfc}\left((t/\tau)^{1/2}\right). \tag{B5}$$

We note in passing that the Mittag-Leffler function is the solution of the fractional relaxation equation [84]

$$\frac{df(t)}{dt} = -\tau^{-\alpha}\,_0D_t^{1-\alpha}f(t). \tag{B6}$$

The Mittag-Leffler function interpolates between the initial stretched exponential form

$$E_\alpha(-(t/\tau)^\alpha) \sim \exp\left(-\frac{(t/\tau)^\alpha}{\Gamma(1+\alpha)}\right) \tag{B7}$$

and the long-time inverse power-law behavior (B3). For $\alpha > 1$, the Mittag-Leffler function shows oscillations [44–46].

References

1. R. Brown, *Philos. Mag.* **4**, 161 (1828); *Ann. Phys. Chem.* **14**, 294 (1828); M. Gouy, *J. Phys. (Paris)* **7**, 561 (1888); F. M. Exner, *Ann. Phys.* **2**, 843 (1900). Actually, "Brownian" motion was already observed by Dutch physician Jan Ingenhousz in 1785.

2. J. Perrin, *Comptes Rendus (Paris)* **146**, 967 (1908); *Ann. Chim. Phys.* **VIII 18**, 5 (1909); A. Westgren, *Z. Phys. Chem.* **83**, 151 (1913); *ibid.* **89**, 63 (1914); *Arch. Mat. Astr. Fysik* **11** (8 & 14) (1916); *Z. anorg. Chem.* **93**, 231 (1915); *ibid.* **95**, 39 (1916); E. Kappler, *Ann. Phys. (Leipzig)* **11**, 233 (1931).

3. A. Fick, *Ann. Phys. (Leipzig)* **170**, 50 (1855). He actually set up his two laws for the temporal spreading of the *concentration* of a tracer substance, not for the probability. The first evolution equation for a probability was the Boltzmann equation [L. Boltzmann *Vorlesungen über Gastheorie* I (J. A. Barth, Leipzig, 1896)], following Maxwell's theory of gas kinetics.

4. K. Pearson, *Nature* **72**, 294, 342 (1905); J. W. Strutt, Lord Rayleigh, *Nature* **72**, 318 (1905); G. Pólya, *Math. Ann.* **83**, 149 (1921).

5. A. Einstein, *Ann. Phys.* **17**, 549 (1905), *ibid.* **19**, 371 (1906); *Albert Einstein, Investigations on the Theory of Brownian Movement*, edited by R. Fürth, Dover, New York, 1956. Einstein coined the name "Brownian motion". Some of Einstein's results were already reported by Bachelier in terms of stock market variations: L. Bachelier, *Ann. Scientif. l'École Norm.* **17**, 21 (1900); *ibid.* **27**, 339 (1910).

6. P. Langevin, *Comptes Rendus* **146**, 530 (1908).

7. N. Wiener, *J. Math. Phys. (MIT)* **2**, 131 (1923); *Bull. Soc. Math. (France)* **52**, 569 (1924); *Acta Math.* **55**, 117 (1930); *Am. J. Math.* **60**, 897 (1938).

8. G. E. Uhlenbeck and L. S. Ornstein, *Phys. Rev.* **36**, 823 (1930).

9. S. Chandrasekhar, *Rev. Mod. Phys.* **15**, 1 (1943).

10. J. L. Doob, *Stochastic Processes*, John Wiley & Sons, New York, 1953.

11. P. Lévy, *Processus stochastiques et mouvement Brownien*, Gauthier–Villars, Paris, 1965.

12. W. Feller, *An Introduction to Probability Theory and Its Applications*, John Wiley & Sons, New York, 1968.

13. N. G. van Kampen, *Stochastic Processes in Physics and Chemistry*, North-Holland, Amsterdam, 1981.

14. H. Risken, *The Fokker–Planck Equation*, Springer-Verlag, Berlin, 1989.

15. G. H. Weiss, *Aspects and Applications of the Random Walk Random Materials and Processes*, series edited by H. E. Stanley and E. Guyon, North-Holland, Amsterdam, 1994.

16. B. D. Hughes, *Random Walks and Random Environments, Vol. 1: Random Walks*, Oxford University Press, Oxford, 1995.

17. A. D. Fokker, *Ann. Phys. (Leipzig)* **43**, 810 (1914).

18. M. Planck, *Sitzber. Preuß. Akad. Wiss.*, p. 324 (1917).

19. M. von Smoluchowski, *Ann. Phys.* **48**, 1103 (1915).

20. J. W. Strutt (Lord Rayleigh), *Philos. Mag.* **32**, 424 (1891).

21. O. Klein, *Arkiv Mat. Astr. Fys.* **16** (5) (1922).

22. H. A. Kramers, *Physica* **7**, 284 (1940).

23. The dimensionless constant A_α and the "internal waiting time constant" τ are chosen for convenience, ensuring the correct dimensionality s^{-1} of $w(t)$.

24. H. Scher, M. F. Shlesinger, and J. T. Bendler, *Phys. Today* **44** (1), 26 (1991).

25. H. Scher and E. W. Montroll, *Phys. Rev.* B **12**, 2455 (1975); G. Pfister and H. Scher, *Phys. Rev.* B **15**, 2062 (1977); *Adv. Phys.* **27**, 747 (1978).

26. Q. Gu, E. A. Schiff, S. Grebner, and R. Schwartz, *Phys. Rev. Lett.* **76**, 3196 (1996).

27. O. Cardoso and P. Tabeling, *Europhys. Lett.* **7**, 225 (1988).

28. H. W. Weber and R. Kimmich, *Macromolecules* **26**, 2597 (1993); E. Fischer, R. Kimmich, and N. Fatkullin, *J. Chem. Phys.* **104**, 9174 (1996); E. Fischer, R. Kimmich, U. Beginn, M. Moeller, and N. Fatkullin, *Phys. Rev. E* **59**, 4079 (1999); R. Kimmich, R.-O. Seitter, U. Beginn, M. Moeller, and N. Fatkullin, *Chem. Phys. Lett.*, in press; K. Binder and W. Paul, *J. Polym. Sci. B Pol. Phys.* **35**, 1 (1997).

29. F. Amblard, A. C. Maggs, B. Yurke, A. N. Pargellis, and S. Leibler, *Phys. Rev. Lett.* **77**, 4470 (1996); *ibid.* **81**, 1135 (1998); E. Barkai and J. Klafter, *ibid.* **81**, 1134 (1998).

30. F. Klammler and R. Kimmich, *Croat. Chem. Acta* **65**, 455 (1992).

31. A. Klemm, H.-P. Müller and R. Kimmich, *Phys. Rev. E* **55**, 4413 (1997); *Physica* **266A**, 242 (1999).

32. S. Liu, L. Bönig, J. Detch, and H. Metiu, *Phys. Rev. Lett.* **74**, 4495 (1995).

33. B. Berkowitz and H. Scher, *Phys. Rev. E* **57**, 5858 (1998); *Phys. Rev. Lett.* **79**, 4038 (1997); B. Berkowitz, H. Scher, and S. E. Silliman, *Water Res. Res.* **36**, 149 (2000).

34. J. Sabelko, J. Ervin, and M. Grubele, *Proc. Natl. Acad. Sci. USA* **96**, 6031 (1999); M. Grubele, *Ann. Rev. Phys. Chem.* **50**, 485 (1999).

35. M. Karplus, *J. Phys. Chem.* **104**, 11 (2000).

36. G. Zumofen and J. Klafter, *Phys. Rev. E* **49**, 4873 (1995); L. Kuznetsov and G. M. Zaslavsky, *ibid.* **58**, 7330 (1998); S. Benkadda, S. Kassibrakis, R. B. White, and G. M. Zaslavsky, *ibid.* **55**, 4909 (1997).

37. M. Kuno, D. P. Fromm, H. F. Hamann, A. Gallagher, and D. J. Nesbitt, *J. Chem. Phys.* **112**, 3117 (2000).

38. S. Schaufler, W. P. Schleich, and V. P. Yakovlev, *Phys. Rev. Lett.* **83**, 3162 (1999); *Europhys. Lett.* **39**, 383 (1997).

39. A. Blumen, J. Klafter, and G. Zumofen, in *Optical Spectroscopy of Glasses*, edited by I. Zschokke, Reidel, Dordrecht, 1986.

40. S. Havlin and D. Ben-Avraham, *Adv. Phys.* **36**, 695 (1987).

41. J.-P. Bouchaud and A. Georges, *Phys. Rep.* **195**, 12 (1990).

42. J. Klafter, M. F. Shlesinger, and G. Zumofen, *Phys. Today* **49**(2), 33 (1996).

43. P. Lévy, *Théorie de l'addition des variables aléatoires*, Gauthier–Villars, Paris, 1954; *Calcul des probabilités*, Gauthier–Villars, Paris, 1925; B. V. Gnedenko and A. N. Kolmogorov, *Limit Distributions for Sums of Random Variables*, Addison–Wesley, Reading, 1954.

44. G. M. Mittag-Leffler, *C. R. Acad. Sci. Paris* **137**, 554 (1903); *R. Acad. dei Lincei, Rendiconti* **13**, 3 (1904); *Acta Math.* **29**, 101 (1905).

45. A. Wiman, *Acta Math.* **29**, 191 (1905); H. Pollard, *Bull. Am. Math. Soc.* **54**, 1115 (1948).

46. *Tables of Integral Transforms*, edited by A. Erdélyi, Bateman Manuscript Project, Vol. I, McGraw–Hill, New York, 1954.

47. R. F. Pawula, *Phys. Rev.* **162**, 186 (1967).

48. M. F. Shlesinger, G. M. Zaslavsky, and J. Klafter, *Nature* **363** 31 (1993).

49. J. Klafter, A. Blumen, and M. F. Shlesinger, *Phys. Rev. A* **35**, 3081 (1987).

50. W. R. Schneider and W. Wyss, *J. Math. Phys.* **30**, 134 (1989).

51. R. Hilfer, *Fractals* **3**, 211 (1995); R. Metzler, W. G. Glöckle, and T. F. Nonnenmacher, *Physica* **211A**, 13 (1994).

52. R. Hilfer and L. Anton, *Phys. Rev. E* **51**, R848 (1995).

53. A. Compte, *Phys. Rev. E* **53**, 4191 (1996).

54. R. Metzler, J. Klafter, and I. Sokolov, *Phys. Rev. E* **58**, 1621 (1998).

55. C. Fox, Trans. *Am. Math. Soc.* **98**, 395 (1961); B. L. J. Braaksma, *Compos. Math.* **15**, 239 (1964).

56. A. M. Mathai and R. K. Saxena, *The H-Function with Applications in Statistics and Other Disciplines*, Wiley Eastern Ltd., New Delhi, 1978; H. M. Srivastava and B. R. K. Kashyap, *Special Functions in Queuing Theory and Related Stochastic Processes* Academic Press, New York, 1982.

57. G. Zumofen and J. Klafter, *Phys. Rev. E* **47**, 851 (1993); J. Klafter and G. Zumofen, *J. Phys. Chem.* **98**, 7366 (1994).

58. R. Metzler, E. Barkai, and J. Klafter, *Europhys. Lett.* **46**, 431 (1999).

59. E. Barkai, R. Metzler, and J. Klafter, *Phys. Rev. E* **61**, 132 (2000).

60. R. Metzler, E. Barkai, and J. Klafter, *Phys. Rev. Lett.* **82**, 3563 (1999).

61. K. B. Oldham and J. Spanier *The fractional calculus*, Academic Press, New York, 1974.

62. R. Metzler and J. Klafter, *Physica* **278A**, 107 (2000).

63. B. I. Henry and S. L. Wearne, *Physica* **276A**, 448 (2000).

64. B. B. Mandelbrot and J. W. van Ness, *SIAM Rev.* **10**, 422 (1968).

65. E. W. Montroll, *J. SIAM* **4**, 241 (1956); E. W. Montroll and G. H. Weiss, *J. Math. Phys.* **6**, 178 (1965); E. W. Montroll and G. H. Weiss, *J. Math. Phys.* **10**, 753 (1969); *The Wonderful World of Stochastics: A Tribute to E. W. Montroll*, Studies in Statistical Mechanics, No. 12, edited by M. F. Shlesinger, North-Holland, Amsterdam, 1985; J. Klafter, A. Blumen, and M. F. Shlesinger, *Phys. Rev. A* **35**, 3081 (1987).

66. B. O'Shaugnessy and I. Procaccia, *Phys. Rev. Lett.* **54**, 455 (1985).

67. K. G. Wang, L. K. Dong, X. F. Wu, F. W. Zhu, and T. Ko, *Physica* **203A**, 53 (1994); K. G. Wang and M. Tokuyama, *Physica* **265A**, 341 (1999).

68. B. J. West and V. Seshadri, *Physica* **113A**, 203 (1982).

69. G. M. Zaslavsky, *Chaos* **4**, 25 (1994); G. M. Zaslavsky, M. Edelman, and B. A. Niyazov, *Chaos*, **7**, 159 (1997); A. I. Saichev and G. M. Zaslavsky, *Chaos* **7**, 753 (1997).

70. H. C. Fogedby, *Phys. Rev. Lett.* **73**, 2517 (1994); *Phys. Rev. E* **50**, 1657 (1994); *ibid.* **58**, 1690 (1998); S. Jespersen, R. Metzler, and H. C. Fogedby, *Phys. Rev. E* **59**, 2736 (1999).

71. D. Kusnezov, A. Bulgac, and G. D. Dang, *Phys. Rev. Lett.* **82**, 1136 (1999).

72. K. M. Kolwankar and A. D. Gangal, *Phys. Rev. Lett.* **80**, 214 (1998).

73. C. Tsallis and D. J. Bukman, *Phys. Rev. E* **54**, R2197 (1996); L. Borland, *Phys. Rev. E* **57**, 6634 (1998).

74. E. Barkai and V. N. Fleurov, *Phys. Rev. E* **58**, 1296 (1998).

75. R. Metzler and J. Klafter, *J. Phys. Chem., B* **104**, 3851 (2000); *Phys. Rev. E,* **61**, 6308 (2000).

76. An equivalent formulation can be found in J. Klafter and G. Zumofen, *J. Phys. Chem.* **98**, 7366 (1994).

77. E. Barkai and R. Silbey, *J. Phys. Chem., B* **104**, 3866 (2000).

78. R. Metzler and J. Klafter, *Europhys. Lett.* **51**, 492 (2000).

79. W. G. Glöckle and T. F. Nonnenmacher, *Macromolecules* **24**, 6426 (1991); *J. Stat. Phys.* **71**, 755 (1993); *Rheol. Acta* **33**, 337 (1994).

80. W. G. Glöckle and T. F. Nonnenmacher, *J. Stat. Phys.* **71**, 741 (1993).

81. K. Weron and M. Kotulski, *Physica* **232A**, 180 (1996).

82. K. S. Cole and R. H. Cole, *J. Chem. Phys.* **9**, 341 (1941).

83. E. Barkai and J. Klafter, *Phys. Rev. Lett.* **81**, 1134 (1998).

84. E. Barkai and V. Fleurov, *J. Chem. Phys.* **212**, 69, (1996); *Phys. Rev. E.* **56**, 6355, (1997); *Phys. Rev. E*, in press.

85. R. Metzler, W. Schick, H.-G. Kilian, and T. F. Nonnenmacher, *J. Chem. Phys.* **103**, 7180 (1995); H. Schiessel, R. Metzler, A. Blumen, and T. F. Nonnenmacher, *J. Phys. A* **28**, 6567 (1995).

86. R. Metzler, J. Klafter, J. Jortner, and M. Volk, *Chem. Phys. Lett.* **293**, 477 (1998).

87. G. H. Weiss, *J. Stat. Phys.* **42**, 1 (1986); V. I. Mel'nikov, *Phys. Rep.* **209**, 1 (1991).

88. R. von Smoluchowski, *Z. Phys. Chem.* **29**, 129 (1917).

89. L. Onsager, *Phys. Rev.* **54**, 554 (1938).

90. L. Pontryagin, A. Andronow, and A. Witt, *Zh. Eksp. Teor. Fiz.* **3**, 172 (1933).

91. L. Landau and E. Teller, *Phys. Z. Sowjetunion* **10**, 34 (1936).

92. E. W. Montroll and K. E. Shuler, *Adv. Chem. Phys.* **1**, 361 (1958).

93. G. H. Weiss, *Adv. Chem. Phys.* **13**, 1 (1967).

94. A. Bar-Haim and J. Klafter, *Phys. Rev. E* **60**, 2554 (1999).

95. O. Benichou, B. Gaveau and M. Moreau, *Phys. Rev. E* **59**, 103 (1999); *J. Chem. Phys.* **110**, 2544 (1999); *ibid.* **111**, 1385 (1999).

96. Y. Abe, D. Boilley, B. G. Giraud, and T. Wada, *Phys. Rev. E* **61**, 1125 (2000).

97. R. F. Grote and J. T. Hynes, *J. Chem. Phys.* **73** (1980) 2715; *ibid.* **74**, 4465 (1981); *ibid.* **75**, 2191 (1981); S. P. Velsko and G. R. Fleming, *ibid.* **76**, 3553 (1982); *Chem. Phys.* **65**, 59 (1982).

98. P. Hänggi and F. Mojtabai, *Phys. Rev. A* **26** (1982) 1168; P. Hänggi, *J. Stat. Phys.* **30**, 401 (1983); *ibid.* **42**, 105 (1986); P. Hänggi, P. Talkner, and M. Borkovec, *Rev. Mod. Phys.* **62**, 251 (1990).

99. A. M. Berezhkovskij et al., *J. Chem. Phys.* **107**, 10539 (1997).

100. C. W. Gardiner, *Handbook of Stochastic Methods*, Springer-Verlag, Berlin, 1983.

101. J. L. Skinner and P. G. Wolynes, *J. Chem. Phys.* **69**, 2143 (1978); *ibid.* **72**, 4913 (1980).

102. D. Bedeaux, K. Lakatos-Lindenberg, and K. E. Shuler, *J. Math. Phys.* **12**, 2116 (1971); *Stochastic Processes in Chemical Physics: The Master Equation*, edited by I. Oppenheim, K. E. Shuler, and G. H. Weiss, MIT Press, Cambridge, MA, 1977.

103. S. Arrhenius, *Z. Phys. Chem.* **4**, 226 (1889).

104. V. I. Mel'nikov and S. V. Meshkov, *J. Chem. Phys.* **85**, 1018 (1986); *Phys. Rev. E* **48**, 3271 (1993).

105. R. Metzler and J. Klafter, *Chem. Phys. Lett.*, **321**, 238 (2000).

106. H. Frauenfelder, P. G. Wolynes, and R. H. Austin, *Rev. Mod. Phys.* **71**, S419 (1999).

107. A. Ansari et al., *Proc. Natl. Acad. Sci. USA* **82**, 5000 (1985).

108. R. H. Austin, K. W. Beeson, L. Eisenstein, H. Frauenfelder, and I. C. Gunsalus, *Biochemistry* **14**, 5355 (1975); H. Frauenfelder, F. Parak, and R. D. Young, *Annu. Rev. Biophys. Chem.* **17**, 451 (1988).

109. I. E. T. Iben, D. Braunstein, W. Doster, H. Frauenfelder, M. K. Hong, J. B. Johnson, S. Luck, P. Ormos, A. Schulte, P. J. Steinbach, A. H. Xie, and R. D. Young, *Phys. Rev. Lett.* **62**, 1916 (1989).

110. W. G. Glöckle and T. F. Nonnenmacher, *Biophys. J.* **68**, 46 (1995).

111. R. Zwanzig, *J. Chem. Phys.* **97**, 3587 (1992).

112. N. Eizenberg and J. Klafter, *Chem. Phys. Lett.* **243**, 9 (1996); *J. Chem. Phys.* **104**, 6796 (1996).

113. M. F. Shlesinger and J. Klafter, *J. Phys. Chem.* **93**, 7023 (1989).

114. V. M. Kenkre, E. W. Montroll and M. F. Shlesinger, *J. Stat. Phys.* **9**, 45 (1973); V. M. Kenkre, *Phys. Lett. A* **65**, 391 (1978).

115. R. W. Davies, *Phys. Rev.* **93**, 1169 (1954).

116. E. Barkai and J. Klafter, in *Chaos, Kinetics and Nonlinear Dynamics in Fluids and Plasmas*, edited by G. M. Zaslavsky and S. Benkadda, Springer-Verlag, Berlin, 1998.

117. B. J. West, P. Grigolini, R. Metzler, and T. F. Nonnenmacher, *Phys. Rev. E* **55**, 99 (1997); R. Metzler and T. F. Nonnenmacher, *Phys. Rev. E* **57**, 6409 (1998).

118. R. Metzler and J. Klafter, *Phys. Rep.* **339**, 1 (2000).

MOMENT FREE ENERGIES FOR POLYDISPERSE SYSTEMS

PETER SOLLICH

Department of Mathematics, King's College, University of London, London, United Kingdom

PATRICK B. WARREN

Unilever Research Port Sunlight, Bebington, Wirral, United Kingdom

MICHAEL E. CATES

Department of Physics and Astronomy, University of Edinburgh, Edinburgh, United Kingdom

CONTENTS

Advances in Chemical Physics, Volume 116, edited by I. Prigogine and Stuart A. Rice.
ISBN 0-471-40541-8 © 2001 John Wiley & Sons, Inc.

I. INTRODUCTION

As a result of the work of Gibbs, the thermodynamics of mixtures of several chemical species is a well-established subject (see, e.g., Ref. 1). But many systems arising in nature and in industry contain, for practical purposes, an infinite number of distinct, though similar, chemical species. Often these can be classified by a parameter, σ, say, which could be the chain length in a polymeric system, or the particle size in a colloid; both are routinely treated as continuous variables. In other cases (see, e.g., Refs. 2–5) σ is instead a parameter distinguishing species of continuously varying chemical properties. In the most general case, several attributes may be required to distinguish the various particle species in the system (such as length and chemical composition in length-polydisperse random copolymers) and σ is then a collection of parameters [6]. The thermodynamics of polydisperse systems (as defined above) is of crucial interest to wide areas of science and technology; it is sometimes also referred to as "continuous thermodynamics" (see, e.g., Ref. 3).

Standard thermodynamic procedures [1] for constructing phase equilibria in a system of volume V containing M different species can be understood geometrically in terms of a free energy surface $f(\rho_j)$ (with $f = F/V$, where F is the Helmoltz free energy) in the M-dimensional space of number densities ρ_j. Tangent planes to f define regions of coexistence, within which the free energy of the system is lowered by phase separation. The volumes of coexisting phases follow from the well-known "lever rule" [1]. Here "surface" and "plane" are used loosely, to denote manifolds of appropriate dimension. This procedure becomes unmanageable, both conceptually and numerically, in the limit $M \to \infty$ which formally defines a polydisperse system. There is now a separate conserved density $\rho(\sigma)$ for each value of σ; $\rho(\sigma)$ is in fact a *density distribution* and the overall number density of particles is written as $\rho = \int d\sigma \rho(\sigma)$. The free energy surface is $f = f[\rho(\sigma)]$ (a functional) which resides in an infinite-dimensional space. Gibbs' rule allows the coexistence of arbitrarily many thermodynamic phases.

To make phase equilibria in polydisperse systems more accessible to both computation and physical intuition, it is clearly helpful to reduce the dimensionality of the problem. Theoretical work in this area has made significant progress in finding simplified forms of the conditions for phase coexistence, thus making these more numerically tractable [3–5, 7–17]. Our aim is to achieve a similar reduction in dimensionality on a higher level, that of the free energy itself. We show that, for a large class of models, it is possible to construct a reduced free energy depending on only a small number of variables. From this, meaningful information on phase equilibria can be extracted by the usual tangent plane procedure, with obvious benefits for a more intuitive understanding of the phase behavior of polydisperse systems. As an important side effect, this procedure also leads to robust algorithms for calculating polydisperse phase equilibria numerically. In particular, such algorithms can handle coexistence of more than two phases with relative ease compared to those used previously [7, 13, 15, 16, 18–25].

A clue to the choice of independent variables for such a reduced free energy comes from the representation of experimental data. We recall first the definition of a "cloud point" (see, e.g., Refs. 3, 7, and 26): This is the point at which, for a system with a given density distribution $\rho(\sigma)$, phase separation first occurs as the temperature T or another external control parameter is varied. The corresponding incipient phase is called the "shadow." Now consider diluting or concentrating the system—that is, varying its overall density ρ while maintaining a fixed "shape" of polydispersity $n(\sigma) = \rho(\sigma)/\rho$. We will find it useful later to refer to the collection of all systems $\rho(\sigma) = \rho n(\sigma)$ which can be obtained by this process as a "dilution line" (in the space of all density distributions $\rho(\sigma)$). Plotting the cloud point temperature T versus density ρ then defines the cloud point curve (CPC), while plotting T versus the density of the shadow gives the shadow curve [27]. The differences in the shapes $n(\sigma)$ of the density distributions in the different phases are hidden in this representation; only the overall densities ρ in each phase are tracked. The density ρ is a particular *moment* of $\rho(\sigma)$ (the zeroth one); higher-order moments would be given by $\int d\sigma \sigma^i \rho(\sigma)$. Generalizing slightly, this suggests that our reduced free energy should depend on several (generalized) *moment densities* $\rho_i = \int d\sigma \, w_i(\sigma)\rho(\sigma)$, defined by appropriate (linearly independent) weight functions $w_i(\sigma)$. Ordinary power-law moment densities are included as the special case $w_i(\sigma) = \sigma^i$. Whatever the choice of moments, we insist on $w_0 = 1$ so that in all cases the zeroth moment density ρ_0 coincides with the overall number density ρ.

As an illustration, consider the simplest imaginable case where the true free energy f already has the required form; that is, it depends only on a finite set of K moment densities of the density distribution:

$$f = f(\rho_i), \qquad i = 1 \ldots K \quad \text{or} \quad i = 0 \ldots K - 1, \tag{1}$$

where the range of i depends on whether $\rho_0 \equiv \rho$ is among the K moment densities on which f depends. (From now on, we use the notation $i = 1 \ldots K$ inclusively, to cover both possibilities.) In coexisting phases, one demands equality of particle chemical potentials, defined as $\mu(\sigma) = \delta f / \delta \rho(\sigma)$, for all σ. Because

$$\mu(\sigma) \equiv \frac{\delta f}{\delta \rho(\sigma)} = \sum_i \frac{\partial f}{\partial \rho_i} w_i(\sigma) = \sum_i \mu_i w_i(\sigma),$$

this implies that all "moment" chemical potentials, $\mu_i \equiv \partial f / \partial \rho_i$, are likewise equal among phases. The second requirement for phase coexistence is that the pressures or osmotic pressures [28], Π, of all phases must also be equal. But from

$$-\Pi = f - \int d\sigma \mu(\sigma) \rho(\sigma) = f - \sum_i \mu_i \rho_i$$

one sees that this again involves only the moment densities ρ_i and their chemical potentials μ_i. Finally, the moment densities also also obey the "lever rule," as follows. Let the overall density distribution of a system of volume V be $\rho^{(0)}(\sigma)$; we call this the "parent" distribution. If (after a lowering of temperature, for example) this parent has split into p coexisting "daughter" phases with σ-distributions $\rho^{(\alpha)}(\sigma)$, each occupying a fraction $v^{(\alpha)}$ of the total volume ($\alpha = 1 \ldots p$), then particle conservation implies the usual lever rule (or material balance) among species: $\sum_\alpha v^{(\alpha)} \rho^{(\alpha)}(\sigma) = \rho^{(0)}(\sigma), \forall \sigma$. Multiplying this by a weight function $w_i(\sigma)$ and integrating over σ shows that the lever rule also holds for the moment densities:

$$\sum_{\alpha=1}^{p} v^{(\alpha)} \rho_i^{(\alpha)} = \rho_i^{(0)}. \tag{2}$$

These results express the fact that any linear combination of conserved densities (a generalized moment density) is itself a conserved density in thermodynamics. We have shown, therefore, that if the free energy of the system depends only on K moment densities $\rho_1 \ldots \rho_K$, we can view these as the densities of K "quasi-species" of particles and can construct the phase diagram via the usual construction of tangencies and the lever rule. Formally this has reduced the problem to finite dimensionality, although this is trivial here because f, by assumption, has no dependence on any variables other than the ρ_i $(i = 1 \ldots K)$.

Of course, it is uncommon for the free energy f to obey (1). In particular, the entropy of an ideal mixture (or, for polymers, the Flory–Huggins entropy term) is definitely not of this form. On the other hand, in very many thermodynamic (especially mean field) models the *excess* (i.e., nonideal) part of the free energy does have the simple form (1). In other words, if we decompose the free energy as (setting $k_B = 1$)

$$f = -Ts_{\mathrm{id}} + \tilde{f}, \qquad s_{\mathrm{id}} = -\int d\sigma\, \rho(\sigma)\, [\ln \rho(\sigma) - 1], \qquad (3)$$

then the excess free energy $\tilde{f}$ is a function of some moment densities only:

$$\tilde{f} = \tilde{f}(\rho_i), \qquad i = 1 \ldots K. \qquad (4)$$

Examples of models of this kind include polydisperse hard spheres (within the generalization by Salacuse and Stell [8] of the BMCSL equation of state [29, 30]), polydisperse homo- and copolymers [2–5, 9], and van der Waals fluids with factorized interaction parameters [10]. With the exception of a brief discussion in Section VI, this chapter discusses models with free energies of the form (3, 4), which we call "truncatable." (This terminology emphasizes that the number K of moment densities appearing in the excess free energy of truncatable models is finite, while for a nontruncatable model the excess free energy depends on all details of $\rho(\sigma)$, corresponding to an *infinite* number of moment densities.) In what follows, we regard each model free energy as given and do *not* discuss the issue of how good a description of the real system it offers, nor how or whether it can be derived from an underlying microscopic Hamiltonian. Whenever we refer to "exact" results, we mean the exact thermodynamics of such a model as specified by its free energy.

Different authors refer in different terms to s_{id} in (3). Throughout this chapter we will call the quantity

$$s_{\mathrm{id}} = -\int d\sigma\, \rho(\sigma)\, [\ln \rho(\sigma) - 1]$$

(which is, up to a factor of $-T$, the ideal part of the free energy density) the "entropy of an ideal mixture." By writing $\rho(\sigma) = \rho n(\sigma)$, where $n(\sigma)$ is the normalized distribution of σ, this can be decomposed as

$$s_{\mathrm{id}} = -\rho(\ln \rho - 1) - \rho \int d\sigma\, n(\sigma) \ln n(\sigma). \qquad (5)$$

The first term on the right-hand side is the "entropy of an ideal gas," while the second term gives the "entropy of mixing." The prefactor ρ reflects the fact that both are expressed per unit volume rather than per particle.

In principle, the entire phase equilibria for any truncatable system [obeying (4)] can be computed exactly by a finite algorithm. Specifically, the spinodal stability criterion involves a K-dimensional square matrix [11–14] whereas calculation of p-phase equilibrium involves solution of $(p-1)(K+1)$ coupled nonlinear equations (see Section III.E). This method has certainly proved useful [2–4, 9, 10, 14], but is cumbersome, particularly if one is interested mainly in cloud and shadow curves, rather than coexisting compositions deep within multiphase regions [2, 4, 9, 26]. Various ways of simplifying the procedure exist [7, 11, 15–17], but there has previously not been a systematic alternative to the full computation. Note also that the nonlinear phase equilibrium equations permit no simple geometrical interpretation or qualitative insight akin to the familiar rules for constructing phase diagrams from the free energy surface of a finite mixture. This further motivates the search for a description of polydisperse phase equilibria in terms of a reduced free energy that depends on only a small number of density variables.

In previous work, the authors originally arrived independently at two definitions of a reduced free energy in terms of moment densities (*moment free energy* for short) [31, 32]. Though based on distinctly different principles, the two approaches led to very similar results. We explain this somewhat surprising fact in the present work; at the same time, we describe the two methods in more detail and explore the relationship between them. We also discuss issues related to practical applications and give a number of example results for simple model systems.

The first route to a moment free energy takes as its starting point the conventional form of the ideal part of the free energy, $f_{id} = T \int d\sigma \, \rho(\sigma) \, [\ln \rho(\sigma) - 1]$, as in (3). This can be thought of as defining a (hyper-)surface, with the "horizontal" coordinate being $\rho(\sigma)$ and the "height" of the surface $f_{id}[\rho(\sigma)]$. As explained in Section II.A, this surface can then be projected geometrically onto one with only a finite set of horizontal coordinates: these are chosen to be the moment densities appearing in the excess free energy. Physically, this corresponds to minimizing the free energy with respect to all degrees of freedom in $\rho(\sigma)$ that do not affect these moment densities. We call this first route the "projection" method.

The second approach rederives the entropy of mixing in the ideal part of the free energy in a form that depends explicitly only on the chosen moment densities. As described in Section II.B, the expression that results is intractable in general because it still contains the full complexity of the problem. However, in situations where there are only infinitesimal amounts of all but one of the phases in the systems, the entropy of mixing can be evaluated in a closed form. Applying this functional form in regimes where it is not strictly valid (i.e., when phases of comparable volumes coexist) generates a moment free energy by this "combinatorial" method.

We show the equivalence of the two approaches in Section II.C. There, we first demonstrate that the general form of the entropy of mixing obtained by the combinatorial method can be transformed to the standard expression $-\int d\sigma\, n(\sigma)\ln n(\sigma)$. Second, we show that the moment free energies arrived at by the two methods are in fact equal, with the projection method being slightly more generally applicable.

In Section III, we then discuss the properties of the moment free energy (as obtained from either the projection or the combinatorial method). By construction, it depends only on a (finite) number of moment densities, thereby achieving the desired reduction in dimensionality. The moment densities can be treated as densities of "quasi-species" of particles, and the standard procedures of the thermodynamics of finite mixtures can be applied to the moment free energy to calculate phase equilibria. This is only useful, however, if the results are faithful to the actual phase behavior of the underlying polydisperse system (as modeled by the given truncatable free energy). We show that this is so; in fact, the construction of our moment free energy is such that exact binodals (cloud-point and shadow curves), critical (and multicritical) points, and spinodals are obtained. Beyond the onset of phase separation, where coexisting phases occupy comparable volumes, the results are not exact, but can be refined arbitrarily by adding extra moment densities. This procedure is necessary also to ensure that, in regions of multiphase coexistence, the correct number of phases is found. In Section IV, we discuss the practical implementation of our method, followed by a number of examples (Section V). Our results are summarized in Section VI, where we also outline perspectives for future work.

Note that throughout this work, we focus on the case of phase coexistence at fixed *volume*. However, as described in Appendix A, the formalism can easily be applied to scenarios where the (mechanical or osmotic) *pressure* is fixed instead.

II.   DERIVATION OF MOMENT FREE ENERGY

A.   Projection Method

The starting point for this method is the decomposition (3) for the free energy of truncatable systems (4):

$$f = T \int d\sigma\, \rho(\sigma)\left[\ln \frac{\rho(\sigma)}{R(\sigma)} - 1\right] + \tilde{f}(\rho_i). \tag{6}$$

As explained in the previous section, truncatable here means that the excess free energy $\tilde{f}$ depends only on K moment densities ρ_i. Note that, in the first (ideal) term of (6), we have included a dimensional factor $R(\sigma)$ inside the logarithm. This is equivalent to subtracting $T \int d\sigma \rho(\sigma) \ln R(\sigma)$ from the free energy. Since this term is *linear* in densities, it has no effect on the exact thermodynamics; it contributes harmless additive constants to the chemical potentials $\mu(\sigma)$. However, in the projection route to a moment free energy, it will play a central role.

We now argue that the most important moment densities to treat correctly are those that actually appear in the excess free energy $\tilde{f}(\rho_i)$. Accordingly, we divide the infinite-dimensional space of density distributions into two complementary subspaces: a "moment subspace," which contains all the degrees of freedom of $\rho(\sigma)$ that contribute to the moment densities ρ_i, and a "transverse subspace," which contains all remaining degrees of freedom (those that can be varied without affecting the chosen moment densities ρ_i) [33]. Physically, it is reasonable to expect that these "leftover" degrees of freedom play a subsidiary role in the phase equilibria of the system, a view justified *a posteriori* below. Accordingly, we now allow violations of the lever rule, so long as these occur *solely in the transverse space*. This means that the phase splits that we calculate using this approach obey the lever rule for the moment densities, but are allowed to violate it in other details of the density distribution $\rho(\sigma)$ (see Fig. 10 in Section V.B.1 for an example). These "transverse" degrees of freedom, instead of obeying the strict particle conservation laws, are chosen so as to minimize the free energy: They are treated as "annealed." If, as assumed above, $\tilde{f} = \tilde{f}(\rho_i)$ only depends on the moment densities retained, this amounts to *maximizing the ideal mixture entropy* (s_{id}) in (6), while holding fixed the values of the moment densities ρ_i. Note that we are allowed, if we wish, to include among the retained densities "redundant" moments on which $\tilde{f}$ has a null dependence. We will have occasion to do this later on.

At this point, the factor $R(\sigma)$ in (6), which is immaterial if all conservation laws are strictly obeyed, becomes central. Indeed, maximizing the entropy over all distributions $\rho(\sigma)$ at fixed moment densities ρ_i yields

$$\rho(\sigma) = R(\sigma) \exp\left(\sum_i \lambda_i w_i(\sigma) \right), \tag{7}$$

where the Lagrange multipliers λ_i are chosen to give the desired moment densities

$$\rho_i = \int d\sigma \, w_i(\sigma) \, R(\sigma) \exp\left(\sum_j \lambda_j w_j(\sigma) \right). \tag{8}$$

The corresponding minimum value of f then defines our projected free energy as a function of the moment densities ρ_i:

$$f_{\mathrm{pr}}(\rho_i) = -Ts_{\mathrm{pr}}(\rho_i) + \tilde{f}(\rho_i), \qquad s_{\mathrm{pr}} = \rho_0 - \sum_i \lambda_i \rho_i. \qquad (9)$$

Here s_{pr} is the projected entropy of an ideal mixture. The first term appearing in it, $\rho_0 = \int d\sigma \rho(\sigma)$, is the "zeroth moment," which is identical to the overall particle density ρ defined previously. If this is among the moment densities used for the projection (or more generally, if it is a linear combination of them), then the term $-T\rho_0$ is simply a linear contribution to the projected free energy $f_{\mathrm{pr}}(\rho_i)$ and can be dropped because it does not affect phase equilibrium calculations. Otherwise, ρ_0 needs to be expressed—via the λ_i—as a function of the ρ_i, and its contribution cannot be ignored. We will see an example of this in Section V.

The projection method yields a free energy $f_{\mathrm{pr}}(\rho_i)$ that only depends on the chosen set of moment densities. These can now be viewed as densities of "quasi-species" of particles, and a finite-dimensional phase diagram can be constructed from f_{pr} according to the usual rules, ignoring the underlying polydisperse nature of the system. Obviously, though, the results now depend on $R(\sigma)$, which is formally a "prior distribution" for the entropy maximization. To understand its role, we recall that the projected free energy $f_{\mathrm{pr}}(\rho_i)$ was constructed as the minimum of $f[\rho(\sigma)]$ at fixed ρ_i; that is, f_{pr} is the lower envelope of the projection of f onto the moment subspace. Crucially, the shape of this envelope depends on how, by choosing a particular prior distribution $R(\sigma)$, we "tilt" the infinite-dimensional free energy surface *before* projecting it. This *geometrical* point of view is illustrated, for a mixture of only two species, in Fig. 1.

To understand the effect of the prior $R(\sigma)$ *physically*, we note that the projected free energy is simply the free energy of phases of a system in which the density distributions $\rho(\sigma)$ are of the form (7). The prior $R(\sigma)$ determines which distributions lie within this "maximum entropy family" (or "family" for short), and it is the properties of phases with these distributions that the projected free energy represents. Typically, one is interested in a system where a fixed overall "parent" (or "feed") distribution $\rho^{(0)}(\sigma)$ becomes subject to separation into various phases. In such circumstances, we should generally choose this parent distribution as our prior, $R(\sigma) = \rho^{(0)}(\sigma)$, thereby guaranteeing that it is contained within the family (7). Having done this, we note that the projection procedure will be *exactly valid*, to whatever extent the density distributions actually arising in the various coexisting phases of the system under study are members of the corresponding family

$$\rho(\sigma) = \rho^{(0)}(\sigma)\exp\left(\sum_i \lambda_i w_i(\sigma)\right). \qquad (10)$$

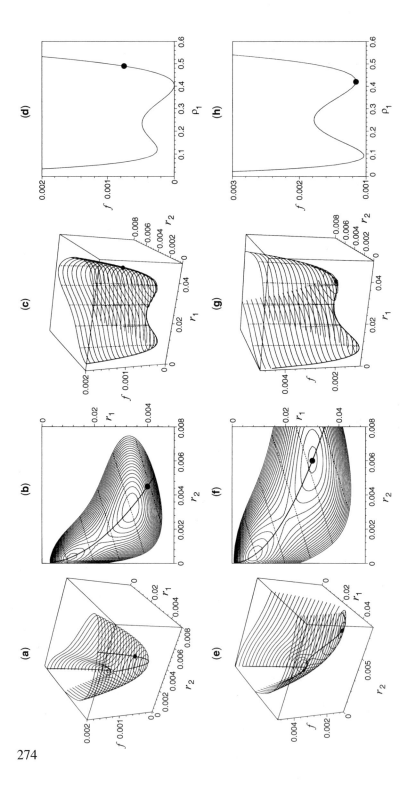

In fact, the condition just described holds whenever all but one of a set of coexisting phases are of infinitesimal volume compared to the majority phase. This is because the density distribution, $\rho^{(0)}(\sigma)$, of the majority phase is negligibly perturbed, whereas that in each minority phase differs from this by a Gibbs–Boltzmann factor, of exactly the form required for (10); we show this formally in Section III. Accordingly, our projection method yields *exact* cloud point and shadow curves. By the same argument, critical points (which in fact lie at the intersection of these two curves) are exactly determined; the same is true for tricritical and higher-order critical points. Finally, spinodals are also found exactly. We defer explicit proofs of these statements to Section III.

The projection method does, however, give only approximate results for coexistences involving finite amounts of different phases. This is because linear combinations of different density distributions from the family (10), corresponding to two (or more) phases arising from the same parent $\rho^{(0)}(\sigma)$, do not necessarily add to recover the parent distribution itself:

$$\sum_\alpha v^{(\alpha)} \rho^{(0)}(\sigma) \exp\left(\sum_i \lambda_i^{(\alpha)} w_i(\sigma)\right) \neq \rho^{(0)}(\sigma) \tag{11}$$

unless all except one of the $v^{(\alpha)}$ are infinitesimal. Moreover, according to Gibbs' phase rule, a projected free energy depending on n moment densities will not normally predict more than $n+1$ coexisting phases, whereas a poly-

Figure 1.   Illustration of free energy projection. The full infinite-dimensional free energy surface $f[\rho(\sigma)]$ for a truly polydisperse system cannot be represented graphically, so a system with only two particle species is used instead. A single phase is then described by its density pair (r_1, r_2). (The projection method remains applicable, although the polydispersity parameter σ is replaced by a discrete species index $a \in \{1, 2\}$, and integrals $\int d\sigma\, \rho(\sigma)\ldots$ are replaced by $\sum_a r_a \ldots$) To illustrate the method we choose a model with an excess free energy depending on a single moment density $\rho_1 = L_1 r_1 + L_2 r_2$ [in fact Flory–Huggins theory for length polydispersity, discussed in more detail in Section V; see Eq. (63)]. Both the transverse and the moment subspace are then one-dimensional. (a) Three-dimensional view of free energy surface. The circle marks the position of the parent $(r_1^{(0)}, r_2^{(0)})$; the thick line traces out the free energy of density pairs (r_1, r_2) in the corresponding family (10). (b) Top view, showing as dashed lines the transverse directions along which the moment density ρ_1 is constant. (c) Side view of the free energy surface, "looking down" the transverse direction. (d) The projected free energy is simply the lower envelope of the free energy surface as seen from this transverse direction. (e–h) The same free energy surface, now tilted differently by using a different parent (or prior). While this does not affect the location of any double tangent planes drawn to the full surface, it *does* produce a different "lower envelope" and hence a different projected free energy. For continuum polydispersity, the role of the density pair (r_1, r_2) is played by a density distribution $\rho(\sigma)$ of a continuous parameter σ.

disperse system can in principle separate into an arbitrary number of phases. We explain in Section III how both of these shortcomings can be overcome by systematically including extra moment densities within the projection procedure. How quickly convergence to the exact results occurs depends on the choice of weight functions for the extra moment densities; we discuss this point further in the context of the examples of Section V.

B.   Combinatorial Method

Now we turn to the second method of constructing a moment free energy. As before, we recognize from the outset that the physics of the problem is contained in the excess free energy, and therefore the most important variables are the moment densities that feature in this. On the other hand, the ideal part of the free energy (essentially the entropy of mixing, up to the trivial ideal gas term) is a bookkeeping device that accounts for the number of ways of partitioning *a priori* a given size distribution between two or more coexisting phases. (To focus the discussion, we identify σ as particle size in this section; the arguments do of course remain valid for a general polydispersity parameter σ.) The entropy of mixing has its origin in $1/N!$ factors in the partition function (where N is a particle number) which are usually derived from the classical limit of quantum statistics. The origin of the $1/N!$ factors in classical statistical mechanics has been the subject of debate ever since the work of Gibbs [34]. The problem is also connected with the question of the extensivity of entropy, along with the Gibbs paradox. Below we present a completely classical derivation of the entropy of mixing which has the additional benefit of indicating the appropriate generalization for moment densities. To start with, follow Gibbs and define a nonextensive free energy F' as a configuration space (Γ) integral

$$e^{-F'/T} = \int d\Gamma e^{-H/T}. \tag{12}$$

No $1/N!$ appears in this because all particles are distinguishable in classical statistical mechanics.

Now consider two phases in coexistence as one joint system. Assume that they occupy volumes $V^{(1)}$ and $V^{(2)}$, respectively, and contain $N^{(1)}$ and $N^{(2)}$ particles. Following literally the prescription in Eq. (12), the free energy of the joint system of $N^{(1)} + N^{(2)} = N$ particles is found from

$$e^{-F'/T} = \sum_{\text{prtns}} e^{-(F'^{(1)}+F'^{(2)})/T}, \tag{13}$$

where the configuration space integral has been done in two parts. First, for each way of partitioning the particles between the two phases, the individual configuration space integrals give a product of the individual partition functions. Second, and crucially, one must sum over the $N!/N^{(1)}!N^{(2)}!$ ways of partitioning the particles between phases. Now define an extensive free energy by reinserting the $1/N!$ as though the particles were indistinguishable:

$$e^{-F_{\text{in}}/T} = \frac{1}{N!}\int d\Gamma e^{-H/T}. \tag{14}$$

Equation (13) can then be written as

$$e^{-F_{\text{in}}/T} = \left\langle e^{-(F^{(1)}_{\text{in}}+F^{(2)}_{\text{in}})/T}\right\rangle_{\text{prtns}}, \tag{15}$$

where the average is taken over all partitions, with equal *a priori* probabilities. This result is the cornerstone of the combinatorial method. If we separate F_{in} into the entropy of an ideal gas and the remaining nonideal (excess) parts, it is written as

$$F_{\text{in}} = NT[\ln\,(N/V) - 1] + \tilde{F}, \tag{16}$$

with $\tilde{F}$ being the excess free energy. Note that F_{in} does not yet contain an entropy of mixing term because it treats the particles as though they are indistinguishable. The entropy of mixing will reappear when we come to do the average over partitions in Eq. (15). Note also that conventional thermodynamics follows from (15) if all particles are identical as far as their mutual interactions are concerned. There is then no entropy of mixing to consider, so that F_{in} is just the conventional free energy F; and since the average over partitions is trivial, one has $F = F^{(1)} + F^{(2)}$. Similarly, as we show in Section II.C, one can recover from Eq. (15) the conventional form of the entropy of mixing as given in Eq. (5).

We now show how the average over partitions in Eq. (15) results in an expression for the entropy of mixing which depends explicitly on moment variables. The key is to note that, by our assumption that we are dealing with a truncatable model, the excess free energy $\tilde{F} = V\tilde{f}(\rho_i)$ depends only on a limited number of moment densities ρ_i (as well as volume V, and temperature T and other state variables that we suppress below). If the density ρ_0 is among the moment densities ρ_i—which we assume throughout this section—then specifying V and the ρ_i is equivalent to specifying N, V, and the *normalized* moments $m_i = \rho_i/\rho_0$ ($i > 0$). So we can think of $\tilde{F}$ as a function of N, V and the m_i. From Eq. (16), F_{in} then depends on the same set of variables. The choice of the m_i [which are moments of the *normalized* particle size distribution $n(\sigma)$] as independent variables is natural in the present context because we are considering the particle number N and volume V of the co-existing phases to be fixed.

To avoid notational complexity, and without loss of generality, we now specialize to the case where there is only one normalized moment m_i, a generalized mean size

$$m = \int d\sigma \, w(\sigma) n(\sigma).$$

We shall also write $\rho_m = \rho m = \int d\sigma \, w(\sigma) \rho(\sigma)$ for the corresponding moment density and denote by $m^{(1)}$ and $m^{(2)}$ the values of the generalized mean size in the first and second phase. The overall ("parent") size distribution is $n^{(0)}(\sigma)$ with mean $m^{(0)}$. In each of the two phases, in which N and V are held fixed, F_{in} only depends on m, and so the average over partitions becomes

$$\langle e^{-(F_{in}^{(1)} + F_{in}^{(2)})/T} \rangle_{\text{prtns}} \; \rightarrow \; \int dm^{(1)} \, \mathcal{P}(m^{(1)}) e^{-(F_{in}^{(1)} + F_{in}^{(2)})/T}. \qquad (17)$$

Here $\mathcal{P}(m^{(1)})$ is the probability distribution for the generalized mean size in the first phase, taken over partitions with fixed $N^{(1)}$ and $N^{(2)}$, with equal *a priori* probabilities. Note that given $m^{(1)}$, $m^{(2)}$ is fixed in the second phase by the moment equivalent of particle conservation: $N^{(1)}m^{(1)} + N^{(2)}m^{(2)} = Nm^{(0)}$. The integral in (17) can be replaced by the maximum of the integrand in the thermodynamic limit, because $\ln \mathcal{P}(m^{(1)})$ is an extensive quantity. Introducing a Lagrange multiplier μ_m for the above moment constraint then shows that the quantity ρ_m has the same status as the density $\rho \equiv \rho_0$ itself: Both are thermodynamic density variables. This reinforces the discussion in the introduction, where we showed that moment densities can be regarded as densities of "quasi-species" of particles.

In what follows, we only need to refer to $m^{(1)}$ (not $m^{(2)}$) and therefore drop the superscript for brevity. Since $\ln \mathcal{P}(m)$ is extensive, we can write $\ln \mathcal{P}(m) = Ns(m)$, where s is the entropy of mixing *per particle* expressed as a function of the moment m (or, more generally, of the full set of normalized moments). The total free energy of the system then takes the form

$$F_{in} = \min_{m} F_{in}^{(1)} + F_{in}^{(2)} - NTs(m), \qquad (18)$$

and we need to calculate the entropy of mixing $s(m)$. We recognize that this quantity depends not only on the generalized mean size, but also on $x = N^{(1)}/N$, the fraction of particles in the first phase. A formal result for $s(m)$ can be obtained by first calculating the joint probability distribution

$\mathcal{P}(xm, x)$ of xm and x. From this, we can find $\mathcal{P}(m)$ according to [35]

$$\mathcal{P}(m) = \frac{\mathcal{P}(xm, x)}{\int dm' \, \mathcal{P}(xm', x)} = \frac{\mathcal{P}(xm, x)}{\mathcal{P}(x)/x}. \tag{19}$$

Writing the log probabilities in (19) as $\ln \mathcal{P}(xm, x) = Ns(xm, x)$ and $\ln \mathcal{P}(x) = Ns(x)$, we then have in the thermodynamic limit

$$s(m) = s(xm, x) - s(x). \tag{20}$$

The method for calculating $\mathcal{P}(xm, x)$ proceeds by introducing an indicator function ϵ_i for each particle—deemed to be 1 if the particle is in the first phase, and 0 if the particle is in the second [36]. Then $Nx = \sum_{i=1}^{N} \epsilon_i$ and $Nxm = \sum_{i=1}^{N} \epsilon_i w(\sigma_i)$. We now write the *moment-generating function* for $\mathcal{P}(xm, m)$ as follows:

$$\int d(xm) dx \, \mathcal{P}(xm, x) \exp\left[N(\theta x + \lambda xm)\right]$$

$$= \left\langle \exp\left[\theta \sum_{i=1}^{N} \epsilon_i + \lambda \sum_{i=1}^{N} \epsilon_i w(\sigma_i)\right]\right\rangle_{\text{prtns}}$$

$$= \left\langle \prod_{i=1}^{N} \exp\left\{\epsilon_i[\theta + \lambda w(\sigma_i)]\right\}\right\rangle_{\text{prtns}}$$

$$= 2^{-N} \prod_{i=1}^{N} \{1 + \exp[\theta + \lambda w(\sigma_i)]\}$$

$$= 2^{-N} \exp\left(\sum_{i=1}^{N} \ln\left\{1 + \exp[\theta + \lambda w(\sigma_i)]\right\}\right). \tag{21}$$

In the second equality we have used the fact that the ϵ_i are independent discrete random variables taking the values 0 and 1 with *a priori* equal probabilities. Taking logarithms of (21) and dividing by N, we obtain on the right-hand side an average over $i = 1 \ldots N$ of a function of σ_i. Since the σ_i are drawn independently from $n^{(0)}(\sigma)$, the law of large numbers guarantees that this average tends to the corresponding average over $n^{(0)}(\sigma)$ in the limit $N \to \infty$ (compare the discussion in Ref. 37). We therefore have the final result

$$\frac{1}{N} \ln \int d(xm) \, dx \, \exp\{N[s(xm, x) + \theta x + \lambda xm]\}$$

$$= \int d\sigma \, n^{(0)}(\sigma) \ln\left\{1 + \exp[\theta + \lambda w(\sigma)]\right\}, \tag{22}$$

where we have dropped an unimportant additive constant $(-\ln 2)$ and have replaced $\mathcal{P}(xm, x)$ by $\exp[Ns(xm, x)]$.

Now, the left-hand side can be evaluated, by the steepest descent method, in the limit $N \to \infty$. Defining

$$\psi(\theta, \lambda) = \int d\sigma \, n^{(0)}(\sigma) \ln \{1 + \exp[\theta + \lambda w(\sigma)]\}$$

we get $s(xm, x) + \theta x + \lambda xm = \psi(\theta, \lambda)$ at the point where $\partial s/\partial x + \theta = \partial s/\partial(xm) + \lambda = 0$. We recognize that the relationship between $s(x, xm)$ and $\psi(\theta, \lambda)$ is a double Legendre transform. Inverting this shows that

$$s(xm, x) = \psi(\theta, \lambda) - \theta x - \lambda xm, \qquad (23)$$

where

$$x = \partial\psi/\partial\theta, \qquad xm = \partial\psi/\partial\lambda. \qquad (24)$$

Equation (23) is essentially the desired result; to obtain $s(m)$ from Eq. (20), only $s(x)$ remains to be determined. It is obtained from $\mathcal{P}(x)$ via $\exp[Ns(x)] = \mathcal{P}(x) = \int d(xm)\mathcal{P}(xm, x)$ which by inspection of the moment generating function is seen to correspond to the point $\lambda = 0$. At this point, $\psi = \ln(1 + e^{\theta})$ and $x = \partial\psi/\partial\theta = e^{\theta}/(1 + e^{\theta})$. We find that $\theta = \ln[x/(1-x)]$ and after some manipulation,

$$s(x) = \psi - x\theta = -x \ln x - (1-x) \ln(1-x). \qquad (25)$$

This is recognized as the standard entropy of mixing that is lost when a total of N particles is partitioned into $N^{(1)} = xN$ and $N^{(2)} = (1-x)N$ particles in two phases. The simple form of the result shows that the calculation leading to Eq. (23) is essentially a generalization of the Stirling approximation.

Inserting Eqs. (23) and (25) into Eq. (20), we finally have the desired result for $s(m)$:

$$s(m) = \psi(\theta, \lambda) - \theta x - \lambda xm + x \ln x + (1-x) \ln(1-x). \qquad (26)$$

Unfortunately, this is mainly formal because neither the integral defining ψ nor the Legendre transform are likely to be tractable. However, we show in Section II.C that Eq. (26) is equivalent to the more conventional form of the entropy of mixing as given by the second term in Eq. (5). From a conceptual point of view, it should be noted that the conventional form is normally derived by "binning" the distribution of particle sizes σ and taking the number of bins to infinity *after* the thermodynamic limit $N \to \infty$ has been taken (see, e.g., Ref. 8). In our above derivation, on the other hand, we have assumed that even for finite N all particles have different sizes σ_i, drawn randomly from $n^{(0)}(\sigma)$; the "polydisperse limit" is thus taken simultaneously with the thermodynamic limit. The relation between these two approaches—which lead to the same results—has been discussed in detail by Salacuse [37]. Note that the first limit is physically more plausible for many homopolymer systems (where there may only be thousands or millions of species, with many particles of each) whereas the second limit is more natural for colloidal materials (and also some random copolymers) in which no two particles present are exactly alike, even in a sample of macroscopic size.

Although the full result (26) is intractable, progress can be made for $x \to 0$, when the number of particles in one phase is much smaller than in the other. The limit $x \to 0$ implies $\theta \to -\infty$ and hence $\psi(\theta, \lambda) \to \exp[\theta + h(\lambda)]$, where $h = \ln \int d\sigma \, n^{(0)}(\sigma) e^{\lambda w(\sigma)}$ is a generalized *cumulant* generating function for $n^{(0)}(\sigma)$. From this we derive $x = \partial\psi/\partial\theta = \psi$, giving $\theta = \ln x - h$. The generalized mean size is given by $xm = \partial\psi/\partial\lambda = \psi \partial h/\partial\lambda$ and hence $m = \partial h/\partial\lambda$. Inserting these results into Eq. (23) then shows that

$$s(xm, x) = x - x \ln x + x(h - \lambda m), \qquad m = \frac{\partial h}{\partial \lambda},$$

and hence from Eq. (26) we obtain

$$s(m) = x(h - \lambda m) + \mathcal{O}(x^2). \tag{27}$$

Noting that $xN = N^{(1)}$ is the number of particles in the small phase, the total free energy (18) can then be written as

$$F_{\text{in}} = F_{\text{in}}^{(2)} + F_{\text{in}}^{(1)} - N^{(1)}T(h - \lambda m). \tag{28}$$

The term $(h - \lambda m)$, being multiplied by $-N^{(1)}T$, can now be interpreted as the entropy of mixing *per particle in the small phase*. It arises from the deviation of the (generalized) mean size $m \equiv m^{(1)}$ in the small phase from the mean size $m^{(0)}$ of the parent, and it is given by the Legendre transform of the generalized cumulant generating function:

$$s_{\text{mix}}(m) = h - \lambda m, \quad \text{where} \quad m = \frac{\partial h}{\partial \lambda}, \quad h = \ln \int d\sigma \, n^{(0)}(\sigma) e^{\lambda w(\sigma)}. \tag{29}$$

Let us now examine the relation between $s_{\text{mix}}(m)$, thereby defined, and $s(m)$ introduced previously. By construction, $Ns(m)$ is the entropy of mixing of the system as a whole; both phases contribute to this. The result (27), which can also be written as

$$Ns(m) = N^{(1)} s_{\mathrm{mix}}(m), \tag{30}$$

appears to contradict this, because it does not contain a term proportional to $N^{(2)}$. The resolution of this paradox comes from the neglected $\mathcal{O}(x^2)$ terms in Eq. (27). In fact, using the single-phase entropy of mixing defined in Eq. (29)—and reinstating on the right-hand side the superscript on $m \equiv m^{(1)}$—we can write (30) in the more symmetric form

$$Ns(m) = N^{(1)} s_{\mathrm{mix}}(m^{(1)}) + N^{(2)} s_{\mathrm{mix}}(m^{(2)})$$

This is still correct to leading (linear) order in x because the added term is $\mathcal{O}(x^2)$. [To see this, use the fact that $m^{(2)} = m^{(0)} - xm^{(1)}/(1-x)$ and hence $m^{(2)} - m^{(0)} = \mathcal{O}(x)$. From Eq. (29) it can then be deduced that $s_{\mathrm{mix}}(m^{(2)}) \sim (m^{(2)} - m^{(0)})^2 = \mathcal{O}(x^2)$.] Similarly, we can rewrite the total free energy (28) as

$$F_{\mathrm{in}} = [F_{\mathrm{in}}^{(1)} - N^{(1)} T s_{\mathrm{mix}}(m^{(2)})] + [F_{\mathrm{in}}^{(2)} - N^{(2)} T s_{\mathrm{mix}}(m^{(2)})]. \tag{31}$$

This shows that in the limit where one of the two phases is much larger than the other one, the free energy of each of the phases—now *including* the entropy of mixing—is given by

$$F = F_{\mathrm{in}} - NT s_{\mathrm{mix}}(m) = -NT[-\ln(N/V) + 1 + s_{\mathrm{mix}}(m)] + \tilde{F}. \tag{32}$$

This expression, which only depends on the particle size distribution through the generalized mean size m, is our desired moment free energy.

The Legendre transform result (29) for the (single-phase) entropy of mixing is appealingly simple; we will illustrate its application in Section V.A, using polydisperse Flory–Huggins theory as an example. As discussed briefly in Appendix B, Eq. (29) also establishes an interesting connection to large deviation theory. However, the most important aspect of Eqs. (29) and (32) is that—as we have shown—it gives exact results for the limiting case $N^{(1)} \ll N^{(2)}$ (that is, $x \to 0$). This includes two important classes of problems, which are also handled exactly by the projection method (for essentially the same reason). The first is the determination of spinodal curves and critical points. Intuitively, these are exact because they are related to the stability of the system with respect to small variations in its density or composition, which can be probed by allowing fluctuations to take place in a vanishingly small subregion. The second concerns the cloud point and shadow curves. Again intuitively, these are exact because by definition only an infinitesimal amount of a second phase has appeared. Formal proofs of these statements are given in Section III.

If the result given by Eqs. (29), (31), and (32) is applied in the regime where it is no longer strictly valid—that is, where the two coexisting phases contain comparable number of particles (or, equivalently, occupy comparable volumes)—then approximate two-phase coexistences can be calculated. It is also straightforward to show that the analogue of Eqs. (29), (31), and (32) holds in the case where two or more small phases coexist with a much larger phase; application in the regime of comparable phase volumes then provides approximate results for multiphase coexistence.

To end this section, let us state in full the analogue of Eqs. (29) and (32) for the case of several moment densities, restoring the notation used in the previous sections. The square bracket on the right-hand side of Eq. (32) is the moment expression for the entropy of an ideal mixture. If, as in Section II.A, we measure this entropy *per unit volume* (rather than per particle, as previously in the current section) and generalize to several moment densities, we find by the combinatorial approach the following moment free energy:

$$f_{\text{comb}} = -T s_{\text{comb}} + \tilde{f}, \qquad s_{\text{comb}} = -\rho_0 (\ln \rho_0 - 1) + \rho_0 \left(h - \sum_{i \neq 0} \lambda_i m_i \right),$$

$$(33)$$

with

$$h = \ln \int d\sigma \, n^{(0)}(\sigma) \exp \left(\sum_{i \neq 0} \lambda_i w_i(\sigma) \right). \tag{34}$$

Here $\rho_0 = N/V$ is the particle density as before.

C.   Relation Between the Two Methods

Although Eq. (33) looks somewhat different from the corresponding result (9) for f_{pr} obtained by the projection method, the two methods are, mathematically, almost equivalent, as we now show. (For brevity we return to the case of a single moment m and continue to refer to the polydisperse feature σ as "size.") First, consider the exact expression for the entropy of mixing (26) derived within the combinatorial method. This should be equivalent to the conventional result used as the starting point (5) in the projection method. To see this, write the Legendre transform conditions (24) out explicitly:

$$x = \int d\sigma \, n^{(0)}(\sigma) \frac{\exp[\theta + \lambda w(\sigma)]}{1 + \exp[\theta + \lambda w(\sigma)]}, \tag{35a}$$

$$x m^{(1)} = \int d\sigma \, w(\sigma) \, n^{(0)}(\sigma) \frac{\exp[\theta + \lambda w(\sigma)]}{1 + \exp[\theta + \lambda w(\sigma)]}. \tag{35b}$$

Here we have reinstated the superscript to show that $m^{(1)}$ is the value in phase one of the (generalized) moment m:

$$m^{(1)} = \int d\sigma \, w(\sigma) n^{(1)}(\sigma).$$

Comparing with (35), we can identify the (normalized) particle size distribution in that phase as [38]

$$n^{(1)}(\sigma) = \frac{n^{(0)}(\sigma)}{x} \frac{\exp[\theta + \lambda w(\sigma)]}{1 + \exp[\theta + \lambda w(\sigma)]}.$$

Particle conservation $x n^{(1)}(\sigma) + (1 - x) n^{(2)}(\sigma) = n^{(0)}(\sigma)$ implies a similar form for the size distribution in phase two:

$$n^{(2)}(\sigma) = \frac{n^{(0)}(\sigma)}{1 - x} \frac{1}{1 + \exp[\theta + \lambda w(\sigma)]}.$$

It then takes only a few lines of algebra to show that the entropy of mixing (26) can be written (for any x) as

$$
\begin{aligned}
s(m) = &-x \int d\sigma \, n^{(1)}(\sigma) \ln \frac{n^{(1)}(\sigma)}{n^{(0)}(\sigma)} - (1 - x) \int d\sigma \, n^{(2)}(\sigma) \ln \frac{n^{(2)}(\sigma)}{n^{(0)}(\sigma)} \\
= &-x \int d\sigma \, n^{(1)}(\sigma) \ln n^{(1)}(\sigma) - (1 - x) \int d\sigma \, n^{(2)}(\sigma) \ln n^{(2)}(\sigma) \\
&+ \int d\sigma \, n^{(0)}(\sigma) \ln n^{(0)}(\sigma).
\end{aligned}
\tag{36}
$$

The second equation, with the third (constant) term discarded, is the standard result [compare Eqs. (5), (6) and (A1)]. However, the first expression appears more natural, and is retained, within the combinatorial derivation; it embodies the intuitively reasonable prescription that entropy is best measured relative to the parent distribution $n^{(0)}(\sigma)$. This was also an essential ingredient of the projection method, as described in Section II.A, where the "prior" in the entropy expression was likewise identified with the parent. Retaining the parent as prior also avoids subtleties with the definition of the integrals in Eq. (36) in the case where the phases contain monodisperse components, corresponding to δ-peaks in the density distributions [8].

Having established that the rigorous starting points of the combinatorial and projection method are closely related, we now show that the subsequent approximations also lead to essentially the same results. The relevant approximations are use of (33) for f_{comb} (which adopts the small x form for s_{mix}) in the combinatorial case and use of (9) for f_{pr} (which minimizes over transverse degrees of freedom) in the projection case. First note that the density ρ_0 for phases within the family (10) is given by

$$\rho_0 = \int d\sigma \rho^{(0)}(\sigma) \exp\left(\sum_i \lambda_i w_i(\sigma)\right).$$

Comparing this with Eq. (34) and using the fact that $n^{(0)}(\sigma) = \rho^{(0)}(\sigma)/\rho_0^{(0)}$, one sees that

$$\rho_0 = \int d\sigma \, \rho^{(0)}(\sigma) \exp\left(\sum_i \lambda_i w_i(\sigma)\right) = \rho_0^{(0)} e^{\lambda_0 + h},$$

where the sum over i now includes $i = 0$. Solving for h, one has $h = -\lambda_0 + \ln(\rho_0/\rho_0^{(0)})$. The ideal mixture entropy derived by the combinatorial route, Eq. (33), can thus be rewritten as

$$s_{\text{comb}} = -\rho_0(\ln \rho_0 - 1) + \rho_0 \ln(\rho_0/\rho_0^{(0)}) - \lambda_0 \rho_0 - \sum_{i \neq 0} \lambda_i \rho_0 m_i$$

$$= \rho_0 - \sum_i \lambda_i \rho_i - \rho_0 \ln \rho_0^{(0)}.$$

This is identical to the projected entropy s_{pr}, Eq. (9), except for the last term. But by construction, the combinatorial entropy assumes that ρ_0—the overall density—is among the moment densities retained in the moment free energy. The difference $s_{\text{comb}} - s_{\text{pr}} = -\rho_0 \ln \rho_0^{(0)}$ is then linear in this density, and the combinatorial and projection methods therefore predict exactly the same phase behavior.

In summary, we have shown that the projection and combinatorial methods for obtaining moment free energies give equivalent results. The only difference between the two approaches is that within the projection approach, one need not necessarily retain the zeroth moment, which is the overall density $\rho_0 = \rho$, as one of the moment densities on which the moment free energy depends. If ρ_0 does not appear in the excess free energy, this reduces the minimum number of independent variables of the moment free energy by one (see Section V for an example).

III.   PROPERTIES OF THE MOMENT FREE ENERGY

In the previous sections, we have derived by two different routes [namely, Eqs. (9) and (33)] our moment free energy for truncatable polydisperse systems. We now investigate the properties of this moment free energy, which we henceforward denote f_m and write in the form (9)

$$f_m(\rho_i) = -Ts_m(\rho_i) + \tilde{f}(\rho_i), \qquad s_m = \rho_0 - \sum_i \lambda_i \rho_i; \qquad (37)$$

here s_m is the "moment entropy" of an ideal mixture. In particular, we formally compare the phase behavior predicted from this free energy (by treating the ρ_i as densities of "quasi-species" of particles, and applying the usual tangency construction) with that obtained from the exact free energy of the underlying (truncatable) model

$$f[\rho(\sigma)] = T \int d\sigma \, \rho(\sigma)[\ln \rho(\sigma) - 1] + \tilde{f}(\rho_i). \qquad (38)$$

Let us first collect a few simple properties of the moment free energy (37) which will be useful later. Recall that Eq. (37) faithfully represents the free energy density of any phase with density distribution in the family

$$\rho(\sigma) = \rho^{(0)}(\sigma) \exp\left(\sum_i \lambda_i w_i(\sigma)\right), \qquad (39)$$

where $\rho^{(0)}(\sigma)$ is the density distribution of the parent. The moment densities ρ_i are then related to the Lagrange multipliers λ_i by

$$\rho_i = \int d\sigma \, w_i(\sigma) \rho^{(0)}(\sigma) \exp\left(\sum_j \lambda_j w_j(\sigma)\right). \qquad (40)$$

If we regard ρ_0 as a function of the λ_i, then from Eq. (40) we have $\partial \rho_0 / \partial \lambda_i = \rho_i$. Together with Eq. (37), this implies that the moment entropy s_m has the structure of a Legendre transform [39]. For the first derivatives of s_m with respect to the moment densities, this yields

$$\frac{\partial s_m}{\partial \rho_i} = -\lambda_i$$

while the matrix of second derivatives is the negative inverse of the matrix of "second-order moment densities" ρ_{ij} [40]:

$$\frac{\partial^2 s_m}{\partial \rho_i \partial \rho_j} = -(M^{-1})_{ij}, \qquad (41a)$$

$$(M)_{ij} = \frac{\partial^2 \rho_0}{\partial \lambda_i \partial \lambda_j} = \int d\sigma \, w_i(\sigma) w_j(\sigma) \rho^{(0)}(\sigma) \exp\left(\sum_k \lambda_k w_k(\sigma)\right) \equiv \rho_{ij}. \qquad (41b)$$

The chemical potentials μ_i conjugate to the moment densities follow as

$$\mu_i = \frac{\partial f_m}{\partial \rho_i} = T\lambda_i + \frac{\partial \tilde{f}}{\partial \rho_i} = T\lambda_i + \tilde{\mu}_i \qquad (42)$$

and their derivatives with respect to the ρ_i, which give the curvature of the moment free energy, are

$$\frac{\partial \mu_i}{\partial \rho_j} = \frac{\partial^2 f_m}{\partial \rho_i \partial \rho_j} = T(M^{-1})_{ij} + \frac{\partial^2 \tilde{f}}{\partial \rho_i \partial \rho_j}. \qquad (43)$$

The pressure, finally, is given by

$$-\Pi_m = f_m - \sum_i \mu_i \rho_i = -T\rho_0 + \tilde{f} - \sum_i \tilde{\mu}_i \rho_i. \qquad (44)$$

Equations (42)–(44) are all calculated via the moment free energy. The three corresponding quantities obtained from the exact free energy (38) are, first, the chemical potentials conjugate to $\rho(\sigma)$:

$$\mu(\sigma) = \frac{\delta f}{\delta \rho(\sigma)} = T \ln \rho(\sigma) + \sum_i \tilde{\mu}_i w_i(\sigma); \qquad (45)$$

second, their derivatives with respect to $\rho(\sigma)$:

$$\frac{\delta \mu(\sigma)}{\delta \rho(\sigma')} = \frac{\delta^2 f}{\delta \rho(\sigma) \delta \rho(\sigma')} = \frac{T\delta(\sigma - \sigma')}{\rho(\sigma)} + \sum_{i,j} \frac{\partial^2 \tilde{f}}{\partial \rho_i \partial \rho_j} w_i(\sigma) w_j(\sigma'); \qquad (46)$$

and, third, the resulting expression for the pressure:

$$-\Pi = f - \int d\sigma \, \mu(\sigma) \rho(\sigma) = -T\rho_0 + \tilde{f} - \sum_i \tilde{\mu}_i \rho_i. \qquad (47)$$

The last of these is identical to the result (44) derived from the moment free energy.

We note one important consequence of the form of the exact chemical potentials (45) for truncatable models: If two phases $\rho^{(1)}(\sigma)$ and $\rho^{(2)}(\sigma)$ have the same chemical potentials $\mu(\sigma)$, then the ratio of their density distributions can be written as a Gibbs–Boltzmann factor:

$$\frac{\rho^{(1)}(\sigma)}{\rho^{(2)}(\sigma)} = \exp\left[\sum_i \beta \left(\tilde{\mu}_i^{(2)} - \tilde{\mu}_i^{(1)}\right) w_i(\sigma)\right]. \qquad (48)$$

This implies that if one of the density distributions is in the family (39), then so is the other. The same argument obviously applies if there are several phases with equal chemical potentials. Conversely, we have for the chemical potential difference between any two phases in the family (39)

$$\Delta\mu(\sigma) = \sum_i (T\Delta\lambda_i + \Delta\tilde{\mu}_i)w_i(\sigma) = \sum_i \Delta\mu_i w_i(\sigma). \qquad (49)$$

The chemical potentials $\mu(\sigma)$ of two such phases are therefore equal if and only if their *moment* chemical potentials μ_i are equal. Combining this with the fact that the pressure (44) derived from the moment free energy is exact [compare Eq. (47)], we conclude the following: Any set of (two or more) co-existing phases calculated from the *moment* free energy obeys the *exact* phase equilibrium conditions. That is, if the phases were brought into contact, they would genuinely coexist. [However, they will not necessarily obey the lever rule with respect to the parent $\rho^{(0)}(\sigma)$ of the given family, unless all but one of them—which then *coincides* with the parent—are of vanishing volume; see Eq. (11).]

A.   General Criteria for Spinodals and (Multi-) Critical Points

We now demonstrate that, for any truncatable model, the moment free energy gives exact spinodals and (multi-) critical points. By this we mean that, using the moment free energy, the values of external control parameters—such as temperature—at which the parent phase $\rho^{(0)}(\sigma)$ becomes unstable or critical can be exactly determined. Our argument treats spinodals and (multi-) critical points in a unified fashion, using the fact that all of them occur when the difference between phases with equal chemical potentials becomes infinitesimal. Because the truncatable models that we are considering are generally derived from mean-field approaches, we do not concern ourselves with critical point singularities, assuming instead that the free energy is a smooth function of all order parameters (the densities in the system). Likewise, we do not discuss the subtle question of how, beyond mean-field theory, free energies can actually be defined in spinodal and unstable regions [41].

Let us first recap the general criteria for spinodals and critical points in multicomponent systems. In order to treat the criteria derived from the exact and moment free energies simultaneously, we use the common notation ρ for the vector of densities specifying the system: For the exact free energy, the components of ρ are the values $\rho(\sigma)$; for the moment free energy, they are the reduced set ρ_i. We write the corresponding vector of chemical potentials as

$$\mu(\rho) = \nabla f(\rho).$$

This notation emphasizes that the chemical potentials are functions of the densities. The criterion for a spinodal at the parent phase $\rho^{(0)}$ is then that there is an incipient *instability* direction $\delta\rho$ along which the chemical potentials do not change:

$$(\delta\rho \cdot \nabla)\mu(\rho^{(0)}) = (\delta\rho \cdot \nabla)\nabla f(\rho^{(0)}) = \mathbf{0}. \tag{50}$$

As the second form of the criterion shows, an equivalent statement is that the curvature of the free energy along the direction $\delta\rho$ vanishes. Equation (50) can also be written as

$$\mu(\rho^{(0)} + \epsilon\delta\rho) - \mu(\rho^{(0)}) = \mathcal{O}(\epsilon^2), \tag{51}$$

which is closely related to the critical point criterion that we describe next.

Near a critical point, the parent $\rho^{(0)}$ coexists with another phase that is only slightly different; if, as we assume here, the free energy function is smooth, these two phases are separated—in ρ-space—by a "hypothetical phase" which has the same chemical potentials but is (locally) thermodynamically unstable. [This is geometrically obvious even in high dimensions; between any two minima of $f(\rho) - \mu \cdot \rho$, at given μ, there must lie a maximum or a saddle point, which is the required unstable "phase."] Now imagine connecting these three phases by a smooth curve in density space $\rho(\epsilon)$. At the critical point, all three phases collapse, and the variation of the chemical potential around $\rho(\epsilon = 0) = \rho^{(0)}$ must therefore obey

$$\mu(\rho(\epsilon)) - \mu(\rho^{(0)}) = \mathcal{O}(\epsilon^3).$$

Similarly, if n phases coexist, we can connect them and the $n-1$ unstable phases in between them by a curve $\rho(\epsilon)$ [42]. If we define an n-critical point as one where all these phases become simultaneously critical, we obtain the criterion

$$\mu(\rho(\epsilon)) - \mu(\rho^{(0)}) = \mathcal{O}(\epsilon^{2n-1}). \tag{52}$$

This formulation was proposed by Brannock [43]; the cases $n = 2$ and $n = 3$ correspond to ordinary critical and tricritical points, respectively. The spinodal criterion (51) is of the same form as (52) if one chooses the curve $\rho(\epsilon) = \rho^{(0)} + \epsilon\delta\rho$.

In summary, we have that the phase $\rho^{(0)}$ is a spinodal or n-critical point if there is a curve $\rho(\epsilon)$ with $\rho(\epsilon = 0) = \rho^{(0)}$ such that

$$\Delta\mu \equiv \mu(\rho(\epsilon)) - \mu(\rho^{(0)}) = \mathcal{O}(\epsilon^l), \tag{53}$$

where $l = 2$ for a spinodal, and $l = 2n - 1$ for an n-critical point. Brannock [43] has shown that these criteria are equivalent to the determinant criteria introduced by Gibbs [34]. The above forms are more useful for us because they avoid having to define infinite-dimensional determinants [as otherwise required to handle the *exact* free energy $f(\rho)$ of a polydisperse system, whose argument ρ denotes an infinite number of components, even in the truncatable case]. They also show the analogy with the standard criteria for single-species systems (where ρ has only a single component) more clearly.

We can now apply (53) to the exact free energy (38) and show that the resulting criteria are identical to those obtained from the moment free energy. First, note that the curve $\rho(\epsilon) \equiv \rho(\sigma; \epsilon)$ can always be chosen to lie within the family (39). This follows from the derivation of the criterion (53): The curve $\rho(\epsilon)$ is defined as passing through l phases with equal chemical potentials, one of them being the parent $\rho^{(0)}$. As shown in Eq. (48), all these phases are therefore within the family (39). But then Eq. (49) implies that the condition (53) that $\Delta\mu(\sigma)$ must be zero to $\mathcal{O}(\epsilon^l)$ is equivalent to the same requirement for the differences $\Delta\mu_i$ in the *moment* chemical potentials. This proves that the conditions for spinodals and (multi-) critical points derived from the exact and moment free energies are equivalent. Intuitively, one can understand this as follows: Having shown that the spinodal/critical point conditions can be formulated solely in terms of density distributions $\rho(\sigma)$ within the family (39), it is sufficient to know the free energy of those density distributions. This is exactly the moment free energy.

Finally, we note that the exactness of these stability and critical point conditions holds not just for the parent $\rho^{(0)}(\sigma)$, but for all phases within the family (39). This is clear, because any such phase could itself be chosen as the parent without changing the family; the moment free energy would change only by irrelevant terms linear in the moment densities. In general, one will not necessarily be interested in the properties of such "substitute" parents. As an exception, substitute parents that differ from the parent only by a change of the overall density *are* of interest: They lie on the dilution line of distributions from which cloud-point and shadow curves are calculated. The dilution line is included in the family (39) if the overall density ρ_0 [with corresponding weight function $w_0(\sigma) = 1$] is retained in the moment free energy; in the combinatorial derivation, this is automatically the case.

B. Spinodals

For completeness, we now give the explicit form [11, 12, 44] of the spinodal criterion (50) for truncatable systems; see also Ref. 45 for an equivalent derivation using the combinatorial approach. Using Eq. (43) and abbreviating the matrix of second derivatives of the excess free energy as $\tilde{F}$, the spinodal condition (50) becomes

$$|\tilde{F} + TM^{-1}| = 0, \qquad (\tilde{F} + TM^{-1})\delta\rho = 0. \tag{54}$$

As before, the (nonzero) vector $\delta\rho$ with components $\delta\rho_i$ gives the direction of the spinodal instability; the moment densities ρ_i and ρ_{ij} that appear in $\tilde{F}$ and M are to be evaluated for the parent distribution $\rho^{(0)}(\sigma)$ being studied. Note that Eq. (54) is valid for any $\rho^{(0)}(\sigma)$; no specific assumptions about the parent were made in the derivation. We can thus simply drop the "(0)" superscript: For any phase with density distribution $\rho(\sigma)$, the point where Eq. (54) first becomes zero, as external control parameters are varied, locates a spinodal instability.

More convenient forms of Eq. (54) that avoid matrix inversions are obtained after multiplication by the second-order moment matrix M [which is positive definite for linearly independent weight functions $w_i(\sigma)$ and therefore has nonzero determinant]:

$$Y = |1 + \beta M\tilde{F}| = 0, \qquad (1 + \beta M\tilde{F})\delta\rho = 0. \tag{55}$$

Here $\beta = 1/T$ in the standard notation. From our general statements in Section III. A, the spinodal criterion derived from the exact free energy (38) must be identical to this; this is shown explicitly in Appendix C. Note that the spinodal condition depends only on the (first-order) moment densities ρ_i and the second-order moment densities ρ_{ij} of the distribution $\rho(\sigma)$ [given by Eqs. (40) and (41)]; it is independent of any other of its properties. This simplification, which has been pointed out by a number of authors [11, 12], is particularly useful for the case of power-law moments (defined by weight functions $w_i(\sigma) = \sigma^i$): If the excess free energy only depends on the moments of order 0, 1...$K - 1$ of the density distribution, the spinodal condition involves only $2K - 1$ moments [up to order $2(K - 1)$].

The general discussion in Section III.A as well as the explicit calculation in Appendix C show that the spinodal instability direction lies within (or more precisely, is tangential to) the family (39) of density distributions. This fact has a simple geometrical interpretation: By construction, the family lies along a "valley" of the free energy surface (compare Fig. 1). *Away from* the valley floor, any change in $\rho(\sigma)$ increases the free energy, corresponding to a positive curvature; the spinodal direction, for which the curvature vanishes, must therefore be *along* the valley floor.

C. Critical Points

Next, we show the explicit form of the critical point criterion for truncatable systems. The general condition (53) for an ordinary critical point ($n = 2$) was shown by Brannock [43] to be equivalent to

$$(\delta\boldsymbol{\rho} \cdot \nabla)\nabla f(\boldsymbol{\rho}^{(0)}) = \mathbf{0}, \qquad (\delta\boldsymbol{\rho} \cdot \nabla)^3 f(\boldsymbol{\rho}^{(0)}) = 0. \tag{56}$$

The first part of this is simply the spinodal criterion, as expected. To evaluate the second part for the moment free energy (37), we need the third derivative of s_{m} with respect to the moment densities ρ_i. Writing (41) as

$$\sum_j \rho_{ij} \frac{\partial^2 s_{\mathrm{m}}}{\partial\rho_j \partial\rho_k} = -\delta_{ik}$$

and differentiating with respect to one of the moment densities, one finds after a little algebra

$$\frac{\partial^3 s_{\mathrm{m}}}{\partial\rho_i \partial\rho_j \partial\rho_k} = \sum_{lmn} (\boldsymbol{M}^{-1})_{il}(\boldsymbol{M}^{-1})_{jm}(\boldsymbol{M}^{-1})_{kn}\, \rho_{lmn}$$

with third-order moment densities ρ_{lmn} defined in the obvious way. Thus, evaluating Eq. (56) for the moment free energy (37), we find that critical points have to obey

$$\sum_{ijk}\left(-T\, v_i v_j v_k \rho_{ijk} + \delta\rho_i\delta\rho_j\delta\rho_k \frac{\partial^3 \tilde{f}}{\partial\rho_i \partial\rho_j \partial\rho_k}\right) = 0, \qquad v_i = \sum_j (\boldsymbol{M}^{-1})_{ij}\delta\rho_j \tag{57}$$

in addition to the spinodal condition (55). As before, the criterion has to be evaluated for the parent phase $\rho^{(0)}(\sigma)$ under consideration; but because $\rho^{(0)}(\sigma)$ can be chosen arbitrarily, it applies to all density distributions $\rho(\sigma)$.

As expected from the general discussion in Section III.A, the criterion (57) can also be derived from the exact free energy; an alternative form involving the spinodal determinant Y is given in Appendix D. Equation (57) shows that the location of critical points depend only on the moment densities ρ_i, ρ_{ij}, and ρ_{ijk} [11, 46]. For a system with an excess free energy depending only on power-law moments up to order $K - 1$, the critical point condition thus involves power-law moments of the parent only up to order $3(K - 1)$.

D. Onset of Phase Coexistence: Cloud Point and Shadow

So far in this section, we have shown that the moment free energy gives exact results for spinodals and (multi-) critical points. Now we consider the onset of phase coexistence, where (on varying the temperature, for example) a parent phase with density distribution $\rho^{(0)}(\sigma)$ first starts to phase separate. As explained in Section I, this temperature together with the overall density $\rho^{(0)} \equiv \rho_0^{(0)}$ of the parent defines a "cloud point"; the density of the incipient daughter phase gives the "shadow." If the parent begins to coexist with p phases simultaneously ($p = 2$ at a triple point, for example), there will be p such shadows.

At the onset of phase coexistence, one of the coexisting phases is by definition the parent $\rho^{(0)}(\sigma)$; the lever rule does not yet play any role because the daughters $\rho^{(\alpha)}(\sigma)$ ($\alpha = 1 \ldots p$) occupy an infinitesimal fraction of the total volume. It then follows from Eq. (48) that all daughters lie within the family (39) of density distributions. As shown in Eq. (49), the condition for equality of chemical potentials $\mu(\sigma)$ between any two of the coexisting phases (parent and daughters) then becomes

$$\Delta\mu(\sigma) = \sum_i \Delta\mu_i w_i(\sigma) = 0$$

and is satisfied if and only if the *moment* chemical potentials μ_i are equal in all phases. Likewise, the exact and moment free energies give the same condition for equality of pressure Π in all phases, because they yield identical expressions (44) and (47) for Π. In summary, we see that the conditions for the *onset* of phase coexistence are identical for the exact and moment free energies; the moment free energy therefore gives exact cloud points and shadows.

E. Phase Coexistence Beyond Onset

As stated in Section II.A, the moment free energy does not give exact results beyond the onset of phase coexistence—that is, in the regime where the coexisting phases occupy comparable fractions of the total system volume. As shown in Section III.A, the calculated phases will still be in exact thermal equilibrium; but the lever rule will now be violated for the "transverse" degrees of freedom of the density distributions. This is clear from Eq. (11): In general, no linear combination of distributions from this family can match the parent $\rho^{(0)}(\sigma)$ exactly.

A more detailed understanding of the failure of the moment free energy beyond phase coexistence can be gained by comparing with the formal solution of the exact phase coexistence problem. Assume that the parent $\rho^{(0)}(\sigma)$ has separated into p phases numbered by $\alpha = 1 \ldots p$. The condition (48), which follows from equality of the chemical potentials $\mu(\sigma)$ in all phases, implies that we can write their density distributions $\rho^{(\alpha)}(\sigma)$ as

$$\rho^{(\alpha)}(\sigma) = \tilde{R}(\sigma)\exp\left(\sum_i \lambda_i^{(\alpha)} w_i(\sigma)\right) \tag{58}$$

for *some* function $\tilde{R}(\sigma)$. If phase α occupies a fraction $v^{(\alpha)}$ of the system volume, particle conservation $\sum_\alpha v^{(\alpha)} \rho^{(\alpha)}(\sigma) = \rho^{(0)}(\sigma)$ then gives

$$\tilde{R}(\sigma) = \frac{\rho^{(0)}(\sigma)}{\sum_\alpha v^{(\alpha)} \exp\left(\sum_i \lambda_i^{(\alpha)} w_i(\sigma)\right)} \tag{59a}$$

$$\rho^{(\alpha)}(\sigma) = \rho^{(0)}(\sigma) \frac{\exp\left(\sum_i \lambda_i^{(\alpha)} w_i(\sigma)\right)}{\sum_\beta v^{(\beta)} \exp\left(\sum_i \lambda_i^{(\beta)} w_i(\sigma)\right)}. \tag{59b}$$

If there are K moment densities ρ_i, $i = 1 \ldots K$, then this exact solution is parameterized by $(p-1)(K+1)$ independent parameters: $(p-1)K$ parameters $\lambda_i^{(\alpha)}$ (noting that the λ_i of one phase can be fixed arbitrarily) and $p-1$ parameters $v^{(\alpha)}$ (noting that one phase volume is fixed by the constraint $\sum_\alpha v^{(\alpha)} = 1$). Comparing Eq. (58) with Eq. (48), one sees that the K quantities

$$T\lambda_i^{(\alpha)} + \tilde{\mu}_i^{(\alpha)} \qquad (i = 1 \ldots K)$$

must be the same in all phases α; the same is true for the pressures $\Pi^{(\alpha)}$, and this gives the required total number of $(p-1)(K+1)$ constraints. This for-

---

Figure 2. This figure continues the example of Fig. 1, which is a Flory–Huggins polymer + solvent model with two chain lengths present. Now, however, we allow for a change of the interaction parameter χ (equivalent to changing temperature). (a) The moment free energy at the cloud point, that is, the value of χ for which the given parent first begins to phase separate. (b) The tieline at the cloud point is drawn in the (r_1, r_2) plane of all density pairs (thin solid line). It is *exact* and connects the parent (solid circle) with another member of the family (39); the family is indicated by the thick solid line. The dotted line is the "transverse" line of constant $\rho_1 (= \rho_1^{(0)})$ passing through the parent. (c) The moment free energy for a higher value of χ. (d) The corresponding tieline (thin dashed line) in the (r_1, r_2) plane. As before, this connects two members of the parent's family (now the thick dashed line), and the two phases so found are in stable thermodynamic equilibrium with each other. Because they now occupy comparable fractions of the system volume, however, the lever rule is violated: The parent does not lie on the tieline. But the lever rule violation occurs only along the transverse direction (dotted line): The fractional phase volumes calculated from the moment free energy result in a total density pair (indicated by the empty circle) which has the same moment density ρ_1 as the parent. Finally, the exact tieline at this χ is also shown (thin solid line). As required, it passes through the parent. But its endpoints now connect members of a *different* family (thick solid line), which derives from an "effective parent" (or effective prior) $(\tilde{r}_1, \tilde{r}_2)$, which may be chosen anywhere along the thick solid line [compare Eq. (58)]. Note that if a single extra moment density were added in this scenario, the resulting maximum entropy family would in fact cover all possible density pairs (r_1, r_2) and the corresponding two-moment free energy would therefore give exact results in all situations. For a truly polydisperse (rather than bidisperse) system, in which the density pairs of this example become density *distributions* $\rho(\sigma)$, each added moment density allows the accuracy of the calculation to be increased.

mally defines the exact solution of the phase coexistence problem for truncatable systems. Its practical value is limited by the difficulties of finding the solution numerically, as pointed out in Section I; see also Section IV. There is also no interpretation of the result in terms of a free energy depending on a small number of densities.

Nevertheless, Eq. (59) is useful for a comparison with the solution provided by the moment free energy method. From Eq. (58), the exact coexisting phases are all members of a family of density distributions of the general form (7); this is simply a consequence of the requirement of equal chemical potentials (48). This "exact coexistence family" has an "effective" prior [47] $\tilde{R}(\sigma) \neq \rho^{(0)}(\sigma)$ and is therefore different from the "original" family (39). The moment free energy only gives us access to distributions from the original family, not the exact coexistence family, and therefore cannot yield exact solutions for phase coexistence (beyond its onset). Figure 2 illustrates this point explicitly in the simplified context of a bidisperse system.

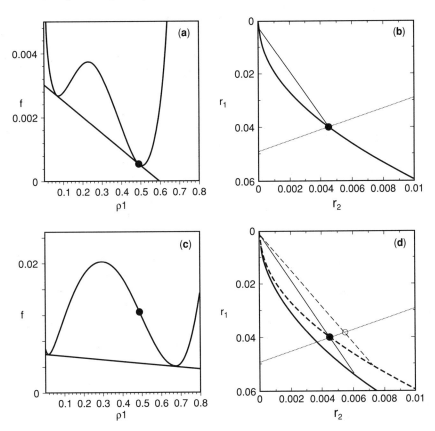

It is now easy to see, however, that—as stated in Section II.A—the exact solution can be approached to arbitrary precision by including extra moment densities in the moment free energy. (This leaves the exactness of spinodals, critical points, cloud points, and shadows unaffected, because none of our arguments excluded a null dependence of $\tilde{f}$ on certain of the ρ_i.) Indeed, by adding further moment densities, one can indefinitely extend the family (39) of density distributions, thereby approaching with increasing precision the actual distributions in all phases present; this yields phase diagrams of ever-refined accuracy.

The roles of the "original" moment densities (those appearing in the excess free energy) and the extra ones are quite different, however. To see this, note from (42) that equality among phases of the moment chemical potentials implies that, for the extra moments only, the corresponding Lagrange multipliers must themselves be *equal in all coexisting phases*. (This is because there is, by construction, no excess part to the "extra" chemical potentials.) We can therefore drop the phase index on these Lagrange multipliers and write the density distribution in phase α as

$$\rho^{(\alpha)}(\sigma) = \rho^{(0)}(\sigma) \exp\left(\sum_{\text{extra } i} \lambda_i w_i(\sigma)\right) \exp\left(\sum_{\text{original } i} \lambda_i^{(\alpha)} w_i(\sigma)\right). \qquad (60)$$

Comparing with Eqs. (7) and (58), we see that the extra Lagrange multipliers can be thought of as providing a "flexible prior" $R(\sigma)$, allowing a better approximation to the effective prior $\tilde{R}(\sigma)$ required by particle conservation. The "tuning" of the prior with these extra Lagrange multipliers also has an important effect on the number of coexisting phases that can be found by the moment method (see Section III.G).

F. Global and Local Stability

Most numerical algorithms for phase coexistence calculations, including ours, initially proceed by finding a solution to the equilibrium conditions of equal chemical potentials and pressures in all phases, rather than by a direct minimization of the total free energy of the system. It is then crucial to verify whether this solution is *stable*, both *locally* (i.e., with respect to small fluctuations in the compositions of the phases) and *globally* with respect to splitting into a larger number (or different) phases (see Fig. 3). Global stability is a particularly important issue in our context because there is in principle no limit on the number of coexisting phases in a polydisperse system.

A useful tool for stability calculations is the "tangent plane distance" [20]. Let us first define this generically for a system with (a vector of) densities ρ, free energy density $f(\rho)$ and chemical potentials $\mu(\rho) = \nabla f(\rho)$; this

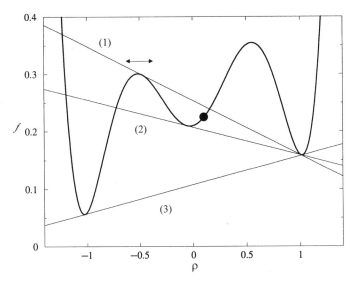

Figure 3. Illustration of local and global stability of solutions of the phase equilibrium conditions. For simplicity, we consider the case of a monodisperse system, characterized by its free energy density f (bold line) as a function of the density ρ. The phase split for a parent with density $\rho^{(0)}$ (indicated by the circle) is to be calculated. Three double tangents to $f(\rho)$ are shown, predicting different two-phase splits of the parent. All obey (by the double tangent property) the conditions of equal chemical potential and pressure in the two phases. But (1) is *locally* unstable: A small fluctuation (indicated by the arrow) of the density of the more dilute coexisting phase lowers the total free energy. (2) is locally stable, but *globally* unstable: The phase split (3) gives a lower total free energy. (3) is globally (and therefore locally) stable.

notation is the same as in Section III.A. Assume that we have found a candidate "phase split"—that is, a collection of p phases $\boldsymbol{\rho}^{(\alpha)}$ ($\alpha = 1 \ldots p$) that satisfy the phase equilibrium conditions of equal chemical potentials $\boldsymbol{\mu} = \boldsymbol{\mu}^{(\alpha)}$ and pressures $\Pi = \Pi^{(\alpha)}$. They occupy fractions $v^{(\alpha)}$ of the total volume, and the overall density distribution is thus $\boldsymbol{\rho}_{\text{tot}} = \sum_\alpha v^{(\alpha)} \boldsymbol{\rho}^{(\alpha)}$. Note that we do not yet assume that the lever rule is satisfied—that is, that $\boldsymbol{\rho}_{\text{tot}} = \boldsymbol{\rho}^{(0)}$; as explained above, this equality will generally not hold for phase splits calculated from the moment free energy. We now define "global tangent plane (TP) stability" for such a phase split as the property that there is no other phase split (i.e., no other tangent plane) that gives a lower total free energy for *the same overall density distribution* $\boldsymbol{\rho}_{\text{tot}}$. In more intuitive language, this means that if we were to put the phases $\boldsymbol{\rho}^{(\alpha)}$ into contact with each other, the resulting system would be thermodynamically stable; neither the composition nor the number of phases would change over time. Note, however, that since in general $\boldsymbol{\rho}_{\text{tot}} = \sum_\alpha v^{(\alpha)} \boldsymbol{\rho}^{(\alpha)}$ is *not* equal to $\boldsymbol{\rho}^{(0)}$, a phase split that is globally TP-stable need not accurately reflect the number and composition of phases into which the parent $\rho^{(0)}$ would actually split under the chosen thermodynamic conditions.

To define the tangent plane distance (TPD), note first that by virtue of the coexistence conditions, all phases $\rho^{(\alpha)}$ lie on a tangent plane to the free energy surface. Points (ρ, f) on this tangent plane obey the equation $f - \boldsymbol{\mu} \cdot \boldsymbol{\rho} + \Pi = 0$, with $\boldsymbol{\mu}$ and Π the chemical potentials and pressure common to all phases. For a generic phase with density distribution ρ and free energy $f(\rho)$, the same expression will have a nonzero value that measures how much "below" or "above" the tangent plane it lies. This defines the TPD

$$t(\boldsymbol{\rho}) = f(\boldsymbol{\rho}) - \boldsymbol{\mu}^{(\alpha)} \cdot \boldsymbol{\rho} + \Pi^{(\alpha)}. \tag{61}$$

Here we have added the superscript α to emphasize that the chemical potential and osmotic pressure used in the calculation of the TPD are those of the calculated phase equilibrium (and hence of any of the participating phases α), rather than those of the test phase ρ. It is then clear intuitively—and can be shown more formally [20]—that the calculated phase coexistence is globally TP-stable if the TPD is nonnegative everywhere. Geometrically, this simply means that no part of the free energy surface must protrude beneath the tangent plane; otherwise the total free energy of the system could be lowered by constructing a new tangent plane that touches the protruding piece. Global TP-stability of course encompasses *local* stability; the latter simply corresponds to the requirement that the TPD be a local minimum (with the value $t = 0$) at each of the phases in the candidate solution.

When verifying global TP-stability, it is obviously sufficient to check the value of the TPD at all of its stationary points. By differentiating (61) with respect to ρ, one sees that at these points, the chemical potentials $\mu(\rho)$ are the same as in the calculated coexisting phases. For our truncatable polydisperse systems, it then follows from Eq. (48) that we only need to consider the TPD for test phases $\rho(\sigma)$ which are in the same family (7) as the calculated coexisting phases. This is a crucial point: Even though global TP-stability is a statement about stability in the infinite-dimensional space of density distributions $\rho(\sigma)$, it can be checked by only considering density distributions from a K-dimensional family. In the (generally hypothetical) case of an exactly calculated phase split, this family would be Eq. (59), with the effective prior $\tilde{R}(\sigma)$ [48]. For a phase split calculated from the moment free energy, it is the family (39) with the parent as prior, with any Lagrange multipliers for extra moments fixed to their values in the calculated coexisting phases. The TPD of such a test phase is [using Eq. (38) for the exact free energy and Eq. (45) for the chemical potentials $\mu^{(\alpha)}(\sigma)$ of the coexisting phases]

$$t[\rho(\sigma)] = f[\rho(\sigma)] - \int d\sigma \mu^{(\alpha)}(\sigma)\rho(\sigma) + \Pi^{(\alpha)}$$

$$= \tilde{f}(\rho_i) + \int d\sigma \rho(\sigma) \left\{ T[\ln\rho(\sigma)-1] - T\ln\rho^{(\alpha)}(\sigma) - \sum_i \tilde{\mu}_i^{(\alpha)} w_i(\sigma) \right\} + \Pi^{(\alpha)}$$

$$= \tilde{f}(\rho_i) + T\int d\sigma\rho(\sigma) \left[\ln\frac{\rho(\sigma)}{\rho^{(0)}(\sigma)} - 1 \right] - \sum_i \left[T\lambda_i^{(\alpha)} + \tilde{\mu}_i^{(\alpha)} \right]\rho_i + \Pi^{(\alpha)}$$

$$= f_{\mathrm{m}}(\rho_i) - \sum_i \mu_i^{(\alpha)}\rho_i + \Pi^{(\alpha)}$$

A comparison with Eq. (61) shows that this is identical to the TPD that one would derive from the moment free energy alone. Here again, the underlying polydisperse nature of the problem can therefore be disregarded once the moment free energy has been obtained. In summary, one can determine whether a phase split calculated from the moment free energy is globally TP-stable (which means that phases of the predicted volumes and compositions, would, if placed in contact, indeed coexist) using only the TPD derived from the moment free energy; the same is trivially true of the weaker requirement of local stability [49].

Recall, however, that the overall density distribution $\rho_{\mathrm{tot}}(\sigma)$ for a phase split found from the moment free energy only has the same moment densities $\rho_i^{(0)}$ as the parent $\rho^{(0)}(\sigma)$, but differs in other details (the transverse degrees of freedom). Global TP-stability thus guarantees that such an approximate phase split is thermodynamically stable, but does not imply that it is identical (in either number or composition of phases) to the exact one. Nevertheless, progressively increasing the number of extra moment densities in the moment free energy will make $\rho_{\mathrm{tot}}(\sigma)$ a progressively better approximation to $\rho^{(0)}(\sigma)$, and so the exact phase split, with the correct number of phases, must eventually be recovered to arbitrary accuracy; see Fig. 4 for an illustration.

G. Geometry in Density Distribution Space

In the above discussion of the properties of the moment free energy, we have focused on obtaining the phase behavior of a system with a given parent distribution $\rho^{(0)}(\sigma)$. This is the point of view most relevant for practical applications of the method, and the rest of this section is not essential for understanding such applications. Nevertheless, from a theoretical angle, it is also interesting to consider the global geometry of the space of all density distributions $\rho(\sigma)$, without reference to a specific parent: The exact free

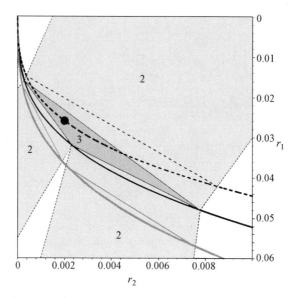

Figure 4. Example of structure of the space of density distributions. We continue the example of Figs. 1 and 2 and consider a bidisperse Flory–Huggins system with a density pair (r_1, r_2). As before, we take the excess free energy of the system to depend on a single moment density, so that the families (7) are one-dimensional (i.e., appear as curves in the graph). In the case sketched here, there is a three-phase region bordered by two-phase regions delineated by dotted lines. The bold (solid, dashed, gray) curves show three families (7) of density pairs; the whole space is partitioned (or "foliated") into an infinite number of such families. All tielines must begin and end within the same family; the thin gray tielines are shown as examples. The moment free energy (without use of extra moments—or in this case, with a single moment density retained) can only access systems from the family passing through the parent (filled circle); it therefore predicts that the parent will separate into the two phases within this family that are connected by a tieline (thin dashed line). This tieline is thermodynamically "real": the two phases at its ends, if put into contact, would remain in stable coexistence. But it does not give the exact phase separation for the given parent because of the lever rule violations allowed by the moment approach (the parent does not lie on the tieline). Retaining one extra moment density in the moment free energy gives, in this simple bidisperse case, the exact result: The parent separates into the three phases at the corners of the three-phase triangle, which lie on a family (bold solid curve) that does not contain the parent.

energy (38) of the underlying (truncatable) model induces two-phase tie lines and multiphase coexistence regions in this space, and one is led to ask how the moment free energy encodes these properties.

As pointed out in Section III.A, the definition of the moment free energy depends on a prior $R(\sigma)$ and represents the properties of systems with density distributions $\rho(\sigma)$ in the corresponding maximum entropy family (7). Instead of identifying $R(\sigma) = \rho^{(0)}(\sigma)$, we now allow a general prior $R(\sigma)$. Conceptually, it then makes sense to associate the moment free energy with the *family* (7) rather than with the specific *prior*. This is because any density distribution from (7) can be chosen as prior, without changing the moment free energy (apart from irrelevant linear terms in the moments ρ_i), or the identity of the remaining family members. The construction of the moment free energy thus partitions the space of all $\rho(\sigma)$ into (an infinite number of) families (7); different families give different moment free energies that describe the thermodynamics "within the family." This procedure gives meaningful results because, as shown at the end of Section III.A, coexisting phases are always members of the *same* family. By considering the entire ensemble of families (7) and their corresponding moment free energies, one can thus in principle recover the exact geometry of the density distribution space and the phase coexistences within it. Note that this includes regions with more than $K + 1$ phases, even when the excess (and thus moment) free energy depends on only K moment densities (see Fig. 4). Consistent with Gibbs' phase rule, however, the families for which this occurs are exceptional: They occupy submanifolds (of measure zero) in the space of all families. Accordingly, to find the families on which such "super-Gibbs" multiphase coexistences occur, the corresponding prior must be very carefully tuned. Indeed, the probability of finding such priors "accidentally," without either adding extra moments to the moment free energy description or solving the exact phase coexistence problem, is zero. Generically, one needs to retain at least n moment densities in the moment free energy to find $n + 1$ phases in coexistence. From Eq. (60), this corresponds to having $n - K$ parameters available for tuning the location of the family (or equivalently, its prior) in density distribution space. But note that, once the extra moments have been introduced, the required tuning need not be done "by hand": It is achieved implicitly by requiring the lever rule to be obeyed not only for the K original moment densities, but also for the $n - K$ extra ones. [The solution of the phase equilibrium conditions thus effectively proceeds in an n-dimensional space; but once a solution has been found, its (global TP-) stability can still be checked by computing the TPD only within a K-dimensional family of distributions. See Section III.F.]

As pointed out above, phase splits calculated from the moment free energy allow violations of the lever rule. In the global view, this fact also has a simple geometric interpretation: The families (7) are (generically) curved. In other words, for two distributions $\rho^{(1)}(\sigma)$ and $\rho^{(2)}(\sigma)$ from the same family, the straight line $\rho(\sigma) = \epsilon\rho^{(1)}(\sigma) + (1 - \epsilon)\rho^{(2)}(\sigma)$ connecting them lies outside the family. More generally, if p coexisting phases $\rho^{(\alpha)}(\sigma)$ have been identified from the moment free energy, then the overall density distribution $\rho_{\text{tot}}(\sigma) = \sum_\alpha v^{(\alpha)}\rho^{(\alpha)}(\sigma)$ is different from the parent $\rho^{(0)}(\sigma)$; geometrically, it lies on the hyperplane that passes through the phases $\rho^{(\alpha)}(\sigma)$.

IV. PRACTICAL IMPLEMENTATION
OF THE MOMENT METHOD

The application of the moment free energy method to the calculation of spinodal and critical points is straighforward using conditions (55), and (57) or (D3), respectively, and is further illustrated in Section V below. Therefore in this section we focus on phase coexistence calculations.

Recall that in the moment approach, each phase α is parameterized by Lagrange multipliers $\lambda_i^{(\alpha)}$ for the original moments (the ones appearing in the excess free energy of the system) and the fraction $v^{(\alpha)}$ of system volume that it occupies. If extra moments are used, there is one additional Lagrange multiplier λ_i for each of them; these are common to all phases. These parameters have to be chosen such that the pressure (44) and the moment chemical potentials μ_i given by Eq. (42) are equal in all phases. Furthermore, the (fractional) phase volumes $v^{(\alpha)}$ have to sum to one, and the lever rule has to be satisfied for all moments (both original and extra):

$$\sum_\alpha v^{(\alpha)} \rho_i^{(\alpha)} = \rho_i^{(0)}.$$

In a system with K original moments that is being studied using an n-moment free energy (i.e., with $n - K$ extra moments), and for p coexisting phases, one has $p(K + 1) + n - K = (p - 1)(K + 1) + n + 1$ parameters and as many equations. Starting from a suitable initial guess, these can, in principle, be solved by a standard algorithm such as Newton–Raphson [50]. Generating an initial point from which such an algorithm will converge, however, is a nontrivial problem, especially when more than two phases coexist.

To simplify this task, we work with a continuous control parameter such as temperature, density $\rho^{(0)}$ of the parent phase, or interaction parameter χ for polymers. Taking the latter case as an example, we start the calculation at a small value of χ where we are sure to be in a single-phase region; thus the parent phase under consideration is stable. The basic strategy is then to increment χ and detect potential new phases as we go along. At each step, the phase equilibrium conditions are solved by Newton–Raphson for the current number of phases. Then we check for local stability of the solution by calculating the Hessian of the TPD around each of the coexisting phases and verifying that it is positive definite. If an instability (i.e., a negative eigenvalue of the Hessian) is found, we search for local minima starting from points displaced either way along the instability direction. If two new local minima are found in this way, we add them to list of phases and delete the unstable phase; if only one new local minimum is uncovered, we add this but retain the old phase. Finally, we check for global (TP-) stability.

As explained in Section III.F, this involves scanning all "test" phases from the same maximum entropy family for possible negative values of the TPD. However, the test phase will have the same extra Lagrange multipliers, and is therefore parameterized in terms of the λ_i for the original moments only. The extra Lagrange multipliers are held fixed, so their number is irrelevant for the computational cost of the stability check. Put differently, within our algorithm the *precision* of the coexistence curves should depend on the *total* number of moments retained, n, rather than the number of moments in the excess free energy, K. *Computational effort*, on the other hand, is dominated by the global stability check, and thus it is mainly sensitive to K (the dimension of the space to be searched for new phases) rather than n.

This is a substantial efficiency gain, but despite it, an exhaustive search for local TPD minima over the K-dimensional space of original Lagrange multipliers is unrealistic except for $K = 1$. We use instead a Monte Carlo-type algorithm [51] to sample the TPD at representative points. If any negative values of the TPD are encountered, we choose the smallest such value, find the nearest local minimum of the TPD and add this point to the list of phases. If any new phases have been found during the (local and global) stability checks, we assign each of them a default phase volume ($v = 0.1$, say), reduce the phase volumes of the old phases accordingly, update the number of phases, and loop back to the Newton–Raphson solution of the phase equilibrium conditions, using the current list of phases as an initial guess. This process is repeated until the calculated solution is found to be stable. Our implementation of this basic scheme also contains some additional elements (such as an adaptive choice of the stepsize for the control parameter χ, and checks for very small phase volumes) that are useful near points where new phases appear or old ones vanish.

Comparing our own approach, which is based on the moment free energy with extra moment densities as just outlined, to (59), which is the exact solution for truncatable systems, we see that the former actually uses more parameters (essentially one phase-independent Lagrange multiplier per extra moment) to represent the solution. At first sight, this may appear counterproductive. However, the Lagrange multipliers $\lambda_i^{(\alpha)}$ and the (fractional) phase volumes $v^{(\alpha)}$ are much less strongly coupled in the moment free energy approach. The phase volumes are only determined by the lever rule, while in the exact solution they "feed back" into the effective prior $\tilde{R}(\sigma)$ and therefore into the equilibrium conditions of equal chemical potentials and pressures. In the examples studied below, we have found that the numerical advantages of this decoupling (in terms of stability, robustness, and convergence of our algorithm) can easily outweigh the larger number of parameters that it requires.

As with any numerical algorithm, it is important to develop robust criteria by which the convergence of the solution can be judged. This has a novel aspect, in the moment method, since violations of the lever rule are allowed: Alongside normal numerical convergence criteria, one needs a method for deciding whether the effect of these violations on the predicted phase behavior is significant. We develop appropriate criteria in Section V.B in the context of a specific example. The basic ideas is that since any state of phase coexistence predicted by the method represents the exact behavior of *some* parent, then so long as this parent is close enough to the true one, the predicted behavior will lie within the range of uncertainty that arises anyway, from not knowing the true parent to arbitrary experimental precision.

V. EXAMPLES

We now illustrate how the moment method is applied and demonstrate its usefulness for several examples. The first two (Flory–Huggins theory for length-polydisperse homopolymers and dense chemically polydisperse copolymers, respectively) contain only a single moment density in the excess free energy and are therefore particularly simple to analyze and visualize. In the third example (chemically polydisperse copolymers in a polymeric solvent), the excess free energy depends on two moment densities, and this will give us the opportunity to discuss the appearance of more complex phenomena such as tricritical points.

A. Homopolymers with Length Polydispersity

Let us start with the simple but well-studied example of polydisperse Flory–Huggins theory [52]. One considers a system of homopolymers with a distribution of chain lengths; the polydisperse feature σ is simply the chain length L—that is, the number of monomers in each chain. (We treat this as a continuous variable.) The density distribution $\rho(L)$ then gives the number density of chains as a function of L. We choose the segment volume a^3 as our unit of volume, making $\rho(L)$ dimensionless. The volume fraction occupied by the polymer is then simply the first moment of this distribution, $\phi \equiv \rho_1 = \int dL\, L\, \rho(L)$. Within Flory–Huggins theory, the free energy density is (in units such that $k_B T = 1$)

$$f = \int dL\, \rho(L)\, [\ln \rho(L) - 1] + (1 - \rho_1) \ln (1 - \rho_1) + \chi \rho_1 (1 - \rho_1), \quad (62)$$

where the Flory χ-parameter plays essentially the role of an inverse temperature. Before analyzing this further, we note that for a bidisperse system with chain lengths L_1 and L_2 and number densities r_1 and r_2, the corresponding expression would be

$$f = r_1(\ln r_1 - 1) + r_2(\ln r_2 - 1) + (1 - \rho_1) \ln (1 - \rho_1) + \chi \rho_1 (1 - \rho_1),$$
$$\rho_1 = L_1 r_1 + L_2 r_2. \tag{63}$$

This free energy, with $L_1 = 10$ and $L_2 = 20$, was used to generate the examples shown in Figs. 1 and 2.

Returning now to (62), we note first that the excess free energy

$$\tilde{f} = (1 - \rho_1) \ln(1 - \rho_1) + \chi \rho_1 (1 - \rho_1) \tag{64}$$

depends only on the momement density ρ_1. If no extra moments are used, the moment free energy is therefore a function of a single density variable, ρ_1. Let us work out its construction explicitly for the case of a parent phase with a Schulz distribution of lengths, given by

$$\rho^{(0)}(L) = \rho_0^{(0)} \frac{1}{\Gamma(a)b^a} L^{a-1} e^{-L/b}. \tag{65}$$

Here a is a parameter that determines how broad or peaked the distribution is; it is conventionally denoted by α, but we choose a different notation here to prevent confusion with the phase index used in the general discussion so far. The chain number density of the parent is $\rho_0^{(0)}$, and the *normalized* first moment $m_1^{(0)} = \rho_1^{(0)}/\rho_0^{(0)}$ is simply the (number) average chain length

$$L_N \equiv m_1^{(0)} \equiv \frac{\rho_1^{(0)}}{\rho_0^{(0)}} = ab. \tag{66}$$

The parent is thus parameterized in terms of a, $\rho_0^{(0)}$, and $\rho_1^{(0)}$ (or b). The density distributions in the family (39) are given by

$$\rho(L) = \rho^{(0)}(L)e^{\lambda_1 L} = \rho_0^{(0)} \frac{1}{\Gamma(a)b^a} L^{a-1} e^{-(b^{-1}-\lambda_1)L} \tag{67}$$

and the zeroth and first moment densities are easily worked out, for members of this family, to be

$$\rho_0 = \rho_0^{(0)} \left(\frac{1}{1 - b\lambda_1}\right)^a, \qquad \rho_1 = \rho_1^{(0)} \left(\frac{1}{1 - b\lambda_1}\right)^{a+1}.$$

Because the family only has a single parameter λ_1, these two moment densities are of course related to one another:

$$\rho_0 = c\rho_1^{a/(a+1)}, \qquad c = \rho_0^{(0)}\left(\rho_1^{(0)}\right)^{-a/(a+1)}. \tag{68}$$

Now we turn to the moment entropy, given by Eq. (37): $s_m = \rho_0 - \lambda_1\rho_1$ and this is to be considered as a function of ρ_1. After a little algebra (and eliminating b in favor of $\rho_1^{(0)}$) one finds

$$s_m = \rho_0^{(0)}\left[(1+a)\left(\frac{\rho_1}{\rho_1^{(0)}}\right)^{a/(a+1)} - a\frac{\rho_1}{\rho_1^{(0)}}\right]. \tag{69}$$

The last term is linear in ρ_1 and can be disregarded for the calculation of phase equilibria. This then gives the following simple result for the moment free energy:

$$f_m = -(a+1)c\rho_1^{a/(a+1)} + \tilde{f}. \tag{70}$$

[In conventional polymer notation, this result would read $f_m = -(\alpha+1)$ $c\phi^{\alpha/(\alpha+1)} + \tilde{f}$.] From this we can now obtain the spinodal condition, for example, which identifies the value of χ where the parent becomes unstable. In our case of a single moment density the general criterion (50) simplifies to

$$\left.\frac{d^2f_m}{d\rho_1^2}\right|_{\rho_1=\rho_1^{(0)}} = 0 \Rightarrow \frac{1}{1-\rho_1^{(0)}} - 2\chi + \frac{a}{a+1}\frac{\rho_0^{(0)}}{(\rho_1^{(0)})^2} = 0. \tag{71}$$

The same condition for a phase with general density distribution $\rho(L)$—rather than our specific Schulz parent $\rho^{(0)}(L)$—follows from Eq. (55) as

$$\frac{1}{1-\rho_1} - 2\chi + \frac{1}{\rho_2} = 0. \tag{72}$$

As expected, this becomes equivalent to Eq. (71) for parents of the Schulz form (65), which obey $\rho_2^{(0)} = \rho_0^{(0)}a(a+1)b^2$ and Eq. (66). Note that in standard polymer notation, Eq. (72) would be written as

$$\frac{1}{1-\phi} - 2\chi + \frac{1}{L_W\phi} = 0, \tag{73}$$

where $L_W = \rho_2/\rho_1$ is the weight average chain length.

The critical point condition is obtained similarly. For a single density variable, Eq. (56) reduces to

$$\left.\frac{d^3 f_m}{d\rho_1^3}\right|_{\rho_1 = \rho_1^{(0)}} = 0 \quad \Rightarrow \quad \frac{1}{(1-\rho_1^{(0)})^2} - \frac{a(a+2)}{(a+1)^2} \frac{\rho_0^{(0)}}{(\rho_0^{(0)})^3} = 0. \tag{74}$$

For a general distribution, the critical point criterion (57) becomes instead

$$-\frac{\rho_3}{\rho_2^3} + \frac{d^3 \tilde{f}}{d\rho_1^3} = 0,$$

and inserting the Flory–Huggins excess free energy (64) gives

$$\frac{1}{(1-\rho_1)^2} - \frac{\rho_3}{\rho_2^3} = 0. \tag{75}$$

Again, this simplifies to Eq. (74) for Schulz parents, for which $\rho_3^{(0)} = \rho_0^{(0)}$ $a(a+1)(a+2)b^3$. The traditional statement of Eq. (75), in terms of $\phi \equiv \rho_1$, L_W, and the Z-average chain length $L_Z = \rho_3/\rho_2$, reads [52]

$$\frac{1}{(1-\phi)^2} - \frac{L_Z}{L_W^2} \frac{1}{\phi^2} = 0. \tag{76}$$

Figure 5 shows the moment free energy (70) for a given parent and different values of χ. The value of χ for which there is a double tangent at the point $\rho_1 = \rho_1^{(0)}$ representing the parent gives the cloud point; the second tangency point gives the polymer volume fraction ρ_1 in the coexisting phase, hence the shadow point. Repeating this procedure for different values of $\rho_1^{(0)}$ (while maintaining $L_N = \rho_1^{(0)}/\rho_0^{(0)}$ and a constant), one obtains the full CPC and shadow curve for a given normalized parent length distribution $n(L) = \rho^{(0)}(L)/\rho_0^{(0)}$.

We now consider the properties of the moment free energy with the (chain number) density ρ_0 retained as an extra moment. This provides additional geometrical insight into the properties of polydisperse chains (while for the numerical determination of the CPC and shadow curve, the above one-moment free energy is preferable). To construct the two-moment free energy, we proceed as before. The family (39) is now

$$\rho(L) = \rho^{(0)}(L)e^{\lambda_0 + \lambda_1 L} = \rho_0^{(0)} \frac{1}{\Gamma(a)b^a} L^{a-1} e^{\lambda_0 - (b^{-1} - \lambda_1)L} \tag{77}$$

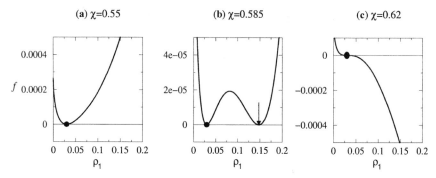

Figure 5. Examples of moment free energy (70) for Flory–Huggins theory of length-polydisperse polymers, with one moment density, ρ_1, retained. The parent is of the Schulz form (65), with $\rho_1^{(0)} = 0.03$, $L_N = 100$ (hence $\rho_0^{(0)} = \rho_1^{(0)}/L_N = 3 \times 10^{-4}$), and $a = 2$ (hence $L_W = 150$); the point $\rho_1 = \rho_1^{(0)}$ is marked by the filled circles. In plot (a), the value of $\chi = 0.55$ is sufficiently small for the parent to be stable: The moment free energy is convex. Plot (b) shows the cloud point, $\chi \approx 0.585$, where the parent lies on one endpoint of a double tangent; the other endpoint gives the polymer volume fraction ρ_1 in the shadow phase. Increasing χ further, the parent eventually becomes spinodally unstable [$\chi \approx 0.62$, plot (c)]. Note that for better visualization, linear terms have been added to all free energies to make the tangent at the parent coincide with the horizontal axis.

with zeroth and first moment densities

$$\rho_0 = \rho_0^{(0)} e^{\lambda_0} \left(\frac{1}{1 - b\lambda_1} \right)^a, \qquad \rho_1 = \rho_1^{(0)} e^{\lambda_0} \left(\frac{1}{1 - b\lambda_1} \right)^{a+1}.$$

The relations can be inverted to express the Lagrange multipliers λ_0 and λ_1 in terms of the moment densities as

$$\lambda_0 = (a+1) \ln \frac{\rho_0}{\rho_0^{(0)}} - a \ln \frac{\rho_1}{\rho_1^{(0)}}$$

$$\lambda_1 = \frac{1}{b} \left(1 - \frac{\rho_0/\rho_0^{(0)}}{\rho_1/\rho_1^{(0)}} \right)$$

and one obtains the moment entropy

$$s_m = \rho_0 - \lambda_0 \rho_0 - \lambda_1 \rho_1$$

$$= -(a+1)\rho_0 \ln \frac{\rho_0}{\rho_0^{(0)}} + a\rho_0 \ln \frac{\rho_1}{\rho_1^{(0)}} + (a+1)\rho_0 - a\frac{\rho_0^{(0)}}{\rho_1^{(0)}}\rho_1. \quad (78)$$

As expected, this reduces to Eq. (69) for systems with density distributions from our earlier one-parameter family (67), where ρ_0 can be expressed as a function of ρ_1 according to Eq. (68). With both ρ_0 and ρ_1 retained as moment densities in the moment free energy, a number of linear terms in Eq. (78) can be dropped, giving the final result

$$f_m = (a+1)\rho_0 \ln \rho_0 - a\rho_0 \ln \rho_1 + \tilde{f}. \tag{79}$$

Note that the dependence on the parent distribution is now only through a. This can be understood from the general discussion in Section III.G: The family (77) of density distributions now contains *all* Schulz distributions with the given a, and the moment free energy is insensitive to which member of this family (specified by $\rho_0^{(0)}$ and $\rho_1^{(0)}$) is used as the parent.

The result (79) can of course also be obtained via the combinatorial method, as follows. The normalized Schulz parent distribution is given by $n^{(0)}(L) = [\Gamma(a)b^a]^{-1} L^{a-1} e^{-bL}$. The cumulant generating function is thus $h(\lambda_1) = -a \ln(1 - b\lambda_1)$, giving $m_1 = \partial h/\partial \lambda_1 = ab/(1 - b\lambda_1)$. Dropping constants and terms linear in the average chain length m_1 (which do not affect the phase behavior), the generalized entropy of mixing per particle, $s_{mix} = h - \lambda_1 m_1$, then becomes $s_{mix} = a \ln m_1$. Adding the ideal gas term and the excess part of the free energy, we thus find the moment free energy density

$$f_m = \rho_0 \ln \rho_0 - a\rho_0 \ln (\rho_1/\rho_0) + \tilde{f}, \tag{80}$$

in agreement with Eq. (79); it is a simple two component free energy. In the conventional notation, Eq. (80) would read

$$f_m = \rho \ln \rho - \alpha\rho \ln (\phi/\rho) + (1 - \phi) \ln (1 - \phi) + \chi\phi(1 - \phi).$$

The spinodal curve (SC) and critical point (CP) condition may now be calculated from Eq. (80), using either the methods outlined in Section III.A or the more traditional determinant conditions [34] (see also Appendix D). One finds

$$\frac{1}{1 - \rho_1} - 2\chi + \frac{a}{a+1}\frac{\rho_0}{\rho_1^2} = 0 \qquad \text{(SC)}, \tag{81}$$

$$\frac{1}{(1 - \rho_1)^2} - \frac{a(a+2)}{(a+1)^2}\frac{\rho_0}{\rho_1^3} = 0 \qquad \text{(CP)}. \tag{82}$$

Evaluated at the parent, these conditions are identical to Eqs. (71) and (74) – which were derived from the one-moment free energy – as they must be. But, as explained at the end of Section III.A, they also hold more generally for all systems with Schulz density distributions with the given value of a; see the remarks after Eq. (79). Of particular interest among these systems are those that differ from the parent only in their overall density while having the same normalized length distribution $n(L) = n^{(0)}(L)$. They form the "dilution line"; their number averaged chain length satisfies $m_1 = m_1^{(0)} = \int dL\, Ln^{(0)}(L)$, which translates to $\rho_1 = L_N\rho_0$ (where $L_N \equiv m_1^{(0)} = ab$ is, as before, the number average chain length of the parent). Inserting this "dilution line constraint" into the above equations, we obtain as the spinodal and critical point conditions for systems on the dilution line

$$\frac{1}{1-\rho_1} - 2\chi + \frac{a}{a+1}\frac{1}{L_N\rho_1} = 0 \qquad \text{(SC)}, \qquad (83)$$

$$\frac{1}{(1-\rho_1)^2} - \frac{a(a+2)}{(a+1)^2}\frac{1}{L_N\rho_1^2} = 0 \qquad \text{(CP)}. \qquad (84)$$

Again, agreement with the general results (73) and (76) is easily verified by noting that for Schulz length distributions, $L_W = \rho_2/\rho_1 = (a+1)b = L_N(a+1)/a$ and $L_Z = \rho_3/\rho_2 = (a+2)b = L_N(a+2)/a$ for the weight and Z-average chain lengths, respectively.

The above discussion focused on spinodals, critical points, and cloud and shadow curves, all of which are found exactly by the moment approach. [As usual, "exact" is subject to the assumed validity of the original, truncatable free energy expression (64).] Within the moment approach, we have also calculated the full phase behavior in the (ρ_1, ρ_0) plane. This was done numerically from the two-moment free energy (80), facilitated by a partial Legendre transformation from ρ_0 to the associated chemical potential μ_0, which can be performed analytically in this case [and effectively brings one back to the one-moment free energy (70)]. Binodal curves, tielines, spinodal curves, and critical points in the (ρ_1, ρ_0) plane are shown in Fig. 6(a)–(c), for three values of χ. These plots give an appealing geometric insight into the problem of phase separation in this system. For instance, the slope of the tielines indicates a size partitioning effect: The more dense phases are enriched in long chains.

The heavy line in Fig. 6(a)–(c) is the dilution line constraint $m_1 = m_1^{(0)}$ or $\rho_1 = L_N\rho_0$. As discussed already, we are interested mainly in systems whose mean composition lies on this line. [With χ as an extra variable, this line becomes a plane $(\rho_1 = L_N\rho_0, \chi)$ in the space (ρ_1, ρ_0, χ).] Not all of the phase behavior shown in Fig. 6 is then accessible. The extremities of phase separation on the $\rho_1 = L_N\rho_0$ line are points where phase separation just starts to occur, and the locus of these points in the $(\rho_1 = L_N\rho_0, \chi)$ plane defines the

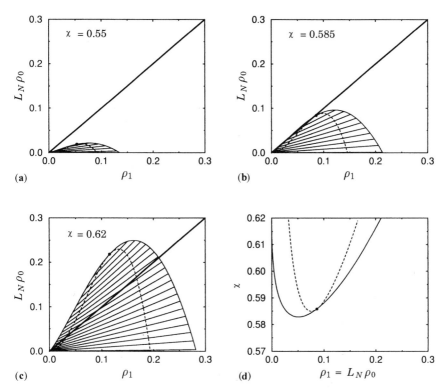

Figure 6. Conventional two-component phase behavior in polydisperse Flory–Huggins theory, shown in the (ρ_1, ρ_0) plane for three values of χ. As in Fig. 5, the parent has $L_N = 100$ and $L_W = 150$ (hence $a = 2$). Along the y-axis, we plot $L_N \rho_0$ rather than ρ_0 so that the dilution line $\rho_1 = L_N \rho_0$, shown as the thick solid line in (a–c), is simply along the diagonal. With χ considered as an additional variable, the dilution line constraint defines a plane $(\rho_1 = L_N \rho_0, \chi)$. The last plot, (d), shows the cut by this plane through the phase behavior in (a–c); the solid line is the cloud point curve, and the dashed line is the spinodal stability condition.

cloud curve, which bounds the phase separation region. This is shown in Fig. 6(d). This plot also includes the spinodal stability curve and the critical point from Eqs. (83) and (84). The spinodal curve and the cloud curve touch at the critical point, which no longer lies at the minimum of either. This distorted behavior is a well-known feature of polydisperse systems. Here it is seen to be due to the way that *regular* phase behavior in (ρ_1, ρ_0, χ) space is cut through by the dilution line constraint. Note that only two moments are needed to understand this qualitative effect, and that Fig. 6 gives direct geometrical insight into it, which would be very hard to extract from any exact solution based on Eqs. (58) and (59).

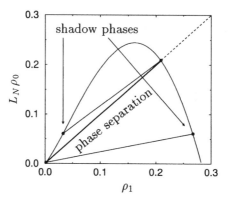

Figure 7. The phase separation region, and compositions of the shadow phases corresponding to the two cloud points, for $\chi = 0.62$ [compare Fig. 6(c)].

The compositions of the phases that just start to appear as the phase separation region is entered do not in general lie on the dilution line $\rho_1 = L_N\rho_0$. These phases are the "shadows"; they lie at the other ends of the cloud point tielines in Fig. 6(a)–(c). This is illustrated for $\chi = 0.62$ in Fig. 7. The shadow phase compositions may be projected onto the $(\rho_1 = L_N\rho_0, \chi)$ plane to give shadow curves, but the projection is not unique. The shadow curve obtained by ignoring the value of ρ_0 and projecting onto the ρ_1-axis, for example, is shown in Fig. 8(a). That obtained by ignoring the value of ρ_1 and projecting onto the ρ_0-axis is shown in Fig. 8(b). The different shapes are due to the fact that the locus of shadow phase compositions in general only intersects the $(\rho_1 = L_N\rho_0, \chi)$ plane at a single point—the critical point.

In conclusion, we emphasize again that all the curves shown in Fig. 6(d) and Fig. 8 are exact, even though they have been constructed from the phase behavior shown in Fig. 6(a)–(c), which is itself a projection of the true phase behavior of the fully polydisperse system.

B. Copolymer with Chemical Polydispersity

Next, we apply the moment free energy formalism to random AB copolymers in the melt or in solution [2–4]. For simplicity, we assume that the copolymer chain *lengths* are monodisperse, so that each chain contains an identical number L of monomers. Even with this simplification, the problem is quite challenging because, as we shall see, a given parent can split into an arbitrarily large number of phases as the interaction strength is increased.

For this problem we choose as our polydisperse feature, σ, the difference between the fractions of A and B monomers on a chain. This takes values in

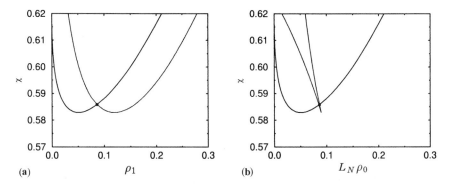

Figure 8. Phase behavior in $(\rho_1 = L_N\rho_0, \chi)$ plane showing (a) ρ_1-projected shadow curve and (b) ρ_0-projected shadow curve.

the range $\sigma = -1\ldots 1$, with the extreme values ± 1 corresponding to pure A and B chains, respectively. We now measure volumes in units of the chain volume (which is L times larger than the monomeric volume used in the previous example). The total density $\rho_0 = \int d\sigma\, \rho(\sigma)$ is then just the copolymer volume fraction ϕ. (Note that for the case of length polydispersity, ϕ was related to the *first* moment of the *length* distribution; here it is the *zeroth* moment of the distribution of *chain compositions* σ.) Within Flory–Huggins theory, the free energy density (in units of $k_B T$ per chain volume) is

$$f = \int d\sigma \rho(\sigma)[\ln \rho(\sigma) - 1] + \frac{L}{L_s}(1 - \rho_0)\ln(1 - \rho_0)$$
$$+ L\tilde{\chi}\rho_0(1 - \rho_0) - L\tilde{\chi}'\rho_1^2. \tag{85}$$

The interaction between A and B monomers enters through $\rho_1 = \int d\sigma \sigma \rho(\sigma)$. In Eq. (85), we have specialized to the case where the interactions of the A and B monomers with each other and with the solvent are symmetric in A and B; otherwise, there would be an additional term proportional to $\rho_0\rho_1$. We have also allowed the solvent to be a polymer of length L_s; the case of a monomeric solvent is recovered for $L_s = 1$. In the following, we absorb factors of L into the interaction parameters by setting

$$\chi = L\tilde{\chi}, \qquad \chi' = L\tilde{\chi}'$$

Defining also $r = L/L_s$, the excess free energy takes the form

$$\tilde{f} = r(1 - \rho_0)\ln(1 - \rho_0) + \chi\rho_0(1 - \rho_0) - \chi'\rho_1^2. \tag{86}$$

The moment free energy (37), with only ρ_0 and ρ_1 retained, is then

$$f_{\mathrm{m}} = \lambda_0 \rho_0 + \lambda_1 \rho_1 - \rho_0 + r(1 - \rho_0) \ln(1 - \rho_0)$$
$$+ \chi \rho_0 (1 - \rho_0) - \chi' \rho_1^2. \tag{87}$$

If extra moment densities with weight functions $w_i(\sigma)$ are included, these simply add a term $\lambda_i \rho_i$ each.

1. Copolymer without Solvent

We first consider the special case of a copolymer melt, for which the overall density (i.e., copolymer volume fraction) is constrained to take everywhere its maximum value $\rho_0 = 1$. The free energy (87) then simplifies to

$$f_{\mathrm{m}} = \lambda_0 + \lambda_1 \rho_1 - \chi' \rho_1^2 \tag{88}$$

up to irrelevant constants [53]. Note that the same expression for the excess free energy of the *dense* system is obtained for a model in which the interaction energy between two species σ, σ' varies as $(\sigma - \sigma')^2$:

$$\tilde{f} = \frac{1}{2} \chi' \int d\sigma\, d\sigma'\, \rho(\sigma)\rho(\sigma')(\sigma - \sigma')^2 = -\chi' \rho_1^2 + \chi' \int d\sigma\, \rho(\sigma)\sigma^2. \tag{89}$$

The second term on the right-hand side is linear in $\rho(\sigma)$ and can be discarded for the purposes of phase equilibrium calculations. The model (88) can therefore also be viewed as a simplified model of chemical fractionation, with σ being related to aromaticity, for example, or another smoothly varying property of the different polymers. Equation (89) suggests that the model should show fractionation into an ever-increasing number of phases as χ' is increased: only the entropy of mixing prevents each value of σ from forming a phase by itself, which would minimize this form of $\tilde{f}$. It is therefore an interesting test case for the moment free energy approach (and the method of adding further moment densities), yet simple enough for exact phase equilibrium calculations to remain feasible [2–4], allowing detailed comparisons to be made.

As a concrete example, we now consider phase separation from parent distributions of the form $\rho^{(0)}(\sigma) \propto \exp(\gamma \sigma)$ (for $-1 \leq \sigma \leq 1$, otherwise zero). The shape parameter γ is then a fixed function of the parental first moment density $\rho_1^{(0)} = \int d\sigma\, \sigma \rho^{(0)}(\sigma)$. Figure 9 shows the exact coexistence curve for $\rho_1^{(0)} = 0.2$, along with the predictions from our moment free energy with n moment densities ($\rho_{i} = \int d\sigma\, \sigma^i \rho(\sigma), i = 1 \ldots n$) retained. Comparable results are found for other $\rho_1^{(0)}$. Even for the minimal set of moment densities ($n = 1$)

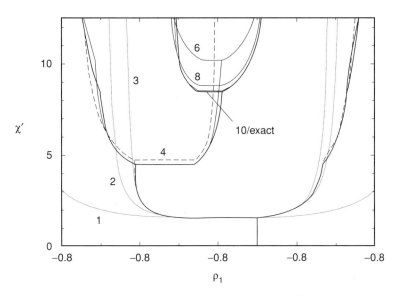

Figure 9. Coexistence curves for a parent distribution with $\rho_1^{(0)} = 0.2$. Shown are the values of ρ_1 of the coexisting phases; horizontal lines guide the eye where new phases appear. Curves are labeled by n, the number of moment densities retained in the moment free energy. Predictions for $n = 10$ are indistinguishable from an exact calculation (in bold).

the point where phase separation first occurs on increasing χ' is predicted correctly (this is the cloud point for the given parent). As more moments are added, the calculated coexistence curves approach the exact one to higher and higher precision. As expected, the precision decreases at high χ', where fractionated phases proliferate; in this region, the number of coexisting phases predicted by the moment method increases with n. However, it is not always equal to $n + 1$, as one might expect from a naive use of Gibbs' phase rule; three-phase coexistence, for example, is first predicted for $n = 4$. In fact, for fractionation problems such as this (but not more generally), study of the low temperature limit (large χ') suggests that to obtain $n + 1$ phases one may have to include up to $2n$ moments. We show below that using localized weight functions (rather than powers of σ) for the extra moments can reduce this number, but also gives less accurate coexistence curves.

In Fig. 10, we show how, for a given value of χ', the lever rule is satisfied more and more accurately as the number of moment densities n in the moment free energy is increased: The total density distribution $\rho_{\text{tot}}(\sigma) = \sum_\alpha v^{(\alpha)} \rho^{(\alpha)}(\sigma)$ approaches the parent $\rho^{(0)}(\sigma)$. This is by design, because the moment method forces the moment densities $\rho_i = \int d\sigma\, w_i(\sigma) \rho_{\text{tot}}(\sigma)$ of

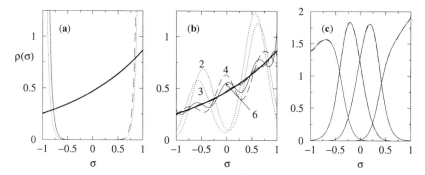

Figure 10. The lever rule in phase coexistences calculated from the moment free energy. We consider the same scenario as in Fig. 9 and take a horizontal cut through that figure, varying n, the number of moment densities retained in the moment free energy, at fixed $\chi' = 10$. (a) Solid and dashed lines: The two daughter phases calculated for $n = 1$ (both are exponential, but with different signs for the exponent). Dotted line: The corresponding total density distribution $\rho_{\text{tot}}(\sigma) = \sum_{\alpha=1,2} v^{(\alpha)} \rho^{(\alpha)}(\sigma)$. This deviates (rather drastically, in this case) from the parent $\rho^{(0)}(\sigma)$ (in bold); the moment calculation with $n = 1$ enforces only that the means ρ_1 of the two distributions be equal. (Their normalizations ρ_0 are trivially equal, because we are considering the dense limit $\rho_0 = 1$.) (b) Total density distributions $\rho_{\text{tot}}(\sigma)$ for increasing values of n. Curves are labeled by the value of n; the curves for $n = 8$ and $n = 10$ (not labeled) are almost indistinguishable from the parent (in bold). As n increases, $\rho_{\text{tot}}(\sigma)$ converges to the parent, being forced to agree with it in an increasing number of its moments. (c) Density distributions of the four daughter phases at $\chi' = 10$. For each, both the exact result and that of the moment free energy calculation with $n = 10$ are shown; they cannot be distinguished on the scale of the plot.

$\rho_{\text{tot}}(\sigma)$ to be equal to those of the parent, for $i = 1 \ldots n$; as n is increased, therefore, $\rho_{\text{tot}}(\sigma)$ approaches $\rho^{(0)}(\sigma)$.

The above results for the dense copolymer model raise two general questions. First, how large does n (the number of moment densities retained) have to be for the moment method to give reliable results for the coexisting phases (and, in particular, the correct number of phases)? And second, how should the weight functions for the extra moment densities be chosen? Regarding the first question, note first that the theoretical framework developed so far only says that the exact results (and therefore the correct number of phases) will be approached as n gets large; it does not say for which value of n a reasonable approximation is obtained. In fact it is clear that no "universal" (problem-independent) value of n can exist. The model studied in this section already provides a counterexample: As χ' is increased, the (exact) number p of coexisting phases increases without limit; therefore the minimum value of n ($n = p - 1$) required to obtain this correct number of phases from a moment free energy calculation can also be arbitrarily large.

Nevertheless, we see already from Fig. 9 that, for the current example, the robustness of the results to the addition of extra moment densities provides a reasonable qualitative check of the accuracy of the coexistence curves. To study the convergence of the moment free energy results to the exact ones more quantitatively, especially in cases where the latter are not available for comparison, we need a measure of the deviation between $\rho_{\text{tot}}(\sigma)$ and $\rho^{(0)}(\sigma)$. To obtain a dimensionless quantity that does not scale with the overall density of the parent, we consider the normalized distributions $n_{\text{tot}}(\sigma) = \rho_{\text{tot}}(\sigma)/\rho_{\text{tot}}$ and $n^{(0)}(\sigma) = \rho^{(0)}(\sigma)/\rho^{(0)}$ and define as our measure of deviation the "log-error"

$$\delta = \int d\sigma\, n^{(0)}(\sigma) \left(\ln \frac{n_{\text{tot}}(\sigma)}{n^{(0)}(\sigma)} \right)^2 . \tag{90}$$

This becomes zero only when $n_{\text{tot}}(\sigma) = n^{(0)}(\sigma)$; otherwise it is positive. When the deviations between the two distributions are small, the logarithm becomes to leading order the relative deviation $n_{\text{tot}}(\sigma)/n^{(0)}(\sigma) - 1$ and we can identify $\delta^{1/2}$ as the root-mean-square relative deviation. We work with the logarithm rather than directly with the relative deviation because the former gives more sensible behavior for larger deviations; in particular, isolated points where $n_{\text{tot}}(\sigma)$ is nonzero while $n^{(0)}(\sigma)$ is close to zero do not lead to divergences of δ [54].

Figure 11 shows the n-dependence of the log-error δ for three fixed values of χ', with the same parent as in Fig 9. [55]. The correct number of phases (3, 4, 5, respectively) is reached at values of $\delta \approx 4 \times 10^{-3}$, 5×10^{-4}, and

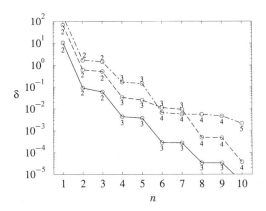

Figure 11. Long-error δ as a function of n for three fixed values of $\chi' = 5, 10, 15$ (bottom to top), for the same scenario as in Fig. 9. The number of phases is indicated next to the curves. Note that in all three cases the correct number of phases is reached for δ of order 10^{-3}.

2×10^{-3} for $\chi = 5, 10,$ and $15,$ respectively. This suggests that the log-error δ might generally be a useful heuristic criterion for guiding the choice of n, with values of δ around 10^{-4} [corresponding to an average deviation between $n_{\text{tot}}(\sigma)$ and $n^{(0)}(\sigma)$ of around 1%] ensuring that the correct number of phases has been detected. Note that although for the current example we did know the correct number of phases (because the simplicity of the problem made an exact calculation feasible), the definition of δ does *not* require us to know in advance any properties of the exact solution of the coexistence problem. Also, note that in real physical materials it is extremely unlikely for the parent or "feedstock" distribution $\rho^{(0)}(\sigma)$ to be known to better than 1% accuracy. Hence even for systems, such as polymers, where a mean-field (and thus truncatable) free energy model is thought to be highly accurate, deviations from the lever rule, at about the 1% level, will almost never be the main source of uncertainty in phase diagram prediction.

We now turn to the second general question, regarding the choice of weight functions for the extra moment densities. Comparing Eq. (60) with the formally exact solution (59) of the coexistence problem tells us at least in principle what is required: The log ratio $\ln \tilde{R}(\sigma)/\rho^{(0)}(\sigma)$ of the "effective prior" and the parent needs to be well approximated by a linear combination of the weight functions of the extra moment densities. However, the effective prior is unknown (otherwise we would already have the exact solution of the phase coexistence problem), and so this criterion is of little use [56].

To make progress, let us try to gain some intuition from the example at hand (dense random copolymers). So far, we have simply taken increasing powers of σ for the extra weight functions. On the other hand, looking at Fig. 10 one might suspect that *localized* weight functions might be more suitable for capturing the fractionating of the parent into the various daughter phases. We therefore tried the simple "triangle" and "bell" weight functions shown in Fig. 12(a). For a given n, there are $n - 1$ extra moment densities (the first moment density, with weight function $w_1(\sigma) = \sigma$, is prescribed by the excess free energy) and we take these to be evenly spaced across the range $\sigma = -1 \ldots 1$, with overlaps as indicated in Fig 12(a). [For the case of triangular weight functions, the log ratio $\ln \rho(\sigma)/\rho^{(0)}(\sigma) = \Sigma_i \lambda_i w_i(\sigma)$ between the density distribution of any phase and that of the parent is thus approximated by a "trapezoidal" function across n equal intervals spanning $-1 \ldots 1$.] Figure 12(b) shows the corresponding coexistence curves for the triangle case with $n = 1, 2, 3, 4$. Note that multiphase coexistence is generally detected for smaller n than for the case of power-law weight functions (Fig. 9), in agreement with the expectations outlined above. However, the predicted number of phases does not vary monotonically with n: In the range of χ' shown in Fig. 12(b), up to four phases are predicted for $n = 3$, whereas $n = 4$ gives no more than three. This effect illustrates a significant difference

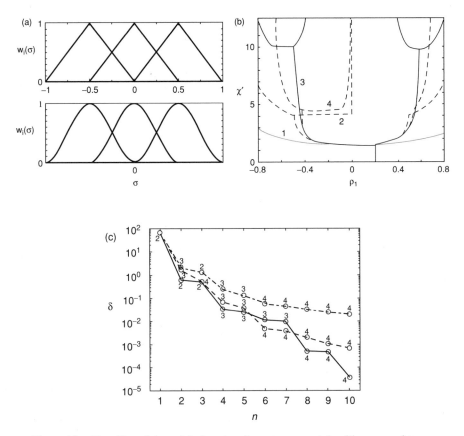

Figure 12.   The effect of the weight functions for extra moment densities on coexistence calculations. (a) The "triangle" and "bell" weight functions, for the case $n - 1 = 3$. (b) Coexistence curves for the triangle weight functions, for the same parent as in Fig. 9. Note that multiphase coexistence is generally detected for smaller n than in Fig. 9, where power-law weight functions were used for the extra moments. On the other hand, the predicted number of phases now no longer varies monotonically with n. (c) Dependence of log-error δ on n for fixed $\chi' = 10$ and three choices of weight functions for the extra moment densities: Power law (solid line), triangle (dashed), bell (dot–dashed). The number of phases is marked next to the curves.

between power-law and triangle weight functions. The former, but not the latter, make up an *incremental set*: Increasing n by one corresponds to adding a new weight function while leaving the existing set of weight functions unchanged. The family (39) of accessible density distributions $\rho(\sigma)$ is thus enlarged without losing any of its former members. Intuitively, one then expects that the number of phases cannot decrease as n is increased. But for nonincremental sets, such as triangle weight functions (and bells, which give qualitatively similar results to triangles), each increase in n creates a larger family but completely discards the previous one, so that there is then no reason for the results to be monotonic in n.

We regard the fact that the number of phases can go up and down with increasing n as a disadvantage of nonincremental weight functions. In the current example, we also find that they generally give higher values of the log-error than power-law weight functions; see Fig. 12(c). Finally, incremental sets of weight functions also have the benefit that the results for n can be used to initialize the calculation for $n + 1$, thus helping to speed up the numerics (though we have not yet exploited this property). On the basis of these observations, we would generally recommend the use of incremental sets of weight functions for extra moment densities. Beyond this, the optimal choice of extra weight functions remains largely an open problem. One avenue worth exploring in future work might be the possibility of choosing extra weight functions *adaptively*. One could imagine monitoring the log-error in a phase coexistence calculation, and then increasing the number of extra moments by one every time it becomes larger than a certain threshold (10^{-4}, say). The new weight function could then be chosen as a best fit to the current lever-rule violation $\ln \rho_{\text{tot}}(\sigma)/\rho^{(0)}(\sigma)$ (for example, as a linear combination of weight functions from some predefined "pool"). We are planning to investigate this method in future work.

To conclude this section on the dense random copolymer model, we briefly discuss the spinodal criterion and ask whether critical or multicritical points can exist for a general parent distribution $\rho^{(0)}(\sigma)$ (with $\rho_0^{(0)} = 1$). The criterion (53), applied to our one-moment free energy $f_{\text{m}}(\rho_1)$, becomes simply

$$f_{\text{m}}(\rho_1) = \mathcal{O}\left((\rho_1 - \rho_1^{(0)})^{l+1}\right), \tag{91}$$

where $l = 2$ for a spinodal and $l = 2n - 1$ for an n-critical point. From Eq. (88), we have that

$$f_{\text{m}} = \lambda_0 + \lambda_1 \rho_1 - \chi' \rho_1^2 = -s_{\text{m}} - \chi' \rho_1^2. \tag{92}$$

The excess part $-\chi' \rho_1^2$ is quadratic in ρ_1 and therefore only enters the spinodal criterion; the additional conditions for critical points only depend on the moment entropy $s_{\text{m}} = -\lambda_0 - \lambda_1 \rho_1$. Because of the dense limit constraint $\rho_1 = 1$, λ_0 is related to λ_1 by

$$-\lambda_0 = \ln \int d\sigma\, \rho^{(0)}(\sigma) e^{\lambda_1 \sigma} \equiv h(\lambda_1).$$

As expected from the general discussion in Section II.C, $-\lambda_0$ is thus the cumulant generating function $h(\lambda_1)$ of $\rho^{(0)}(\sigma)(\equiv n^{(0)}(\sigma))$ in the dense limit we are considering), introduced in Eq. (34) within the combinatorial approach. The moment entropy $s_{\mathrm{m}} = h - \lambda_1\rho_1$ is its Legendre transform; its behavior for $\rho_1 \approx \rho_1^{(0)}$ reflects that of h for small λ_1. We can thus use the expansion $h = \sum_{j=1}^{\infty}(c_j/j!)\lambda_1^j$, where the c_j are the cumulants of $\rho^{(0)}(\sigma)$; $c_1 \equiv \rho_1^{(0)}$ is its mean, c_2 is its variance, and so on. Using the properties of the Legendre transform, it then follows that the leading term of the moment entropy is $s_{\mathrm{m}} = (\rho_1 - \rho_1^{(0)})^2/(2c_2)$. Inserting this into Eq. (92), we see immediately that the spinodal condition becomes

$$2\chi'c_2 = 2\chi'\left(\rho_2^{(0)} - (\rho_1^{(0)})^2\right) = 1. \tag{93}$$

To find the additional conditions for critical points, consider what happens if the cumulants $c_3, \ldots, c_l$ are zero, hence $h = \rho_1^{(0)}\lambda_1 + c_2\lambda_1^2/2 + \mathcal{O}(\lambda_1^{l+1})$. It can then easily be shown that the moment entropy behaves as $s_{\mathrm{m}} = (\rho_1 - \rho_1^{(0)})^2/(2c_2) + \mathcal{O}((\rho_1 - \rho_1^{(0)})^{l+1})$; conversely, one can show that this behavior of s_{m} implies that the cumulants $c_3, \ldots, c_l$ are zero. We thus have the simple condition that for an n-critical point to be observed, the cumulants $c_3, \ldots, c_{2n-1}$ of the parent distribution have to be zero; the value of χ' at this critical point is given by the spinodal condition (93). (For an ordinary critical point, $n = 2$, the condition $c_3 = 0$ is well known in this context [2].) We therefore come to the somewhat surprising conclusion that even in a very simple polydisperse system such as this—with a single moment density occurring in the excess free energy—multicritical points of arbitrary order can occur, at least in principle. In practice, the required fine-tuning of the parent will of course make experimental study of these points difficult.

2. Copolymer with Solvent

We now consider the random copolymer model in the presence of solvent— that is, for a copolymer volume fraction $\rho_0 < 1$. We are not aware of previous work on this model in the literature, but will briefly discuss below the link to models of homopolymer/copolymer mixtures [57]. The excess free energy (86) then depends on two moment densities, rather than just one as in all previous examples. For simplicity, we restrict ourselves to the case of a neutral solvent that does not in itself induce phase separation; this corresponds to $\chi = 0$, making the excess free energy

$$\tilde{f} = r(1 - \rho_0)\ln(1 - \rho_0) - \chi'\rho_1^2. \tag{94}$$

This is still sufficiently simple that the exact spinodal condition (55)—for a system with a general density distribution $\rho(\sigma)$—can be worked out analytically; one finds

$$2\chi' = \frac{r\rho_0 + 1 - \rho_0}{r(\rho_2\rho_0 - \rho_1^2) + \rho_2(1 - \rho_0)}. \tag{95}$$

In the dense limit $\rho_0 \to 1$, the earlier result (93) is recovered. To study critical and multicritical points, we use the moment free energy corresponding to (94),

$$f_m = \lambda_0\rho_0 + \lambda_1\rho_1 - \rho_0 + r(1 - \rho_0)\ln(1 - \rho_0) - \chi'\rho_1^2, \tag{96}$$

and start from the criterion (53). The moment chemical potentials are given by

$$\begin{aligned} \mu_0 &= \lambda_0 - r[\ln(1 - \rho_0) + 1] \\ \mu_1 &= \lambda_1 - 2\chi'\rho_1, \end{aligned} \tag{97}$$

and the curve $\rho(\epsilon)$ referred to in Eq. (53)—along which the critical phases merge—is simply parameterized in terms of $\lambda_0(\epsilon)$ and $\lambda_1(\epsilon)$. The restriction that the curve must pass through the parent at $\epsilon = 0$ means that $\lambda_0(0) = \lambda_1(0) = 0$, so to get the small ϵ behavior of the chemical potentials (97) we can expand for small λ_0 and λ_1. To simplify the algebra, we restrict ourselves to parent distributions $\rho^{(0)}(\sigma)$ that are symmetric about $\sigma = 0$. Using this symmetry and the fact that ϵ is simply a dummy parameter that can be reparameterized arbitrarily, we can then set $\lambda_1 = \epsilon$ and $\lambda_0 = a_2\epsilon^2 + a_4\epsilon^4 + \mathcal{O}(\epsilon^6)$.

To have a spinodal, the moment chemical potentials (97) must differ from those of the parent by no more than $\mathcal{O}(\epsilon^2)$. We find that this gives the condition $2\chi' = 1/\rho_2^{(0)}$, in agreement with (95): Symmetric parents have $\rho_1^{(0)} = 0$. For a critical point, the chemical potential difference must be decreased to $\mathcal{O}(\epsilon^3)$. This gives no new condition (but determines a_2), implying that parents with symmetric distributions are automatically critical at their spinodal. This result has its origin in the symmetry $\sigma \to -\sigma$ of the excess free energy, and can be understood by considering the ρ_0, ρ_1, χ' phase diagram of the moment free energy (96). In this phase diagram there is a two-dimensional spinodal surface, and within this surface lies a line of critical points. The symmetry $\sigma \to -\sigma$ (corresponding to $\rho_1 \to -\rho_1, \lambda_1 \to -\lambda_1$) forces this critical line to lie within the plane $\rho_1 = 0$ of symmetrical distributions, implying that for such distributions spinodal and critical points coincide.

For a tricritical point, finally, where the chemical potential difference must be further reduced to $\mathcal{O}(\epsilon^3)$, we find after some algebra the condition

$$\rho_4^{(0)} = \frac{3r(\rho_2^{(0)})^2}{r\rho_0^{(0)} + 1 - \rho_0^{(0)}}.$$

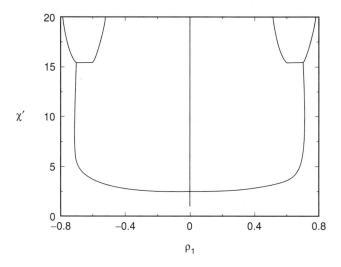

Figure 13. Coexistence curve for copolymer with solvent; the parent has a uniform density distribution $\rho^{(0)}(\sigma) = \mathrm{const.} = \rho_0^{(0)}/2$ with overall density (i.e., copolymer volume fraction) $\rho_0^{(0)} = 0.6$. The results were obtained from the moment free energy with $n = 11$ moment densities retained [weight functions $w_i(\sigma) = \sigma^i$, $i = 0\ldots 10$]. Note the tricritical point at $\chi' = 2.5$: The parent splits continuously into three phases.

As an example, consider a uniform parent distribution $\rho^{(0)}(\sigma) = \mathrm{const}$, which can be written as $\rho^{(0)}(\sigma) = \rho_0^{(0)}/2$, using the fact that $\rho_0^{(0)} = \int_{-1}^{1} d\sigma\, \rho^{(0)}(\sigma)$. Then $\rho_2^{(0)} = \rho_0^{(0)}/3$ and $\rho_4^{(0)} = \rho_0^{(0)}/5$ and so a tricritical point occurs if the overall density (i.e., the copolymer volume fraction) is $\rho_0^{(0)} = 3/(2r+3)$. Figure 13 shows the coexistence curve calculated for this parent (with $r = 1$), which clearly shows the tricritical point at the predicted value $\chi' = 1/(2\rho_2^{(0)}) = r + 3/2 = 2.5$. Our numerical implementation manages to locate the tricritical point and follow the three coexisting phases without problems; we take that as a signature of its robustness [58]. Note that the tricritical point that we found is closely analogous to that studied by Leibler [57] for a symmetric blend of two homopolymers and a symmetric random copolymer that is, nonetheless, chemically monodisperse (in the sense that $\sigma = 0$ for all copolymers present). In fact, in our notation, the scenario of Ref. 57 simply corresponds to a parent density of the form $\rho^{(0)}(\sigma) \sim \delta(\sigma - 1) + \delta(\sigma + 1)$, with the copolymer ($\sigma = 0$) now playing the role of the neutral solvent.

We conclude this section with another illustration of the geometrical intuition that the moment free energy can provide. In Fig. 14, we plot the moment free energy $f_m(\rho_0, \rho_1)$, Eq. (96), for the uniform parent $\rho^{(0)}(\sigma)$ studied above.

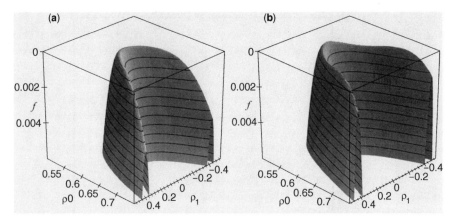

Figure 14. Moment free energy $f_m(\rho_0, \rho_1)$ near the tricritical point, for the same parent as in Fig. 13. For ease of visualization, the free energy surface is shown *upside down*, with f_m increasing in the downward direction. (a) At the tricritical point $\chi' = 2.5$, the parent is stable (note that linear terms have been added to the free energy to make the tangent plane at the parent horizontal). (b) As χ' is increased into the three-phase region (here $\chi' = 2.65$), the parent splits continuously into three phases.

Two values of χ' are shown: the tricritical value ($\chi' = 2.5$) and a value just in the three-phase region ($\chi' = 2.65$); the developing three minima of the free energy can clearly be seen. Even though the underlying polydisperse system has an infinite number of degrees of freedom $\rho(\sigma)$, the moment method thus gives us a tool for understanding and visualizing the occurrence of the tricritical point in terms of a simple free energy surface with two independent variables.

VI. CONCLUSION

Polydisperse systems contain particles with an attribute σ that takes values in a continuous range (such as particle size in colloids or chain length in polymers). They therefore have an infinite number of conserved densities, corresponding to a density *distribution* $\rho(\sigma)$. The fact that the free energy can depend on all details of $\rho(\sigma)$ makes the analysis of the phase behavior of such systems highly nontrivial. However, in many (especially mean-field) models the *excess* free energy only depends on a finite number of moment densities—that is, (generalized) moments of $\rho(\sigma)$; the only dependence on other details of $\rho(\sigma)$ is through the ideal part of the free energy. For such models, which we call truncatable, we showed how a *moment free energy* can be constructed, which only depends on the moment densities appearing in the excess free energy. Two possible approaches for the construction of the moment free energy were described in Section II. The first, which we call the projection method (Section II.A), is primarily geometrical; the second is mainly combinatorial (Section II.B) and starts from first principles to derive an expression for the ideal part of the free energy in terms of moment densities. In Section II.C, we showed that these two approaches give essentially equivalent results.

In Section III, we explored the properties of the moment free energy in detail. The moment free energy depends only on a finite number of moment densities, and it can be used to predict phase behavior using the conventional methods known from the thermodynamics of finite mixtures. We showed that, for any truncatable model, this procedure yields the same spinodals, critical points, and multicritical points as the original free energy, even though the latter depends on all details of the density distribution $\rho(\sigma)$. It also correctly predicts the onset of phase coexistence—that is, the cloud points and corresponding shadows. Beyond the onset of phase coexistence (*i.e.*, when one or more coexisting phases occupy comparable volumes), the results from the moment free energy are approximate, but can be made arbitrarily accurate by retaining additional moment densities in the moment free energy. We also explained how to check the local and global stability of a phase coexistence calculated from the moment free energy, and how the moment free energy reflects overall phase behavior in the space of density distributions $\rho(\sigma)$.

We discussed aspects of the numerical implementation of the moment free energy method in Section IV, highlighting its many advantages in terms of robustness, efficiency, and stability. The method allows violations of the lever rule, but then ensures that these be kept small by adding $n - K$ extra moments (where K is the number of moment densities actually present in the excess free energy of the underlying truncatable model). The solution of the phase equilibrium conditions thus proceeds in an n-dimensional space; but global (tangent plane) stability can always be verified by minimimimization (of the tangent plane distance) in a space of dimension K. For $K > 1$ exact minimization is not practicable, but standard search methods can instead be employed; we chose a Monte Carlo-type algorithm for this.

We then surveyed several sample applications in Section V. For length-polydisperse homopolymers (treated within Flory–Huggins theory), we showed explicitly how the moment free energy with only a single moment density (the volume fraction of polymer) gives the correct spinodals and critical points. By retaining the chain number density as an additional moment density, we could rationalize the behavior of cloud point and shadow curves as cuts through conventional two-species phase diagrams. As a second example, the case of copolymers with chemical polydispersity was considered, both without and with solvent. This gave us the opportunity to test the performance of the moment free energy method for multiphase coexistence, and in particular to study how the accuracy of the predictions in the coexistence region increases as more and more moment densities are retained. In the presence of solvent, the same theory predicts a tricritical point under appropriate conditions, which was detected without problems, along with the associated three-phase region, by our numerical algorithm. A plot of the two-moment free energy gave a simple geometrical picture of the origin of this tricritical point.

The results for the copolymer model showed that the log-error (90) can be used to decide when enough moment densities have been included to locate the correct number of phases and their compositions. For general purposes we suggest a log-error tolerance of order 10^{-4}; this corresponds to limiting the rms lever-rule violations at the predicted phase coexistence (found by global minimization of the moment free energy) to 1% or so of the parent distribution. Given that in practice the parent is itself not known to such high precision, these violations lie well within an acceptable range of error. Indeed, every state of coexistence actually predicted by our algorithm represents the *exact* phase equilibrium of *some or other* parent, indistinguishable from the real one to within experimental accuracy. Here, as elsewhere in this paper, "exact" means exactly as predicted by a truncatable model free energy; such models are, of course, themselves approximate, which is another reason not to expend resources reducing the log-error below the level suggested above.

In summary, with exact results for cloud point and shadow curves, critical points, and spinodals, as well as refinably accurate coexistence curves and multiphase regions, the moment free energy method allows rapid and accurate computation of the phase behavior of many polydisperse systems. Moreover, by establishing the link to a projected free energy $f_m(\rho_i)$ which is a function of a finite set of conserved moment densities ρ_i, it restores to the problem much of the geometrical interpretation and insight (as well as the computational methodology) associated with the phase behavior of finite mixtures. This contrasts with many procedures previously in use for truncatable polydisperse systems [2–5, 9, 10]. Some previous approximations to the problem have used (generalized) moments as thermodynamic coordinates (see, e.g., Refs. 11, 14, 15, and 17). The moment free energy method provides a rational basis for these methods and, so long as the generalized moments are correctly chosen, guarantees that many properties of interest are found exactly. Note also that the commonplace method of "binning" the σ-distribution into discrete "pseudocomponents" can be seen as a special case of the moment method in which each weight function is zero outside the corresponding bin; but since these moments do not, in general, coincide with those contained in the excess free energy, the advantages described above are then lost.

We finish with some remarks about extensions to the moment method, within and outside the sphere of phase diagram prediction. First, methods of this type may extend to nontruncatable models, for which the excess free energy *cannot* be written directly in terms of a finite number of moments as in Eq. (6). For example, many mean-field theories correspond to a variational minimization of the free energy: $F \le \langle E \rangle_0 - TS_0$, where the subscript 0 refers to a trial Hamiltonian [59]. In such a case, one might choose to *first* make a physically motivated decision about which (and how many) moment densities ρ_i to keep, and then include among the variational parameters the "transverse" degrees of freedom of the density distribution $\rho(\sigma)$. We have not pursued this approach, though it may form a promising basis for future developments. Second, we note that physical insight based on moment free energies may increasingly play a part in understanding kinetic problems. For example, one of us [60] has argued that in many systems the zeroth moment (mean particle density) can relax more rapidly (by collective diffusion) than can the higher moments (by interdiffusion of different species), and that this may lead to novel kinetics in certain areas of the phase diagram. And finally, we point to recent work [61] in which the moment method has been extended, via density functional theory, to allow study of inhomogeneous states; this opens the way to studying the effect of polydispersity on interfacial tensions and other interfacial thermodynamic properties.

Acknowledgments

We thank P. Bladon, N. Clarke, R. M. L. Evans, T. McLeish, P. Olmsted, I. Pagonabarraga, W. C. K. Poon, and R. Sear for helpful discussions. PS is grateful to the Royal Society for financial support. This work was also funded in part under EPSRC GR/M29696 and GR/L81185.

APPENDIX A: MOMENT (GIBBS) FREE ENERGY FOR FIXED PRESSURE

In this appendix, we explain how the construction of the moment free energy is extended to scenarios where the pressure (rather than the overall system volume) is held constant. One then describes the system by the particle number distribution $N(\sigma) = V\rho(\sigma)$; this is normalized so that its integral gives the total number of particles, $N = \int d\sigma N(\sigma)$. The relevant thermodynamic potential is the Gibbs free energy $G = \min_V F + \Pi V$, which we now construct. For a truncatable system, whose free energy *density* obeys Eqs. (3) and (4) by definition, the Helmholtz free energy can be written as

$$F = T \int d\sigma N(\sigma) \left[\ln \frac{N(\sigma)}{V} - 1 \right] + \tilde{F},$$

where the excess free energy $\tilde{F}(N_i, V) = V\tilde{f}(\rho_i)$ depends on volume V and on the moments $N_i = V\rho_i = \int d\sigma N(\sigma) w_i(\sigma)$ of the particle number distribution. If $\tilde{F} = 0$, a Legendre transform gives the Gibbs free energy of an ideal mixture:

$$
G_{id} = \min_V F_{id} + \Pi V = T \int d\sigma\, N(\sigma) \ln \frac{\Pi N(\sigma)}{NT}
$$

$$
= NT \int d\sigma\, n(\sigma) \ln n(\sigma) + NT \ln \frac{\Pi}{T}.
$$

Here $n(\sigma) = N(\sigma)/N = \rho(\sigma)/\rho$ is the normalized particle distribution. In the general case, we split off the ideal part

$$
G = G_{id} + \tilde{G} = \min_V F + \Pi V
$$

and obtain for the excess Gibbs free energy

$$
\tilde{G} = \min_V NT \left(\ln \frac{NT}{\Pi V} - 1 \right) + \tilde{F}(N_i, V) + \Pi V.
$$

This is a function of N and the N_i (as well as the fixed control parameters Π, T, which we suppress in our notation). We can therefore write the excess Gibbs free energy per particle, $\tilde{G}(N_i, N)/N = \tilde{g}(m_i)$, as a function of the moments $m_i = N_i/N$ $(i = 1 \ldots K)$ of the normalized particle distribution $n(\sigma)$. Altogether, the Gibbs free energy per particle of a truncatable system becomes

$$
g[n(\sigma)] = G/N = T \int d\sigma\, n(\sigma) \ln n(\sigma) + T \ln \frac{\Pi}{T} + \tilde{g}(m_i). \qquad \text{(A1)}
$$

This has again a truncatable structure; the excess part $\tilde{g}$ "inherits" its moment structure from $\tilde{f}$. Note that because of the normalization of the m_i, $\tilde{g}$ does not depend on the density $\rho \equiv \rho_0$. It is therefore normally a function of one less variable than $\tilde{f}$, unless $\tilde{f}$ is already independent of ρ [as is the case in Flory–Huggins theory, Eq. (64), for example]. The chemical potentials $\mu(\sigma)$ follow from Eq. (A1) as

$$
\mu(\sigma) = \frac{\delta}{\delta N(\sigma)} Ng[N(\sigma)/N] = \frac{\delta g}{\delta n(\sigma)} + g - \int d\sigma'\, n(\sigma') \frac{\delta g}{\delta n(\sigma')}
$$

$$
= T \ln n(\sigma) + \sum_i w_i(\sigma) \frac{\partial \tilde{g}}{\partial m_i} + \left[T \ln \frac{\Pi}{T} + \tilde{g} - \sum_i m_i \frac{\partial \tilde{g}}{\partial m_i} \right]. \qquad \text{(A2)}
$$

Multiplying by $n(\sigma)$ and integrating over σ, one has $g = \int d\sigma \mu(\sigma) n(\sigma)$ as it should be [compare Eq. (A1)]. Comparing Eq. (A2) with Eq. (45), one deduces that

$$\frac{\partial \tilde{g}}{\partial m_i} = \tilde{\mu}_i$$

and that the density ρ of a phase with moments m_i is given by

$$T \ln \left(\frac{T \rho}{\Pi} \right) = \tilde{g} - \sum_i m_i \tilde{\mu}_i.$$

To construct the moment Gibbs free energy, one can now proceed as in the constant volume case: In the ideal contribution, we replace $\ln n(\sigma) \to \ln[n(\sigma)/n^{(0)}(\sigma)]$, with a normalized parent distribution $n^{(0)}(\sigma)$, and then minimize g at fixed values of the m_i. The minimum occurs for distributions $n(\sigma)$ from the family

$$n(\sigma) = n^{(0)}(\sigma) \exp \left(\lambda_0 + \sum_i \lambda_i w_i(\sigma) \right). \tag{A3}$$

The extra Lagrange multiplier λ_0 comes from the normalization condition $m_0 = \int d\sigma\, n(\sigma) = 1$. Inserting into Eq. (A1), one obtains the moment Gibbs free energy

$$g_m(m_i) = T \left(\lambda_0 + \sum_i \lambda_i m_i \right) + T \ln \frac{\Pi}{T} + \tilde{g}(m_i), \tag{A4}$$

where the m_i are given by

$$m_i = \int d\sigma\, w_i(\sigma) n^{(0)}(\sigma) \exp \left(\lambda_0 + \sum_i \lambda_i w_i(\sigma) \right)$$

$$= \frac{\int d\sigma\, w_i(\sigma) n^{(0)}(\sigma) \exp \left(\sum_i \lambda_i w_i(\sigma) \right)}{\int d\sigma\, n^{(0)}(\sigma) \exp \left(\sum_i \lambda_i w_i(\sigma) \right)}.$$

The derivatives of g_m can again be identified as moment chemical potentials and obey

$$\mu_i \equiv \frac{\partial g_m}{\partial m_i} = T \lambda_i + \frac{\partial \tilde{g}}{\partial m_i} = T \lambda_i + \tilde{\mu}_i. \tag{A5}$$

The constraint $m_0 = 1$ does not affect this result because the partial derivative is taken with the values of all moments other than m_i fixed anyway. For the same reason, the second derivatives of the ideal part of g_m are (up to a factor T) given by the inverse of the matrix of second-order normalized moments m_{ij}, by analogy with (43). (Note that the first row and column of the second-order moment matrix, corresponding to $i = 0$ or $j = 0$, need to be retained; they can only be discarded *after* the inverse has been taken.)

From Eq. (A2), one deduces immediately that if one of a number of coexisting phases is in the family (A3), then so are all others. Equality of the exact chemical potentials $\mu(\sigma)$ in all phases is then equivalent to equality of the quantities

$$\tilde{\mu}_i + T\lambda_i = \mu_i \quad (i = 1 \dots K), \qquad T\lambda_0 + \tilde{g} - \sum_i m_i \tilde{\mu}_i. \qquad (A6)$$

But from Eqs. (A4) and (A5), these are simply the moment chemical potentials $\mu_i = \partial g_m / \partial m_i$ and (up to an unimportant additive constant) the associated Legendre transform $g_m - \sum_i m_i \mu_i$. Therefore, phase equilibria can be constructed by applying the usual tangent plane construction to the moment Gibbs free energy, in analogy with the constant volume case. Note, however, that this tangent plane construction now takes place in the space of the normalized moments m_i, rather than that of the unnormalized moment densities ρ_i. As pointed out above, this generically reduces the dimension by one. For the trivial case of a monodisperse system, $\tilde{f}$ is a function of ρ only and $\tilde{g}$ does not depend on any density variables at all, so that the tangent plane condition becomes void as expected.

The exactness statements in Section III can also be directly translated to the constant pressure case. The arguments above imply directly that the onset of phase coexistence is found exactly from the moment Gibbs free energy: All phases are in the family (A3), because one of them (the parent) is, and the requirement of equal chemical potentials is satisfied. Spinodals and (multi-) critical points are also found exactly. Arguing as in Section III. A and using the vector notation of Eq. (53), the criterion for such points is found as

$$\Delta\boldsymbol{\mu} \equiv \boldsymbol{\mu}(\boldsymbol{n}(\epsilon)) - \boldsymbol{\mu}(\boldsymbol{n}^{(0)}) = \mathcal{O}(\epsilon^l), \qquad (A7)$$

where $l = 2$ for a spinodal and $l = 2n - 1$ for an n-critical point. Again, the curve $\boldsymbol{n}(\epsilon)$ only has to pass through phases with equal chemical potentials and can therefore be chosen to lie within the family (A3). As discussed above [Eq. (A6)], the chemical potential difference on the left-hand side of Eq. (A7) is then zero to the required order in ϵ if the same holds for the $\Delta\mu_i$ and Δh, where $h = g_m - \sum_i m_i \mu_i$. It may look as if the constraint on h gives one more condition here than in the constant volume case; but in fact $\Delta \mu_i = \mathcal{O}(\epsilon^l)$ for all i already implies that $\Delta g_m = \mathcal{O}(\epsilon^{l+1})$ and hence $\Delta h = \mathcal{O}(\epsilon^l)$. The exact condition (A7) for spinodals and critical points is therefore again equivalent to the same condition obtained from the moment Gibbs free energy.

APPENDIX B: MOMENT ENTROPY OF MIXING AND LARGE DEVIATION THEORY

In this appendix, we discuss some interesting properties of the Legendre transform result (29) for the moment entropy of mixing, in particular its relation to large deviation theory (LDT).

We first note that Eq. (29), which we derived by taking the limit $x \to 0$ of the result (26) for general x, can also be obtained by a more direct route. In the limit $x \to 0$, the sizes of particles in the smaller system become independent random variables drawn from $n^{(0)}(\sigma)$; the second phase can be viewed as a *reservoir* to which the small phase is connected. One writes the moment generating function for $\mathcal{P}(m)$ in the small phase as a product of xN independent moment generating functions of $n^{(0)}(\sigma)$ and then evaluates the integral over $\mathcal{P}(m)$ by a saddle point method [36].

One can also view (29) as a generalization of Cramér's LDT [62]. Recall that this theory treats the "wings" of a distribution like $\mathcal{P}(m)$ correctly, to which the central limit theorem (CLT) does not apply [63]. However, the CLT approximation in this problem has quite a beautiful interpretation and is worth describing separately. It can be derived by assuming that $n^{(0)}(\sigma)$ is itself a Gaussian in Eq. (29), or just written down directly:

$$\mathcal{P}(m) \sim \exp\left[-N \frac{(m - m^{(0)})^2}{2\, v^{(0)}}\right],$$

where $v^{(0)}$ is the variance of the parent distribution. We see that this gives a term in the free energy of the form

$$\Delta f = -T \rho s = \rho T \frac{(m - m^{(0)})^2}{2 v^{(0)}}.$$

This result states that there is an *entropic spring* term in the free energy that penalizes deviations of the mean size m from the mean size in the parent distribution. The spring constant is inversely proportional to the variance in the parent distribution; thus if the parent is narrower, it is harder to move away from the parental mean size [64]. However, as indicated above, the CLT is not sufficiently accurate for our purposes. For instance it gives a finite weight to $m < 0$ even if $w(\sigma)$ is strictly positive. That said, the CLT may be attractive for narrow size distributions, and one might make a connection with the recent results of Evans et al. [65] (even though the latter do not seem to rely on a specific shape of the distribution). We have not fully explored this avenue.

APPENDIX C: SPINODAL CRITERION FROM EXACT FREE ENERGY

In this appendix, we apply the spinodal criterion (50) to the exact free energy (38) and show that it can be expressed in a form identical to Eq. (55). This result has been given by a number of authors [11, 12, 44], but we include it here for the sake of completeness.

Choosing for ρ the vector $\rho(\sigma)$ [whose "components" are the values of $\rho(\sigma)$ for all σ], the spinodal criterion (50) applied to Eq. (38) becomes

$$\int d\sigma' \frac{\delta^2 f}{\delta\rho(\sigma)\delta\rho(\sigma')} \delta\rho(\sigma') = \sum_{i,j} \frac{\partial^2 \tilde{f}}{\partial\rho_i\partial\rho_j} w_i(\sigma)\delta\rho_j + T\frac{\delta\rho(\sigma)}{\rho(\sigma)} = 0. \quad \text{(C1)}$$

The change of $\ln\rho(\sigma)$ along the instability direction [which is $\delta\rho(\sigma)/\rho(\sigma)$] is therefore a linear combination of the weight functions $w_i(\sigma)$. This means that the instability direction is within the family (39), consistent with our general discussion in Section III.A. One can now rewrite (C1) in the form

$$\delta\rho(\sigma) = -\beta\sum_{i,j} w_i(\sigma)\,\rho(\sigma)\frac{\partial^2 \tilde{f}}{\partial\rho_i\partial\rho_j}\,\delta\rho_j \quad \text{(C2)}$$

and take the moment with the kth weight function $w_k(\sigma)$ to get

$$\delta\rho_k + \beta\sum_{i,j} \rho_{ki}\frac{\partial^2 \tilde{f}}{\partial\rho_i\partial\rho_j}\delta\rho_j = 0.$$

As promised, this is identical to the spinodal condition (55) derived from the moment free energy, with the matrix multiplications written out explicitly.

APPENDIX D: DETERMINANT FORM OF CRITICAL POINT CRITERION

In this appendix, we give the form of the critical point criterion (57) that uses the spinodal determinant Y from Eq. (55) [34]. At a critical point, the instability direction connects two neighboring points on the spinodal. The first-order variation of the spinodal determinant Y along this direction must therefore vanish. This gives

$$\delta Y = \sum_i \frac{\partial Y}{\partial \rho_i} \delta \rho_i + \sum_{i \leq j} \frac{\partial Y}{\partial \rho_{ij}} \delta \rho_{ij} = 0. \tag{D1}$$

Here the $\delta \rho_{ij}$ are the changes in the second-order moment densities along the instability direction. We can express these as a function of the $\delta \rho_i$ via the change in the Lagrange multipliers λ_i: From the definition (40), a change in the λ_i changes the first- and second-order moment densities according to

$$\frac{\partial \rho_i}{\partial \lambda_j} = \rho_{ij} = M_{ij}, \qquad \frac{\partial \rho_{ij}}{\partial \lambda_k} = \rho_{ijk}. \tag{D2}$$

Hence changes in second- and first-order moment densities are related by

$$\delta \rho_{kl} = \sum_i \rho_{kli} \delta \lambda_i = \sum_{ij} \rho_{kli} (M^{-1})_{ij} \delta \rho_j.$$

Inserting this into (D1) gives the determinant form of the critical point condition

$$\sum_i \frac{\partial Y}{\partial \rho_i} \delta \rho_i + \sum_{ij} \sum_{k \leq l} \frac{\partial Y}{\partial \rho_{kl}} \rho_{kli} (M^{-1})_{ij} \delta \rho_j = 0. \tag{D3}$$

Of course, if in an application of the moment free energy method one succeeds in calculating Y explicitly as a function of the ρ_i alone (rather than as a function of the ρ_i and the ρ_{ij}), then one can determine critical points simply from the condition

$$\delta Y = \sum_i \frac{dY}{d\rho_i} \delta \rho_i = 0.$$

We have written a total derivative sign here to indicate that the dependence of Y on the ρ_{ij} is accounted for implicitly.

References

1. R. T. DeHoff, *Thermodynamics in Material Science*, McGraw-Hill, New York, 1992.
2. B. J. Bauer, *Polymer Eng. Sci.* **25**, 1081 (1985).
3. M. T. Rätzsch and C. Wohlfarth, *Adv. Polymer Sci.* **98**, 49 (1991).

4. A. Nesarikar, M. Olvera de la Cruz, and B. Crist, *J. Chem. Phys.* **98**, 7385 (1993).

5. K. Solc, *Makromol. Chem.-Macromol. Symp.* **70-1**, 93 (1993).

6. Some components of σ can also be discrete. For example, in a blend of two homopolymers A and B, both of which are length-polydisperse, one would choose $\sigma = (L, a)$, where L is the chain length and $a = 1, 2$ labels polymer A and B, respectively. All integrations $\int d\sigma$ are then to be read as the combined sum and integral $\int dL \sum_{a=1,2}$.

7. P. A. Irvine and J. W. Kennedy, *Macromolecules* **15**, 473 (1982).

8. J. J. Salacuse and G. Stell, *J. Chem. Phys.* **77**, 3714 (1982).

9. K. Solc and R. Koningsveld, *Coll. Czech. Chem. Comm.* **60**, 1689 (1995).

10. J. A. Gualtieri, J. M. Kincaid, and G. Morrison, *J. Chem. Phys.* **77**, 521 (1982).

11. P. Irvine and M. Gordon, *Proc. R. Soc. London A* **375**, 397 (1981).

12. S. Beerbaum, J. Bergmann, H. Kehlen, and M. T. Rätzsch, *Proc. R. Soc. London A* **406**, 63 (1986).

13. E. M. Hendriks, *Ind. Eng. Chem. Res.* **27**, 1728 (1988). Results in this paper formalize and summarize many earlier studies (and were rediscovered in several later ones).

14. E. M. Hendriks and A. R. D. Vanbergen, *Fluid Phase Eq.* **74**, 17 (1992).

15. R. L. Cotterman and J. M. Prausnitz, *Ind. Eng. Chem. Proc. Des. Dev.* **24**, 434 (1985).

16. S. K. Shibata, S. I. Sandler, and R. A. Behrens, *Chem. Eng. Sci.* **42**, 1977 (1987).

17. P. Bartlett, *J. Chem. Phys.* **107**, 188 (1997).

18. J. M. Kincaid, K. B. Shon, and G. Fescos, *J. Stat. Phys.* **57**, 937 (1989).

19. M. L. Michelsen, *Fluid Phase Eq.* **33**, 13 (1987).

20. M. L. Michelsen, *Fluid Phase Eq.* **9**, 1 (1982).

21. R. L. Cotterman, R. Bender, and J. M. Prausnitz, *Ind. Eng. Chem. Proc. Des. Dev.* **24**, 194 (1985).

22. M. L. Michelsen, *Ind. Eng. Chem. Proc. Des. Dev.* **25**, 184 (1986).

23. M. L. Michelsen, *Fluid Phase Eq.* **30**, 15 (1986).

24. M. L. Michelsen, *Computers Chem. Eng.* **18**, 545 (1994).

25. M. L. Michelsen, *Fluid Phase Eq.* **98**, 1 (1994).

26. N. Clarke, T. C. B. McLeish, and S. D. Jenkins, *Macromolecules* **28**, 4650 (1995).

27. The parts of the CPC where the incipient phase has higher or lower density than the original phase are sometimes referred to as dew and bubble curve, respectively.

28. In systems containing solvent whose degrees of freedom have been eliminated (integrated out) from the free energy, this pressure is the osmotic pressure.

29. T. Boublik, *J. Chem. Phys.* **53**, 471 (1970).

30. G. A. Mansoori, N. F. Carnahan, K. E. Starling, and T. W. Leland, Jr., *J. Chem. Phys.* **54**, 1523 (1971).

31. P. Sollich and M. E. Cates, *Phys. Rev. Lett.* **80**, 1365 (1998).

32. P. B. Warren, *Phys. Rev. Lett.* **80**, 1369 (1998).

33. Formally, one can partition the space of all density distributions $\rho(\sigma)$ into equivalence classes, defined by having the same values for the moment densities ρ_i. The moment subspace is then the space of these equivalence classes; the transverse subspace corresponds to variations of $\rho(\sigma)$ within a given equivalence class.

34. J. W. Gibbs, *The Collected Works of J. Willard Gibbs; The Scientific Papers of J. Willard Gibbs*, reprinted, Dover, New York, 1960.

35. In the standard notation of probability theory, the quantity $\mathcal{P}(m)$ should really be written as $\mathcal{P}(m|x)$—that is, the distribution of m, given the fixed value of x. Equation (19) is then recognized as Bayes' theorem.

36. An alternative development may be to express the desired quantities in terms of contour integrals in the complex plane, using Cauchy's theory of residues, as described in E. Schrödinger, *Statistical Thermodynamics*, reprinted, Dover, New York, 1989. We have not explored this predominantly formal development.

37. J. J. Salacuse, *J. Chem. Phys.* **81**, 2468 (1984).

38. As expected, this is of the form required for an exact solution of the phase coexistence problem; compare Eq. (59).

39. The derivation of the moment entropy using the combinatorial method already suggests this property. The only difference is that we now treat the *moment densities* ρ_i as conjugate to the Lagrange multipliers λ_i, whereas in the combinatorial derivation—which works with the normalized particle distribution $n(\sigma)$—this role is played by the *normalized* moments m_i. Compare Eq.(29).

40. Note that if an analytic expression for s_m in terms of the ρ_i can be obtained, then derivatives such as $\partial^2 s_m/\partial\rho_i\partial\rho_j$ can of course be evaluated directly in terms of the ρ_i. The second-order moments ρ_{ij} then never appear explicitly in the calculation.

41. M. E. Fisher and S. Zinn, *J. Phys. A* **31**, L629 (1998).

42. To simplify matters, we make the common assumption that there is only a single instability direction. This implies that the n coexisting phases have to meet along a line (curve). For the more general case, where two or more instability directions appear simultaneously, see Ref. 43.

43. G. R. Brannock, *J. Chem. Phys.* **95**, 612 (1991).

44. J. A. Cuesta, *Europhys. Lett.* **46**, 197 (1999).

45. P. B. Warren, *Europhys. Lett.* **46**, 295 (1999).

46. S. Beerbaum, J. Bergmann, H. Kehlen, and M. T. Rätzsch, *Proc. R. Soc. London A* **414**, 103 (1987).

47. Of course, this effective "prior" is not known in advance (so the name is a slight misnomer); it can only be determined by an exact solution of the phase coexistence problem. As a consequence, it also changes as control parameters such as temperature are varied.

48. If we had a solution of the exact phase coexistence problem, this means that we can investigate its stability by checking only the TPD of a K-dimensional family of density distributions $\rho(\sigma)$ (where K is the number of moment densities in the excess free energy). It might then seem that the TPD does in fact play the role of a "low-dimensional" (depending only on K variables) free energy surface, in apparent contradiction to our statement that the exact phase coexistence solution does not permit such a low-dimensional interpretation. This paradox is resolved by noting that the TPD only "comes into existence" once the full phase coexistence problem has been solved; unlike the moment free energy, it is *not* a given free energy surface from which the phase coexistence is derived by a tangent plane construction.

49. More explicitly, the exact free energy and TPD are convex with respect to the transverse degrees of freedom of $\rho(\sigma)$; local stability therefore only needs to be checked for the moment degrees of freedom, where again the moment free energy is sufficient.

50. W. H. Press, S. A. Teukolsky, W. T. Vetterling, and B. P. Flannery, *Numerical Recipes in C*, 2nd ed., Cambridge University Press, Cambridge, 1992.

51. B. Hesselbo and R. B. Stinchcombe, *Phys. Rev. Lett.* **74**, 2151 (1995).

52. P. J. Flory, *Principles of Polymer Chemistry*, Cornell University Press, Ithaca, NY, 1953.

53. This result can also be obtained by considering a system *with* solvent at fixed osmotic pressure Π (see Appendix A) and then taking the limit $\Pi \to \infty$. Note that the normalized moment $m_1 = \rho_1/\rho_0$ and the (unnormalized) moment density ρ_1 are identical for this model (and in our chosen units), because the dense limit corresponds to $\rho_0 = 1$.

54. One might also have chosen the cross-entropy as an error measure; this corresponds to leaving out the square in Eq. (90), thus allowing cancellations between positive and negative values of the logarithm. In our numerical experiments, this tended to give misleadingly small values of δ.

55. One may expect intuitively that, starting from $\delta = 0$ at the cloud point, δ should increase as one moves deeper into the coexistence region by increasing χ'. We find that this is so, but only as an overall trend: When new phases are detected as χ' is increased, δ may temporarily decrease before increasing again. This explains the crossing of the curves for $\chi' = 10$ and $\chi' = 15$ in Fig. 11.

56. One exception might be the case where the polydispersity variable σ can assume values from an infinite range (as in the length-polydisperse polymer problem treated in Section V.A, where there is no upper limit on $\sigma \equiv L$). The form of the formally exact solution (59) can then give information about the asymptotic behavior of the extra weight functions. We leave this issue for future work.

57. L. Leibler, *Makromol. Chem. Rapid Commun.* **2**, 393 (1981).

58. Note that our algorithm does not make use of the symmetry $\sigma \to -\sigma$ which could in principle be used to reduce the complexity of the three-phase coexistence problem to that of ordinary two-phase coexistence. The scenario therefore provides a genuine test of the multiphase capabilities of the algorithm.

59. R. P. Feynman, *Statistical Mechanics*, Benjamin, Reading, MA, 1972.

60. P. B. Warren, *Phys. Chem. Chem. Phys* **1**, 2197 (1999).

61. I. Pagonabarraga, M. E. Cates, and G. J. Ackland, *Phys. Rev. Lett.* **84**, 911 (2000).

62. H. Cramér, *Act. Sci. Indust.* **736**, 5 (1938); see also R. S. Ellis, *Entropy, Large Deviations and Statistical Mechanics*, Springer, New York, 1985; for an accessible summary see U. Frisch and D. Sornette, *J. Physique*, **7**, 1755 (1997).

63. The CLT only claims that $\mathcal{P}(m)$ converges to a Gaussian if one simultaneously rescales $m - m^{(0)}$ by $N^{1/2}$, and is thus strictly limited to the region $|m - m^{(0)}| \lesssim N^{-1/2}$. LDT, on the other hand, addresses the asymptotic behavior of the Cramér "rate function" $\ln \mathcal{P}$ for all m.

64. In polymer theory, the LDT result corresponds to the theory of large chain extensions: P. J. Flory, *Statistical Mechanics of Chain Molecules*, Interscience, New York, 1969. Another mapping exists onto Debye's theory for dielectric properties of molecules with permanent dipoles: P. Debye, *Polar Molecules*, reprinted, Dover, New York, 1958.

65. R. M. L. Evans, D. J. Fairhurst, and W. C. K. Poon, *Phys. Rev. Lett.* **81**, 1326 (1998).

CHEMICAL PHYSICS OF THE ELECTRODE–ELECTROLYTE INTERFACE

J. W. HALLEY

School of Physics and Astronomy, University of Minnesota Minneapolis, MN

S. WALBRAN

Forschungszentrum Jülich GmbH, Institut fuer Werkstoffe und Verfahren der Energietechnik (IVW-3), Jülich, Germany

D. L. PRICE*

Physics Department, University of Memphis, Memphis, TN

CONTENTS

I. INTRODUCTION

Electrochemistry might be said to have originated with the publication of physicist A. Volta's report on the discovery of the battery to the British Royal Society 200 years ago [1]. Major contributions by Michael Faraday [2] and Helmholtz [3] continued this involvement by physicists in its developments

*Deceased.

Advances in Chemical Physics, Volume 116, edited by I. Prigogine and Stuart A. Rice.
ISBN 0-471-40541-2 © 2001 John Wiley & Sons, Inc.

in the nineteenth century. During the twentieth century, however, involvement by chemical physicists in understanding the fundamental processes and structures associated with electrode–electrolyte interfaces was not very widespread. Electrochemists exploited the phenomena as analytical tools, and the field remained of great importance to such important technological subjects as batteries, fuel cells, corrosion, electrochemical etching, and many aspects of biology. However, the systems were widely regarded as too complex to serve as quantitative test beds for quantum mechanics or statistical mechanics. The neglect of theorists is understandable because the theory of liquids has lagged the theory of solids, making a convincing account of the details of this solid, polar liquid interface difficult to achieve. In this contribution we will report recent work to model the electrode–electrolyte interface at the microscopic level. The goal of this activity is to develop models that are eventually capable of predicting the macroscopic properties of interfaces, as manifest in measurements of differential capacitance and voltammagrams, for example, from a chemical description of the constituents of the interface. At the present we are quite far from achieving this goal, but significant progress has been made. A theme of the presentation is that to eventually achieve it, it will not be necessary to have a complete description at the electronic structure level of every aspect of the system. Instead it will often be possible to achieve predictive models by use judicious use of electronic structure, atomic dynamics and continuum methods, with careful attention to the connections between them.

Recent progress in fundamental understanding has arisen from advances in computational power as well as with improvements in experimental technique. It should be said that the latter, experimental advances are in some respects far in advance of the theoretical and modeling progress, though the emphasis of this article, in keeping with the authors' interest and contributions, is mainly on the former. Insights and discoveries arising particularly from *in situ* scanning tunneling microscopy [4], grazing incidence x-ray diffraction [5], and electron diffraction [6, 7] provide challenges that have not yet been met by theorists and modelers. Other techniques, such as surface enhanced Raman spectroscopy, second harmonic generation [8], and extended x-ray absorption fine structure, provide a different kind of theoretical challenge, not discussed here, since a more complete knowledge of the experimental method itself is needed in order to fully interpret the results.

In this contribution, we will first provide an overview of the nature of the systems and phenomena and the modeling and computational challenge which they represent. In the following two sections we describe calculations of the electronic and atomic structure of the interface and of electron transfer at the interface. In each case we present some details of our own results involving copper–water interfaces and electron transfer from a copper ion in water to a copper electrode, but we have made an attempt to refer to all the similar calculations of other workers of which we are aware. Finally we summarize and discuss possible directions for future work. We are interested in developing quantitative, predictive theories and models, not just in using modeling for insight (though new insights sometimes results). For this reason we stress quantitative connections with experiment throughout.

II. LENGTH AND TIME SCALES AT THE ELECTRODE–ELECTROLYTE INTERFACE

The modern theory of the electrode–electrolyte interface can be said to have reached an important early milestone with the publication of D. Grahame's review [9] of his experiments on the differential capacitance of the mercury–aqueous electrolyte interface 50 years ago. Fifteen years ago, there was no consensus on the meaning of the results, though the reliability of the data had withstood (and continues to withstand) the test of time. The aspect of the Grahame data which resisted explanation was the inner layer part of the capacitance, obtained by subtracting off a term associated with the ionic screening of the electric field by the the electrolyte and modeled with the Gouy–Chapman theory [10]. It was strikingly asymmetric about the point of zero charge (PZC). (This data are shown, with other later data on other electrodes, in Fig. 3.)

Before considering the quantitative aspects of the current understanding of this phenomenon, it is useful to review how this simple experiment on the differential capacitance of an electrochemical interface probes several of the length scales of interest in the problem. The capacitance per unit area of a layer is inversely proportional to the distance by which it separates the charges and is proportional to the effective dielectric constant so that $1/C \propto d/\epsilon$. For a preliminary discussion we may regard the electrochemical interface as consisting of a region near the interface in which the behavior is dominated by the behavior of electrons associated with the solid electrode, a region in which the solvent dominates the behavior as envisioned by Stern [11] and a region in which screening by ions of the electrolyte dominates the electrostatics. (See Fig. 1).

Roughly, these layers provide contributions to the capacitance which add inversely, as capacitances in series. For a metal–aqueous electrolyte interface, α for the electronic structure contribution is approximately λ_{TF}, the Thomas Fermi screening length, which is characteristically $\approx 10^{-1}$ Å and $\epsilon \approx 1$. For the inner water layer, $d \approx 1$ Å and ϵ is around 10. The screening length of the ionic liquid depends on the ionic strength and the potential drop across the interface but is of order 10 Å at the ionic strengths of interest in the Grahame experiments and when the potential drop across the interface is at the value at

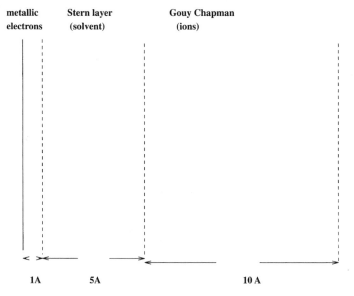

Figure 1.   Microscopic scales near the interface.

which the charge per unit area on the electrode is zero (PZC). Thus, near the PZC the capacitance is dominated by the properties of the ionic liquid. However, as the potential drop across the interface is moved away from the PZC by the experimenter, this Debye–Hückel / Gouy–Chapman [10] screening length shrinks and the capacitance begins to depend on the properties of the inner layers associated with the solvent near the electrode and the electronic structure of the electrode. Our estimates show that both of the latter two contributions could be of comparable significance, as does appear to often be the case. (For a semiconductor electrode, the same type of order of magnitude analysis may be applied to understand the differential capacitance, but the conclusions are quite different because the relevant length in the semiconductor is the width of the depletion layer which is a great deal larger than the screening length in the electrolyte. In fact for most experiments the capacitance is completely dominated by the electronic properties of the semiconductor and is commonly used to diagnose of the properties of the electrode [12, 13].)

This qualitative analysis of length scales can also guide discussion of appropriate modeling scales and techniques. For the properties nearest the interface, details of the electronic structure are required, whereas to describe the solvent near the interface, an atomic scale description is needed. Finally for

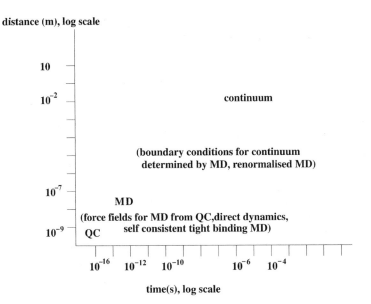

distance (m), log scale

10

10^{-2} continuum

                    (boundary conditions for continuum
                    determined by MD, renormalised MD)

10^{-7}
             MD
        (force fields for MD from QC,direct dynamics,
10^{-9} QC   self consistent tight binding MD)

10^{-16} 10^{-12} 10^{-10} 10^{-6} 10^{-4}

time(s), log scale

Figure 2. Illustration of simulation techniques available at various time and length scales. QC means first principles, quantum chemical methods. MD refers to classical molecular dynamics methods. (Monte Carlo methods are useful in roughly the same range of time and distance.) Methods for connecting QC, MD, and continuum methods are indicated in parentheses.

the ionic liquid (in the absence of adsorption of ions on the surface), a continuum description is adequate. We illustrate with a plot of length and time scales, currently popular in discussing materials science modeling challenges, in Fig. 2. Methods that provide the most microscopic details are constrained by limitations of computational power to the shortest lengths and times. In electrochemical applications we require *ab initio* techniques closest to the interface, molecular dynamics and Monte Carlo methods in the solvent near the surface, and reaction diffusion descriptions in the bulk of the electrolyte. Because all these methods are, on the whole, well-established, the main methodological challenge is to link them quantitatively in order to provide a unified description of various experimental phenomena.

An overview of the approaches that have been taken to linking different theoretical and computational modeling descriptions is also provided in Fig. 2: The first principles (QC) descriptions are based on the Schrödinger equation and the Born Oppenheimer approximation as realized in most chemical applications by density functional [14] or Hartree–Fock [15] methods. Molecular dynamics (MD) methods [16], based on classical Newtonian mechanics, are commonly parameterized by determining force fields for the equations of motion by quantum chemical calculations on small clusters. We have taken this approach in the modeling of electron transfer reactions described below. This approach, generically related to the idea of "parameter passing" between models, has some significant limitations in the electrochemical context, particularly in describing metals and the more complex forms of charge transfer. In a more general approach, sometimes known as direct dynamics pioneered by Car and Parrinello [17], an approximate solution to the Schrödinger equation is found for each atomic configuration and the atoms are moved (classically) in response to the resulting forces. This is much more computationally expensive than the parameter passing approach. We provide some examples of its applications to the study of the structure and capacitance of simple metal–water interfaces below. Those examples reveal the importance of going beyond parameter passing for electrochemical problems but also the magnitude of the computational challenge of trying to describe many of the phenomena of interest in electrochemistry by direct dynamics methods in the forseeable future. For these reasons, we also briefly describe a third approach to the linking of electronic structure to atomic dynamics, in which a self-consistent tight binding model is fit to *ab initio* results and is then used to describe the atomic dynamics of the electrochemical system near the interface. In the case of kinetics, we are aware of very few studies using these methods and we will describe instead a study of electron transfer which essentially depends on parameter passing to make connection of the atomic dynamics model to first principles electronic structure.

The problem of linking atomic scale descriptions to continuum descriptions is also a nontrivial one. We will emphasize here that the problem cannot be solved by heroic extensions of the size of molecular dynamics simulations to millions of particles and that this is actually unnecessary. Here we will describe the use of atomic scale calculations for fixing boundary conditions for continuum descriptions in the context of the modeling of static structure (capacitance) and "outer shell" electron transfer. Though we believe that more can be done with these approaches, several kinds of electrochemical problems—for example, those associated with corrosion phenomena and both inorganic and biological polymers—will require approaches that take into account further intermediate mesoscopic scales. There is less progress to report here, and our discussion will be brief.

III. STATIC STRUCTURE OF THE INTERFACE

Here we describe first and mainly the structure of metal aqueous electrolyte interfaces and refer only briefly and toward the end to the complications associated with the very common existence of oxides and other films and adsorbate layers between metal and electrolyte in experimental systems. This is because progress in theoretical modeling of the latter systems remains quite rudimentary, despite their tremendous importance from a technical point of view. Even within the class of metal–aqueous electrolyte systems, we will only briefly mention the large class of phenomena associated with surface reconstruction and phase transitions in adsorbed monolayers which experimentalists have discovered. Some progress has been made in phenomenologically modeling these phenomena using Monte Carlo methods [18], particularly in the case of adsorbed monolayers. For the most part these studies have not made connection with underlying electronic structure in order to calculate the parameters in the Monte Carlo models so that the theories become quantitatively predictive rather than phenomenological. In principle this is certainly possible and is likely to be a fruitful area for work in the future.

Though X-ray and scanning tunneling microscope data are available for many surfaces, we begin the discussion with some collected, and rather old, differential capacitance data on sp-metal electrodes as shown [19] in Fig. 3. To obtain such data, one must study the charge as a function of overpotential (potential–potential of zero charge) in a range of overpotential in which no charge transfer processes (electron transfer, dissolution, adsorption, or desorption) are occurring so that any currents are transient and purely capacitive. The figure shows inverse "inner layer" differential capacitances obtained by subtracting the inverse of the Gouy–Chapman capacitance, describing the contribution of the ions of the electrolyte, from the inverse of the total capacitance. This subtraction is itself a delicate matter, mainly because it requires an estimate of the "roughness" of the surface, defined as the ratio of the area associated with the Gouy–Chapman capacitance to the macroscopic geometrical area. In practice, these roughness factors are obtained by requiring that the resulting inner layer capacitance be independent of electrolyte ionic strength. The roughness factor is assumed to be potential-independent, which is not strictly correct [20]. Despite these caveats, it is striking that the data for sp metal electrodes shown in Fig. 3 show a very definite trend. Over a significant range of overpotential around the PZC, the inverse inner layer differential capacitance is nearly linear, varying from a smaller value of inverse capacitance at positive overpotentials to a larger one at negative overpotentials and the slopes are of the same order of magnitude for all the metals. In the early 1980s we suggested that this trend might be qualitatively understood as arising from changes in the electronic structure of the interface with changes in the overpotential. The idea is illustrated in the cartoon in Fig. 4: If the electronic charge is assumed (consistent with Ref. 21) to move with potential in a block without change of shape, carrying the solvent and electrolyte with it, then one can obtain [19, 22] a value of the slope

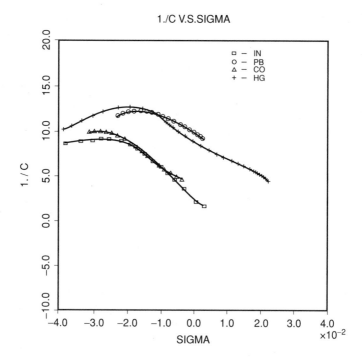

Figure 3. Inverse of the inner-layer capacitance of some *sp*-metal–water interfaces as a function of surface charge (atomic units) as deduced from experimental data. From Ref. 29 by permission.

of the right order of magnitude and sign from this picture. (The general idea that the electronic structure of the metal is a significant contributor to the structure of the capacitance had been anticipated, unknown to us, by many years by O. K Rice [23]. Below we cite the work of others who published studies similar to ours at about the same time). Most earlier attempts [24–26] to account for the observed asymmetry of the inner layer capacitance about the PZC were based on elaborations of the Stern model [11] which attributed the inner layer capacitance to a layer of tightly packed water at the surface of the metal electrode.

Such a Stern layer surely exists, but attempts to model its effects by use of relatively simple statistical mechanical models of a layer of dipoles did not easily yield the observed asymmetry [24–26]. Instead, these models predicted a maximum in the inner-layer capacitance at the PZC, arising from the saturation of the dielectric response of the water layer at potentials far from the PZC. Such a maximum did appear at lower temperatures in

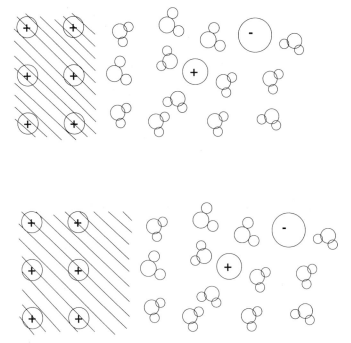

Figure 4. Illustration of the mechanism to which we attribute the assymmetry shown in Figure 3. When the electrode (left) is positively charged (top figure), then the electrons of the metal (shading) retreat into the metal, followed by the solvent and ionic screening charge, with the net result that the charge separation is reduced relative to its value at the point of zero charge and the capacitance is larger. Conversely, when the electrode is negatively charged (bottom figure), then the electrons of the metal are pushed out from the positive ion cores of the metal near the surface, and the solvent and ionic screening charge are pushed away from the metal surface, resulting in a lower capacitance.

Grahame's data, though the main feature is the asymmetry already noted. Later, when reliable data on solid noble metal electrodes appeared [27, 28], however, the predicted peak was even more in evidence, still accompanied, however, by an asymmetry that was not easily understood within the Stern models. In the early 1980s, several groups [22, 29–33] attempted to calculate the contribution of the metal electrons to the capacitance using jellium-like models. They always found an asymmetry like the one suggested by the simple picture in Fig. 4, as long as they allowed the separation between the model of the solvent and the metal to vary in order to minimize the total energy at fixed electrode charge. Results for the inner layer capacitance for a number [22, 29, 30] of those earlier, jellium-like models are summarized in Fig. 5, where they are compared with experimental results for the inner layer

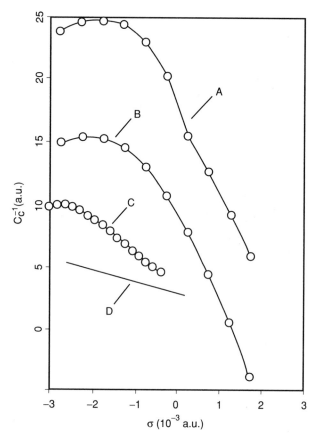

Figure 5. Theoretical results from Refs. 22(D), 29(A), and 30(B) compared with experimental result for cadmium (C) as reported in Ref. 56. Reproduced from Ref. 30 by permission.

capacitance of cadmium. The sign and order of magnitude of the effect are consistently correct, but these theories are clearly not quantitative.

In fact these calculations did not treat the time scales correctly because they generally fixed most features of the atomic structure of solvent and then calculated the resulting electronic structure, for fixed potential drop across the interface. (A recent calculation [34] that takes more detailed account of the electronic structure of the electrode than these early calculations also suffers from this defect.) In fact, of course, in the Born–Oppenheimer approximation, the electronic structure should be recalculated for each atomic configuration in an ensemble of atomic configurations that follow the Born–Oppenheimer surface. This became possible with Car–Parrinello

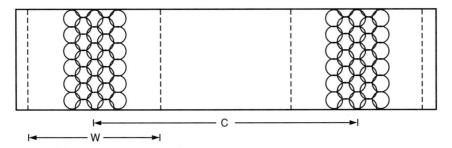

Figure 6. Slab configuration of atoms of the electrode for the calculation described in the text. From Ref. 35 by permission.

[17] and related techniques. We illustrate the possibilities with a description of a computational model of the metal–water interface [35, 36] which we recently used for a relatively detailed study of the differential capacitance of the copper–water interface. Similar plane wave calculations have been carried out by the group of Strauss and Voth [37] (silicon–water interface), by M. Philpott and coworkers [38] (platinum–water interface), and by M. Yamamoto et al. [39] (platinum water interface).

Our calculations [35] were carried out with a unit cell (Fig. 6) having dimensions $42.3 \text{ Å} \times 15.3 \text{ Å} \times 15.3 \text{ Å}$. Each cell contained a thin, five-layer slab of metal atoms, arranged in an FCC lattice. The exposed faces of the slabs were (001) surfaces of the FCC structure, with 36 atoms per layer, for a total of 180 metal atoms per unit cell. Each unit cell contained two metal surfaces and a large empty region between them, which was filled in some of the calculations with 245 H_2O molecules represented by pseudopotentials for the electronic wave functions of the electrons associated with the metal, and moving in response to forces which included classical water–water interactions as well as forces arising from the presence of the electrons of the metal. The cell was periodically continued in all three dimensions. We could apply a net drop in electric potential between neighboring slabs to induce net surface charges on the metal surfaces. Only the metal's valence electron wave functions were calculated; the water molecules were taken to be closed-shell systems with fixed electron densities, as were the core electrons of the metal. In the model for copper, only one electron per copper atom (an "s-like" electron) is treated explicitly.

The valence electron wave functions were taken to be zero outside of a "near-metal region" (see Fig. 6). The wave functions were expanded in a basis of plane waves. One-electron wave functions were computed through the following well-known [40] iterative process: Compute the $(I + 1)$th iterate as

$$\psi_{I+1} = \psi_I - \alpha H \psi_I \tag{1}$$

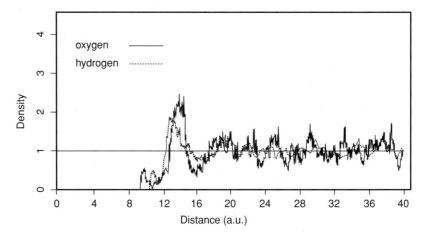

Figure 7. Oxygen and hydrogen density at the copper surface from the calculation of the 001 copper–water interface with zero charge on the electrode as described in the text. From Ref. 35 by permission.

(α is a small positive real constant.); orthogonalize to all the one-electron eigenstates already found; then iterate again. The effective one-electron potential in the Schrödinger equation contained three contributions: $V_{\text{eff}} = V_{\text{ps}} + V_{\text{cl}} + V_{\text{xc}}$. V_{ps} arises from the pseudopotentials of the atoms and ions (including long-range electrostatic potentials), V_{cl} is the classical electrostatic potential generated by the electron density, and V_{xc} is the exchange and correlation potential.

For the copper-like model, the pseudopotentials were initially derived from all electron calculations on individual ions in standard ways but were simplified (local, s-like) and were adjusted to give approximately the right work function for copper metal. After obtaining self-consistent wave functions, the atoms are moved with the Verlet algorithm [41] using forces in the form $F_i = F_{i,\text{atoms}} + F_{i,\text{electrons}}$. $F_{i,\text{atoms}}$ were taken from water–water interactions from the central force model and electrostatic hydrogen–metal ion and oxygen–metal ion interactions. $F_{i,\text{electrons}}$ is the force on hydrogen, oxygen, or metal atom i arising from the calculated valence electron density and is obtained from $-\nabla_{Ri} \int d\vec{r}\, V_{\text{ps}}(\vec{r} - \vec{R}_i)\rho(\vec{r})$ by Fourier transform, where $\rho(\vec{r})$ is the electron density. Details concerning the pseudopotentials used appear in Ref. 35.

Using this model and calculational method, a calculation with just one water molecule in the space between the simulated copper slabs gave an on-top site (oxygen down) as the lowest energy water configuration on a 100 surface with a binding energy of about 0.4 eV. Then we did a calculation [35] with 245 molecules in the space between the planes and found the oxy-

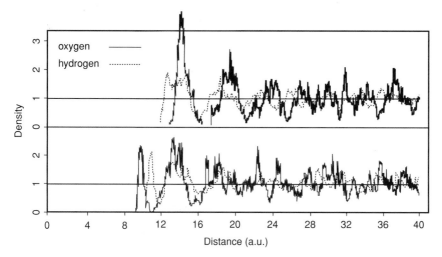

Figure 8.   Oxygen and hydrogen density at the copper surface from the calculation of the 001 copper–water interface with positive charge (top) and negative charge (bottom) on the electrode as described in the text. From Ref. 35, by permission.

gen and hydrogen densities between the planes shown in Fig. 7. In this figure, water molecules captured in the on-top sites found to be the lowest energy binding sites for single water molecules are represented by the small oxygen peak at a distance 9 a.u. from the center of the slab. This small peak contains much less than a monolayer of water molecules captured in these on-top sites so that it is evident that of all the on-top sites, only a few are occupied. At positive potentials, this peak disappeared in the calculation, while at negative ones, it grew to describe a full monolayer (Fig. 8).

From this result and other results of the calculation presented in Ref. 35, we concluded that the metal is causing a repulsion between the water molecules near the surface in this model. This was a quantum mechanical effect not arising from classical image forces, which were present in classical simulations not showing the effect. Recently, quantum chemical all-electron Hartree–Fock calculations on quite large (14 atom) clusters of copper atoms, with various numbers of adsorbed water molecules on a (001) surface, were carried out by Larry Curtiss and coworkers [42] and have confirmed that this "through metal" repulsion of water molecules on a copper surface occurs in a much more complete model of the electronic structure. In these Hartree–Fock calculations (which cannot explore such a large number of atomic configurations as we could in Ref. 35), a single water molecule was found to be bound in an on-top site by 0.72 eV (compared to about 0.4 eV in the plane wave calculation just described). We show two atomic configura-

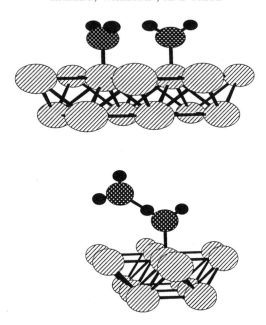

Figure 9. Configurations of atoms used for the Hartree–Fock calculations of the energy of pairs of water molecules at a Cu(001) surface as described in the text.

tions for which Hartree–Fock calculations were also made in Fig. 9, each involving two water molecules on a 14-atom cluster representing the Cu(100) surface. In the first configuration, in which two neighboring waters are in on-top sites, the binding energy of the second water molecule was only 0.03 eV, indicating a strong repulsion of a pair of near-neighbor on-top water molecules due to the "through metal" interaction, consistent with the results in Ref. 35. On the other hand, in the configuration in which a second water was hydrogen-bonded to the first water molecule, which was in turn in an on-top site of the copper surface, the second water was bound by 0.38 eV in the Hartree–Fock calculation. Reference 42 reports energetic calculations for monolayers of water on (001) copper surfaces with similar qualitative conclusions.

These results show that including quantum mechanical electronic rearrangement in dynamics calculations of the configurations of water on a metal surface can reveal effects that are not present in classical models of the water metal interface which treat the interaction of water with the surface as a static, classical potential energy function. For example, in classical calculations of the behavior of models of water at a paladium surface the interaction with one water molecule with the surface had a similar on-top binding site, a classical simulation predicted a full monolayer of water molecules in the on-top site when water at liquid densities was present [43].

To further explore the validity of our simplified Car–Parinello-like model of the copper–water interface, we used it to calculate the capacitance observed in experiments [36]. We describe some details of the calculation of Ref. [36] because they illustrate a connection of microscopic (direct dynamics) calculations to a continuum theory (in this case the Gouy–Chapman theory [10]) by matching boundary conditions in the fashion described schematically in Fig. 2. To calculate the full capacitance of the interface, we need to take account of the screening the charge on the metal by the ions of the electrolyte. By imposing a field across the slab of the direct dynamics model [35] of the metal–water interface, we could calculae the charge on each side of the interface as a function of the total potential drop across the simulated interface (which did not contain screening by the ions of the electrolyte). Results for the (001) and (110) surfaces are shown in Fig. 10.

(Details concerning the model, which, for the water, was slightly different from that of Ref. 35, appear in Refs. 36 and 44.) These results show the electrostatic response of the first two, metallic and Stern, layers in Fig. 1. within the direct dynamics model. At distances of more than about 10 Å, the ions

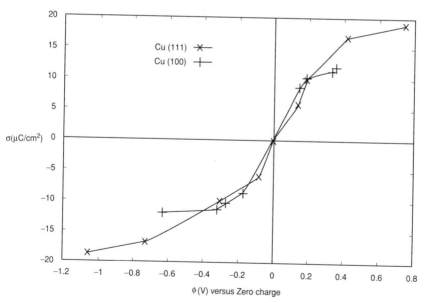

Figure 10. Charge on the electrode as a function of the potential across half the simulation box (containing solvent but no ions) for two crystal faces as described in the text. From Ref. 36 by permission.

farther out in the solution eventually screen the field which is present in the solvent in these calculations. It is *not* practical to include these ions directly in the molecular dynamics simulation. At typical solution strengths there are about 100 water molecules per ion, and we need around 100 ions to represent the double layer. The current Ewald methods scale as N^2 so the needed calculation would be approximately 10^8 times as expensive. Note further that the simulation must be done many times for different field strengths. Furthermore, for dilute solutions and no specific adsorption, direct simulation of the ions is totally unnecessary, because the continuum Gouy–Chapman model works very well [45] for describing the distribution of ions in the double layer at low molar concentrations as long as there is no specific adsorption of ions to the surface. The remaining problem is to correctly couple the continuum Gouy–Chapman theory to the simulation of the "inner-layer" metal and water parts of the simulation. We did this by calculating the macroscopic electric field to provide a boundary condition for the continuum theory at the plane $z = z_c$ at which the model begins to represent the system by the Gouy–Chapman model. Given the macroscopic electric field $E(z)$ for a given charge σ on the electrode and assuming that in the continuum theory the solvent responds linearly to the field, we have, on the electrolyte side of z_c, that $d(\epsilon E(z)(z)/dz = 4\rho((z)$, where $\rho(z)$ is the ionic charge density in the electrolyte and we have allowed for the possibility that the local dielectric constant of the solvent is different from the bulk one, but have assumed a local response function. Some details of the procedure by which the boundary condition is fixed are omitted here. A detailed account appears in Ref. 44. Using a simple model for $\epsilon(z)$ (to which the results are not sensitive), we numerically solved the Poisson–Boltzmann equation for $z > z_c$:

$$-d(\epsilon(z)d\phi_{GC}(z)/dz)/dz = 4\pi\rho(z), \qquad (2)$$

where ϕ_{GC} is the total electrostatic potential in the continuum model on the electrolyte side of z_c with the charge density $\rho(z) = 2n_0q_0 \sinh((q_0\phi_{GC}(z))$. This yielded the net potential drop from inside the electrode to the bulk of the solution as

$$\phi = (\phi_{MD}(z \rightarrow -\infty - \phi_{MD}(z = z_c)) + (\phi_{GC}(z = z_c) - \phi_{GC}(z \rightarrow \infty)). \quad (3)$$

$\phi_{MD}(z)$ is the electrostatic potential obtained from the simulation. $z \rightarrow -\infty$ is replaced by a point inside the model metal slab. Some values of the macroscopic electric field, obtained by Gaussian average of the local field obtained in the Car–Parinello-like mode of Ref. 35, are shown in Fig. 11. An interesting feature is that the macroscopic field is usually smaller near the electrode than it is in the bulk of the solvent, even though the dielectric constant of the

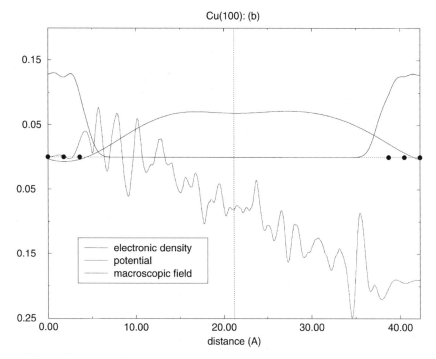

Figure 11.   Values of the macroscopic electric field as calculated from the simulation for one overpotential. From Ref. 36, by permission.

water (around 80) is large so that there is substantial dielectric screening in the bulk solvent. This is because the metal electrons screen the field very effectively. This screening effect is absent in models of the electrode which do not explicitly take account of electronic structure. To illustrate this point, in Fig. 12 we reproduce the result of a classical calculation [46], of the potential distribution near a model electrochemical interface. In the model that led to Fig. 12 the electrodes were described as "ideal metal" walls (infinite dielectric constant and no explicit electronic structure) and the same model of water used in the model which gave the results in Fig. 11. One sees that there is a very large field near the metal-electrolyte interface in the model. This large field has frequently been obtained by classical modeling studies of the interface (e.g., Refs. 47 and 48) and has sometimes been mentioned in a context which suggests that it might produce extraordinary effects [49]. In fact, we are finding that in our models in which the electrons near the interface are allowed to screen this field, they do so very effectively, reducing it in many cases to a value even less than the dielectrically screened value expected (and calculated) deep in the solvent.

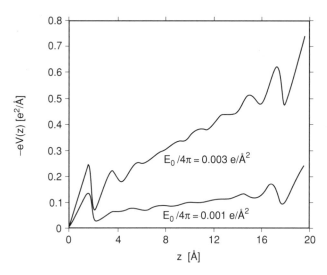

Figure 12. Values of the calculated electrostatic potential in a *classical* simulation containing a similar number of water molecules between two charged electrodes, but no explicit account of screening by the electrons of the metal. Note the large slopes corresponding to big fields near the electrodes. From Ref. 46, by permission.

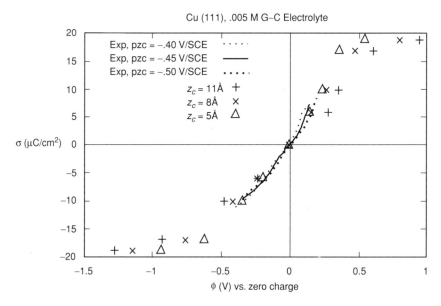

Figure 13. Calculated charge as a function of potential for the water–copper surface in contact with an electrolyte, compared with experimental values. From Ref. 36, by permission.

Using calculated values of the macroscopic field and Eq. (13), we computed the total overpotential as a function of charge for direct comparison with the macroscopic capacitance experiment. One set of results is shown in Fig. 13. Actually, Fig. 13 shows a favorable case. Reported experiments are not all in agreement. We refer to Ref. 36 for further details.

Though this calculation of a model copper–water interface gave some interesting and possibly useful results and insights, it also has some definite limitations. These mainly arise from the limitation to explicit calculation of only the s-like conduction electrons in the copper. Among the liabilities arising from this limitation are the fact that it was not possible to allow the copper atoms to move in the simulations just described. The s-like electrons alone did not yield realistic forces between the copper atoms, and the metal was not stable. Another symptom of the inadequacy of the model became manifest when we tried to describe adsorption of a chloride ion to a copper surface using it [50]. Many features of the chloride adsorption were correctly described, but the binding energy was almost an order of magnitude larger than the experimental one.

Removing the constraint and treating the d electrons explicitly, however, is extremely expensive. The main reason for this [51] is that there are 5 d orbitals. The calculational cost increases as a power of the number of orbitals, which is between 2 and 3, giving increases in computational costs of order 10^2, which is beyond practical reach at the present. To avoid this problem, one can focus on sp metals, work with smaller systems, or develop new or improved methods. We next describe efforts in each direction.

Recently, one of us (D. L. P.) has made [52] a detailed calculation for a cadmium interface which takes s- and p-like bands into full account. This is a very very nearly *ab initio* calculation of the molecular and electronic distributions at the interface of the (001) surface of hcp cadmium and liquid water. In cadmium, unlike copper, the d electrons are not expected to make a significant contribution to the interaction of the electrode with the water, but because Cd is divalent, a study of Cd which includes nonlocality in the pseudopotential tests our ability to make a less phenomenological model in a system with more electrons per ion using these methods in a way that is computationally affordable.

The method employed was similar to that of Ref. 35, but with several improvements. *ab initio*, norm-conserving, nonlocal pseudopotential were used to represent the metal ions. This capability enables reliably realistic representation of the metal's electronic structure. Thus the cadmium pseudopotential was able, for example, to reproduce the experimental cadmium–vacuum work function using no adjustable parameters (unlike the procedure followed in Ref. 35). Pseudopotentials of the Troullier and Martins form [53] were used with the Kleinman–Bylander [54] separable form, and a real space projection technique was used. The second improvement was a new water pseudopotential, which, while not norm-conserving, was constructed in an *ab initio* manner and included proper treatment of the nonlinear exchange and correlation interactions [55]. This pseudopotential was constructed by solving the all-electron water problem (in the local density approximation), obtaining the electron density of an isolated water molecule, and then calculating (using a full-potential LMTO method) the all-electron water-molecule/Cd slab energy versus distance curves, using a three-layer Cd slab. The water pseudopotential was then constructed by (i) fitting the molecular all-electron density to a site-centered function which was smooth in the interior regions of the water molecule, but matched the all-electron density in the outer regions of the molecule and (ii) obtaining a Coulomb potential which was smooth in the interior and matched the all-electron Coulomb potential in the outer regions. Details are described in Ref. 52. Other changes relative to the earlier application of this method include a general shape two-dimensional unit-cell [permitting the modeling of a hcp (001) surface used in the present study] and an associated spherical cutoff for the various reciprocal lattice expansions, improved electrostatic (Ewald) calculations, and incorporation of a interpolation method for evaluating atom–atom force laws.

The calculation of the (001) cadmium-surface–water interface used a hexagonal unit cell with a *c* lattice constant of 135 a.u., and an *a* lattice constant of 33.788 a.u., corresponding to 36 cadmium atoms in the surface layer. The Cd slabs were 5 layers thick (perpendicular to the *c*-axis), for a total of 180 cadmium atoms per unit cell. The region between the slabs was filled with 525 water molecules. The plane wave expansion of the electron eigenstates had a cutoff of 9 Ry, giving 20,209 basis functions. Calculational details are described in reference [52].

Figure 14 shows a cross-section of the unit cell at one time step, with zero applied potential, and it clearly shows marked differences from the similar

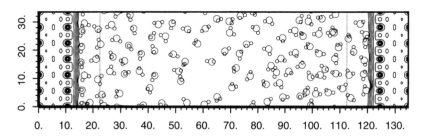

Figure 14.   Contour plot of the electron density at one time step during the zero-voltage molecular dynamics run. The circles indicate the simultaneous positions of those water molecules with the oxygen atom within 3 a.u. of the plane of the contour plot. From Ref. 52, by permission.

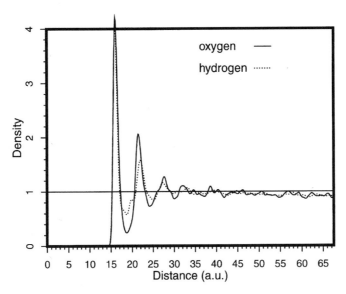

Figure 15. Transverse and time average of the oxygen and hydrogen density, as a function of distance from the center of the metal slab, with the "left" and "right" sides of the unit cell averaged together. The horizontal line indicates the density that would correspond to bulk water (with $1\,g/cm^3$). From Ref. 52, by permission.

calculation for the singly valent, "Cu-like" metal–water interface discussed in Ref. 35 and above. In particular, there are no tightly bound, oxygen-in-water molecules attached to the surface. The water molecules in contact with the surface appear to be loosely bound, effectively "floating" on the metal's valence charge. This figure also suggests that the surface layer water molecules are predominantly oriented flat on the surface, with the hydrogen and oxygen atoms equally distant from the cadmium surface. This appearance is supported by Fig. 15, which shows the time-averaged oxygen and hydrogen densities as a function of distance from the center of the cadmium slab. The density shows at least three peaks in the oscillations of the density near the surface, with a distinct and narrow surface layer. The surface layers of oxygen and hydrogen are quite nearly coincident with each other, with both peaks at the same distance from the cadmium surface. The last layer of Cd ions is at 10.62 a.u. from the center of the slab, confirming that the surface water's are primarily flat. The figure shows, however, that a small fraction of the surface water molecules are oriented with some in a hydrogen-out configuration.

A view of the two-dimensional structure of this surface layer is given in Fig. 16, which shows the positions, at one time step, of the surface layer water

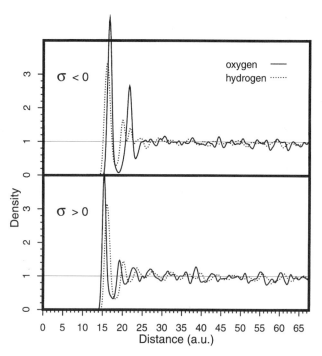

Figure 16. The surface layer of water, in one unit cell, at one instant during the zero applied voltage calculation. The large, light blue spheres represent the surface layer of the cadmium slab. From Ref. 52, by permission.

molecules in relation to the surface layer of cadmium ions. Again, it can be seen that a majority of the water molecules lie flat on the surface, apparently arranged in a two-dimensional, hydrogen-bonded structure that seems to require that a small fraction of the surface molecules be oriented with one hydrogen oriented away from the surface in order to maintain the network of bonds.

The arrangement of surface molecules appears to have little correlation with the lateral positions of the surface cadmium ions. Inducing charge on the metal surface through application of a potential across the unit cell has some significant effects upon this structure. Figure 17 shows the hydrogen and oxygen densities associated with the most negative and most positive charge induced upon the Cd surface. Negative charge on the metal surface enhances the density variations of the fluid. The relative motion of the hydrogen and oxygen densities gives rise to the polarization of the liquid, but an absolute outward displacement of the oxygen distribution with negative charge induced on the metal surface, and an inward displacement of the oxygen distribution for positive induced charge is especially evident. This

Figure 17. Transverse and time average of the hydrogen and oxygen densities for (top) $\sigma < 0$ and (bottom) $\sigma > 0$, as a function of distance from the center of the slab. From Ref. 52, by permission.

is quite consistent with the qualititative picture that guided our earlier models (Fig. 4 and the accompanying discussion.).

To get some idea of the capacitance of the cadmium–water interface implied by this calculation without carrying out a detailed analysis like that in Ref. 36, the following expression [22] for the transverse average of the potential at the metal–water interface was used: $V(x) = 4\pi\sigma(\bar{x}_e - \bar{x}_s) - \frac{4\pi\sigma}{\epsilon}$ $(x - \bar{x}_s)$ Here σ is the net charge per unit area induced on the metal surface, $\bar{x}_e$ is the center of mass of the induced electronic charge, and $\bar{x}_s$ is the center of mass of the induced solvent (polarization) charge. If ϵ were infinite, the expression would be that for parallel plate capacitor with plates at the electronic and solvent-induced charge centers of mass, $\bar{x}_e$ and $\bar{x}_s$. With the large dielectric constant of water (80), over 98% of an electrolyte's counter charge is accounted for by the liquid solvent's polarization surface charge and the compact capacitance may be identified with the first term in the potential drop, $4\pi\sigma(\bar{x}_e - \bar{x}_s)$, which is determined by the detailed nature of the transition from metal to bulk water, and characterized by the gap between $\bar{x}_e(\sigma)$ and $\bar{x}_s(\sigma)$. In the cadmium calculation, the calculated center of mass of electronic induced charge $\bar{x}_e(\sigma)$ was reasonably well-behaved, the center of mass of induced polarization charge $\bar{x}_s(\sigma)$ was not. The oscillations of density of the polarized fluid, and a systematic error in the polarization charge density resulted in calculated values for $\bar{x}_s$ which contain noise and uncertainties at

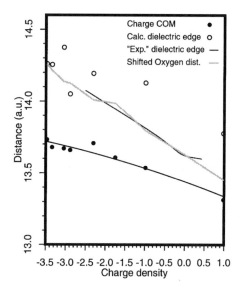

Figure 18. The centers of mass of the induced electronic charge $(\bar{x}_e)$, and induced polarization charge $(\bar{x}_s)$ as a function of the amount of induced surface charge, in units of 10^{-3} e/(a.u.)^3. The black filled circles show the calculated values of $\bar{x}_e$, and the solid black line is a quadratic fit to these values. The open circles indicate the calculated center of mass of the $\bar{x}_s$ (equal to the edge position of an equivalent, classical uniform dielectric), and the line labeled ""Exp." dielectric edge" indicates where they would need to be in order to reproduce the experimental compact capacity. The line labeled "Shifted Oxygen dist." is the position of the oxygen surface layer as a function of charge, shifted downward by 2.4 a.u. From Ref. 52, by permission.

least on the order of half of an atomic unit. Figure 18 shows calculated values of $\bar{x}_e$ and $\bar{x}_s$ (filtered with exponentials decaying from the surface, with the start of the exponential decay situated to maintain equal and opposite net charge on both sides of the fluid slab) and compares to experiment by indicating what $\bar{x}_s$ is needed in order to reproduce the experimental results. Despite the uncertainty in the calculated value for the induced polarization charge $\bar{x}_s$, it is evident that the calculation has predicted the order of magnitude of the experimental [56] compact capacitance, corresponding to a parallel plate capacitor with a gap of about half of a Bohr radius.

Another advantage of making these direct dynamics calculations on *sp* metals is that, unlike the simplified model of copper, the full *sp*-metal simulations have realistic forces between the metal atoms, so that surface relaxations and, eventually, reconstructions can be studied. This was not done in the cadmium study just described, but one of us (SW) has recently explored the possibilities with a calculation of the relaxation of an aluminum surface. The calculation was done on a system consisting of $5 \times 5 \times 5 = 125$ aluminum

TABLE I
Aluminum Slab (Unrelaxed)

Quantity	Simulation (eV)	Experiment (eV)	Reference
Work function (001)	4.49	4.41	75
Cohesive energy	3.65	3.42	89

centers in an FCC lattice exposing the 001 faces. The simulation cell dimensions were $27.07 \times 27.07 \times 45.00$ a.u. Three electrons ($3s^2\ 3p^1$) per aluminum center were treated explicitly. A plane-wave cutoff energy of 16 Ry was used. Troullier–Martins pseudopotentials were used. For computational speed, only the local $3s$ pseudopotential was used. Tests on the Al–Al dimer showed only small differences in binding energy and separation distance compared to the local-$3s$ + nonlocal-$3p$ pseudopotential treatment of the Al centers. We have allowed the atomic structure of this system to relax in the absence of water. The aluminum metal surface appears to be stable. The outermost layer relative to original (bulk) lattice has expanded by about $+0.181\ \text{Å}$ in the direction normal to the 001 face. The calculated work function was 4.08 eV compared to a computed value of 4.49 eV and a reported experimental value of 4.41 eV. These results, together with calculated cohesive energies, appear in Table I. (Note that the work function is $\Phi \equiv V(\infty) - \epsilon_{\text{fermi}}$ and the cohesive energy $E_c \equiv E_{\text{slab}}/N_{\text{atom}} - E_{\text{atom}}$ includes surface energy terms).

We have also begun a study of the aluminum slab in water using the water model described in Ref. 35. We have introduced 180 water molecules into the simulation cell in an ice structure and have allowed them to thermally equilibrate to 300 K with the electronic structure of the aluminum slab frozen.

These results show that the plane wave methods can probably be extended to a rather complete treatment of sp metal electrodes as long as water does not dissociate at their surfaces. To treat all valence sp electrons in water by plane wave methods is certainly possible, but is quite expensive and has so far only been reported for relatively small numbers of water molecules. It appears that use of these methods for electrochemical problems may present obstacles similar to those presented by transition metal electrodes. For these reasons, we have also been working [57, 58] to develop self-consistent tight binding methods for electrochemical modeling. We use quantitative information on the energy level structure of the isolated ions of the material to parameterize the on-site (charge and, in principle, spin-dependent) energies of the model and LDA calculations of band structure and total energy to parameterize the intersite parameters of the model. The approach has some common features with other work [59]. The main elements of the approach were described in Refs. 58 and 57. We fit a Coulomb self-consistent tight binding model to bulk first principles calculations of the total energy and band structure of the material in order to complete the parameterization of the model. We express the energy in terms of eigenfunctions $\psi_\lambda(\vec{r})$ of the one-particle density matrix. In our approach we make use of a tight binding basis and expand the $\psi_\lambda(\vec{r})$ in terms of it. We make a Hartree–Fock approximation for intersite terms and describe on-site energies in terms of a charge- and spin-dependent functional which fits the known properties of the isolated atoms of the material. One-electron equations (like the Kohn–Sham equations) are iterated to self-consistency in the one-particle density matrix. We have also implemented an order N version of the approach in which the energy is minimized directly as a functional of the one particle density matrix.

We published [57] a study of the low index surfaces of rutile TiO_2 using this approach in which we showed that it was possible to obtain correct relaxed structures of low index surfaces using it. More recently we have fitted the model to first principles calculations of the electronic structure of VO_2, RuO_2, and MnO_2 (rutile forms). The latter two systems are metallic, and MnO_2 has a complicated spin structure that preliminary results suggest can be reproduced by this method. To apply the approach to electrochemical systems, we could use a classical model of water, somewhat as in the plane-wave calculations described above. However, at present we are attempting to make a similar description of the water itself, in order to allow dissociation and charge transfer in a natural way. To study very common and important metal–oxide–water interfaces, we also need to couple such a model to the underlying metal of the electrode. Tight binding models of metals have been developed for many years. However, without Coulomb self-consistency they are not useful for interfaces. Preliminary studies of the TiO_2–titanium metal interface using this approach indicate that it may be possible, but more elaborate tight binding models are required in order to treat the metal and oxide on the same footing.

IV. MODELS OF KINETICS

Electron transfer at electrode–electrolyte interfaces is a key step in a wide variety of chemical processes, including, for example, those occurring in stress corrosion cracking [60], in photoelectrochemical reactions of various sorts [61], and in fuel cells and batteries [62]. The general framework for considering such questions was set out by Marcus [63]. The main contribution of this work by Marcus (also independently by some Soviet theorists [64]) was its demonstration of the importance of solvent fluctuations in determining the kinetic barrier both for electrochemical and for homogeneous electron transfer reactions in condensed media. Using a classical model for the dielectric properties of the solvent. Marcus provided a framework for parameterizing a model for the kinetic barrier which was widely used to describe experimental data. Because of its phenomenological character, however, the theory did not have much predictive power and even its basic assumptions were difficult to subject to detailed experimental test. (A brief historical review of early electron transfer theory appears in Ref. 65.)

At the time of Marcus' seminal papers, the phenomenological approach was the only one possible because detailed simulation of solvent behavior was not feasible. More recently, it has become possible to include much more detailed quantum mechanical description of the constituents in simulations and thus to both put the Marcus ideas on a quantitative footing (at least in some simple cases) and to check the Marcus assumptions. Several theorists have approached this problem by using simulation to parameterize a Newns–Anderson one-electron Hamiltonian [66, 67] which describes electron transfer. We have followed a somewhat different route, which utilizes detailed properties of the reactant and product states and which is adequate for electron transfer reactions that occur when the ion accepting an electron is sufficiently far from the electrode ("outer shell"). Formally, our approach is similar to the work of Kuznetsov [68]. In practice, all of this work has treated the electrons of the electrode classically, except to take into account a "coupling constant" that is essentially a quantum mechanical matrix element connecting the reduced and oxidized states of the ion through electron transfer from the electrode. For a study of "inner shell" processes in which the reacting ion is close to the electrode surface during transfer, as well as for related problems of adsorption and dissolution rates, much more detailed accounts of the details of the electronic structure of the electrode, using methods like those described in the last section, will be required. Even for outer shell processes, we and others have been forced in our descriptions to take account only of very simple dependence of the coupling constant on solvent and reactant configurations, whereas a realistic model might reveal qualitative as well as quantitative inadequacies in the assumptions which we have made about these dependencies. Again, direct dynamics methods will ultimately make direct simulation of these aspects possible. However, at the present time, because of the limitations of direct dynamics methods for transition metals, we are mainly limited for kinetics studies to calculations in which the connection to first principles electronic structure is made through parameter passing by determination of force fields for classical molecular dynamics simulation. This is the approach we used in study [69] of the cuprous–cupric electron transfer reviewed in this section. The classical molecular dynamics description of the solvent and the reactant ion does not take full account of the screening effects of the electrons of the metal electrode on the electrostatic potential distribution which were discussed in the last section. Though we argued [69] that this may not significantly affect the barrier heights which we find, it may be the reason that we have not obtained correct transfer coefficients from the calculation. Here the natural remedy is once more to solve the problem of doing direct dynamics for transition metal systems, so that these screening effects can be directly incorporated in a model for the electron transfer.

The cuprous–cupric electron transfer reaction is believed to be the rate-limiting step in the process of stress corrosion cracking in some engineering environments [60]. Experimental studies of the temperature dependence of this rate at a copper electrode were carried out at Argonne. Two remarkable conclusions arise from the study reviewed here [69]: (1) Unlike our previous study of the ferrous–ferric reaction [44], we find the cuprous–cupric electron transfer reaction to be adiabatic, and (2) the free energy barrier to the cuprous cupric reaction is dominated in our interpretation by the energy required to approach the electrode and not, as in the ferrous-ferric case, by solvent rearrangement.

To describe this electron transfer reaction we required a model of the water–copper interface and of the copper ion in solution. We used the Toukan–Rahman [70] flexible central potential molecular dynamics model for water which we used for the previous ferrous–ferric study, with 216 water molecules between slabs of metal. In the context of the general framework of electron transfer theory, it is of some importance to note that this water model reproduces correct dielectric behavior of water in frequency ranges up to near infrared. Simulations were on systems containing two metal electrodes separated by a region of about 20 Å which was filled with water. The metal was modeled as an ideal classical metal (perfect dielectric) with an additional electrode–water interaction that was fit to the results of the calculation of the electronic structure of the copper–water interface described in the previous section [35, 36]. We fit the interaction energy of a single water molecule with the electrode as described in Ref. 35 to a sum of electrostatic images and single-particle potentials, with the image plane 8.5 au from the center of the copper slab. The single-particle potentials used for the fit are of the form

$$U(z) = A/z^9 - Be^{-z/a} \qquad (4)$$

(Values of the parameters are given in Ref. 70.) The results of Ref. 35 showed that the wall induced a significant water–water repulsive force when two particles were simultaneously near the wall. To reproduce this effect in our classical simulation, we included an extra repulsion from the wall for oxygens that are close to one another. The parameters of this multibody force were adjusted so that the classical simulation reproduced the oscillations in the oxygen density which were obtained from the less phenomenological model of Ref. 35 and illustrated in Fig. 8. The multibody potential used in the classical simulation for oxygen centers 1 and 2 was of form

$$U(\vec{r}_1, \vec{r}_2) = Ce^{-r_\perp^2/(r_0/2)^2} \left(e^{-z_1/z_0} + e^{-z_2/z_0} \right). \qquad (5)$$

Here $r_\perp$ is the magnitude of the projection of the vector $\vec{r}_1 - \vec{r}_2$ between oxygens onto a plane parallel to the metal walls, and z_1 and z_2 are the distances of oxygens 1 and 2 from the wall. Values of the parameters appear in Ref. 69, where the resulting oxygen densities from the classical and direct dynamics simulations are compared. While the densities are rather similar, the potential distributions are not, as emphasized above.

The molecular dynamics calculations were carried out on a slab of water of width 18.6 Å between two ideal metal walls separated by 20.1 Å. We used periodic boundary conditions in the directions parallel to these model electrodes and periodically continued the sample in the direction normal to the model electrodes for the purpose of calculating Coulomb potentials including image charges in the electrodes as described in Ref. 71. We take account of charges on the electrodes here by a method similar to the one used in Refs. 44 and 71, imposing a constant field across the electrolyte and deducing from the resulting simulation both the equivalent charge on the electrodes and the potential drop across the interface. However, for the study of the cuprous–cupric electron transfer we added a potential that takes account in a mean field way of the presence of the ions of the background electrolyte as described later.

To describe the interaction between the reacting copper ions and this model of water, our collaborator, Larry Curtiss of Argonne [69], derived pair potentials for Cu^{1+} and Cu^{2+} interacting with a water molecule by fitting energies calculated from *ab initio* molecular orbital theory at the Hartree–Fock level. In previous work [72] on solvation of ferrous and ferric ions, we found that the neglect of the many-body interactions leads to overbinding of the solvated ion. In the calculations on the cuprous–cupric electron transfer, Curtiss used a new method to effectively incorporate many-body effects into the pair potential in order to obtain more accurate solvation energies. This was done in two steps. First, the energies for the $Cu^{n+} \ldots H_2O$ pair interaction were fitted to a potential of the form

$$V = Ae^{-BR_{CuO}} - \frac{C}{R_{CuO}^4} - \frac{D}{R_{CuO}^6} - \frac{F}{R_{CuO}^{12}} + \frac{q_O q_{Cu}}{R_{CuO}} + \frac{q_H q_{Cu}}{R_{CuH}} + \frac{q_{H'} q_{Cu}}{R_{CuH'}}. \qquad (6)$$

Second, *ab initio* calculations were carried out to determine the Cu–O distances, binding energies, and stretching force constants of hexaaquo Cu^{n+} clusters, $Cu^{n+}(H_2O)_6, n = 1, 2$. The A, B, and C parameters in the last equa-

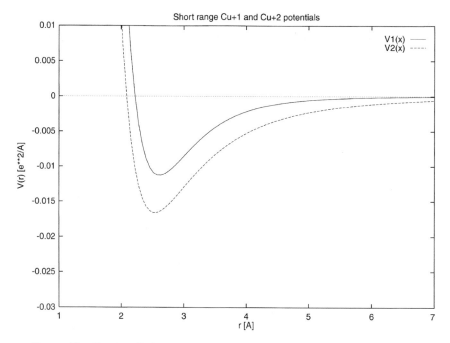

Figure 19. Cuprous–H_2O and Cupric–H_2O pair potentials deduced from *ab initio* electronic structure calculations as described in the text. From Ref. 69, by permission.

tion were then varied to obtain optimal Cu–O distances, binding energies, and breathing force constants of the cluster as close as possible to the *ab initio* results using the pair potential for the Cu–OH_2 interactions and the Toukan–Rahman pair potential for the water–water interactions in the cluster. The parameters for the final pair potentials are given in reference [69] and the potentials (with the Coulomb interactions subtracted) are shown in Fig. 19.

The *ab initio* molecular orbital calculations were done on the ground states of the hydrated copper ion clusters, $Cu^{n+}(H_2O)_6$, having D_{2h} symmetry. For the singly charged clusters, this corresponded to Cu^+ in an ^1S (d^{10}) state. For the doubly charged clusters, this corresponded to Cu^{2+} in an ^2D (d^9) state with the d electron removed from the molecular orbital that gives the ground state of the cluster. In calculations on the octahedral cluster, $Cu^{2+}(H_2O)_6$, the Jahn–Teller effect was not considered; that is, all CuO distances were held fixed. The basis set employed in the calculations consisted of a contracted $[9s3p2d]$ basis set on Cu, a $[3s2p]$ contraction for oxygen, and a $[2s]$ contraction for hydrogen as described in detail in a previous study of Cu^{+1} and Cu^{+2} hydrated clusters [73]. A d function having an exponent of 0.85 is added to oxygen. The HOH angle and OH distances of the H_2O mole-

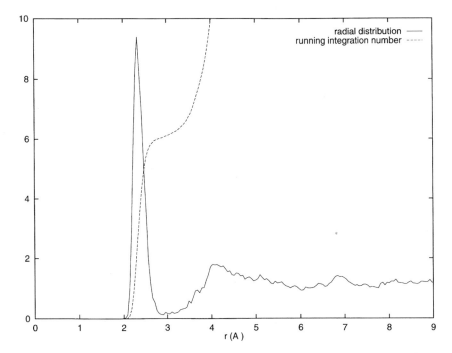

Figure 20.   Calculated oxygen density around the cuprous ion using the potential from first principles calculations as described in the text. From Ref. 69, by permission.

cule were held rigid at their experimental values of $104.5°$ and $0.958 \, \text{Å}$, respectively.

From the interaction potentials just described, we obtained structure of the copper ions in the water of expected form by molecular dynamics simulation, as shown in the radial distribution functions plotted in Figs. 20 and 21, which were obtained with the ion near the center of the simulation cell with zero applied field. The solvation energetics, reported in Table II, are in good agreement with the experimental values, though the uncertainties are significant on the electrochemical scale [74]. Further details concerning the simulations giving these results appear in Ref. 69.

We first used the model described to estimate the equilibrium potential for the electron transfer reaction. We took considerable trouble, some of which is described here with more detail in Ref. 69, to assure that the equilibrium potential was reasonably well-described by the model, because if a simulation model does not correctly reproduce the electrode charge density at which the two ionic charge states are in equilibrium, then the simulation may proceed in a very unrealistic electrostatic environment. This can lead to qualitatively misleading results. Unlike the ferrous–ferric reaction, the cuprous–cupric

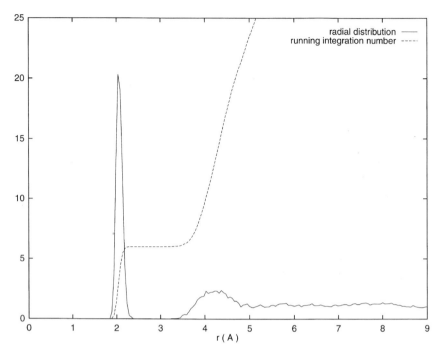

Figure 21. Calculated oxygen density around the cupric ion using the potential from first principles calculations as described in the text. From Ref. 69, by permission.

TABLE II
Calculated and Observed Solvation Energies of Copper Ions

	Simulation (kJ/mol)	Experiment[a] (kJ/mol)
$E_{\text{solvation}}$, Cu^{1+}	580 ± 70	576
$E_{\text{solvation}}$, Cu^{2+}	2085 ± 70	2105
ΔE	1505 ± 70	1529
ΔF	1485 ± 70	1481

[a]Y. Marcus, *Ion Solvation*, John Wiley & Sons, Chichester, 1985.

system is at equilibrium when the charge on the electrode is low, around $10 \, \mu C/cm^2$. The equilibrium is achieved when the concentrations of the Cu^{1+} and Cu^{2+} are given by

$$\frac{[Cu^{2+}]}{[Cu^{1+}]} = e^{-\beta \Delta F}. \tag{7}$$

(More precisely, the experimentally estimated activity ratio is equal to the right-hand side of this equation. However, the difference is insignificant for

TABLE III
Contributions to the Free Energy Difference

Symbol	Meaning	Source	Typical Value (eV)
I	Ionization energy	Experiment	20.3
ΔF_{solv}	Solvation difference	Simulation	15.4
$\Delta\phi^{\sigma}_{H_2O}$	Electrostatic drop	Simulation	0.3
Φ	Work function	Experiment	4.6

the problem considered here.) The components of the free energy difference ΔF are

$$\Delta F = I - \Delta F_{solv} - \Delta\phi^{\sigma}_{H_2O} - \Phi, \tag{8}$$

where I is the ionization potential, ΔF_{solv} is the difference of solvation energies, $\Delta\phi^{\sigma}_{H_2O}$ is the potential energy drop between the electrode and the ion, and Φ is the electrode work function. Approximate values for these quantities in the standard case when $\Delta F = 0$ are shown in Table III.

We have taken a value of the work function of 4.6 eV which is consistent with reported values for some copper electrodes (polycrystalline and Cu(100) from Ref. [75] and Cu(111) from Ref. 76), though reported values fall within a range of ± 0.4 V. One sees that the electrostatic potential drop required to make $\Delta F = 0$ under these assumptions is small (0.3 eV), so that little charge is required on the electrode at equilibrium. Using these values, our classical molecular dynamics model, without taking account of the screening electrolyte, gives an equilibrium field at equal reactant and product concentrations of approximately $E_0/4\pi = 0.00035 \pm 0.00015\, e/\mathring{A}^2$, corresponding to an electrode charge of about $0.006 \pm 0.002\, e/\mathring{A}^2$, or $10 \pm 3\, \mu C/cm^2$. Calculations of the differential capacitance of copper electrodes with low-index crystal faces that we described in the previous section indicate that such a charge corresponds to an electrode potential near 0.25 ± 0.1 V above the potential of zero charge, which is similar to the estimate above. Using the experimentally determined [77] potential of zero charge for the Cu(111) electrode, at -0.45 V/SCE, results in a value for the standard equilibrium potential around -0.2 V/SCE, which is in reasonable agreement with the experimental value of -0.09 V/SCE. But we note that the accuracy of the methods, even assuming the complete validity of the model, is never better than a few tenths of volts. Thus, detailed quantitative simulation of detailed electrochemical quantities which are abundantly available from experiment with precision of hundredths of volts, remains very difficult.

On the other hand, in the experiments on the cuprous–cupric rate which we were trying to understand in Ref. 69, the electrolyte concentration was ~ 1 M and resulted in screening lengths so short that it was unreasonable to ignore them, as in the estimate above. Accordingly, in the calculations of the barrier height for the cuprous–cupric electron transfer, we included the effects of the electrolyte ions through a self-consistent procedure in which a continuum charge density due to the ions of the supporting electrolyte was included in the simulation. This inclusion of ionic screening allowed us to describe the free energy surface for the copper ions relative to a bulklike region in which the field is essentially zero and allowed an unambiguous comparison with experimental results. (Our procedure did not take account of rearrangements of the ions of the electrolyte which are not reacting, due to the presence of the reacting acceptor or donor copper ion, but it did take account of the average electrolyte charge redistribution arising from the average charge on the electrode.) We omit the details of this self-consistent mean field method for taking account of the ionic charge here, referring the reader to Ref. 69. (Note that we did *not* calculate a "Frumkin correction" to the experimentally estimated potential drop from the electrode to acceptor. Such corrections are negligible at high ionic strengths.)

With this molecular dynamics model we first determined the electrode charge and corresponding field associated with equilibrium of the cuprous–cupric reaction as follows. If the ion concentrations of the initial- and final-state ions of the electron transfer process in the bulk of the solution are C_i and C_f, respectively, then the ratio of these concentrations when the electrode is in equilibrium with the bulk so that equal oxidizing and reducing currents flow is given by the condition in Eq. (7). ΔF is the free energy difference between the initial- and final-state ions far from the electrode, but in the presence of a charge on the electrode. We may regard this equation, for a fixed concentration ratio, as determining the equilibrium charge (and, through the capacitance, the equilibrium potential) for the reaction. To calculate ΔF we carry out, as in Ref. 46, simulations of the molecular dynamics model in the mixed potential $V_\lambda = \lambda V_i + (1 - \lambda) V_f$ with $0 < \lambda < 1$. Here V_i and V_f (denoted V_i' and V_f' in Ref. 46) are total potential energies of the classical molecular dynamics system when the ion is in the initial and final charge state, respectively. They contain constants taking account of the ionization energy of the cuprous and cupric ions and of the work function of the copper surface. However, we treat the work function as constant here and given by the vacuum value, for which we use the value 4.57 eV consistent with the (rather wide) range of published experimental values. (The position dependence of the effective work function used in Ref. 46 arose from a double counting error.) Then the free energy difference ΔF in Eq. (7) is obtained from [46]

$$\Delta F = -\int_0^1 d\lambda \langle (V_f - V_i) \rangle \lambda, \qquad (9)$$

where the ion should be in the bulk solution far from the electrode. In the calculations we placed the ion at the center of the simulation box, as far as possible from the model electrodes. The ionization energy was set to the experimental value of $I = 20.29\,$eV [78], and the work function was chosen to be $\Phi = 4.57\,$eV. Using an experimental bulk concentration ratio of 0.0025, close to the one used by the experimenters, we obtained a prediction of $\sigma_{eq} = 5\,\mu C/cm^2$ for the charge on the electrode at equilibrium. This corresponds to a potential drop relative to zero charge of 0.25 V. This may be compared to the difference of 0.5–0.8 eV between the experimental equilibrium potential of 0.05 V/SCE and the potential of zero charge of the interface, which has been reported to be -0.45 V/SCE for the Cu(111) surface and -0.79 V/SCE for the Cu(100) surface [77]. In view of the facts that the experimental values for the point of zero charge are subject to significant uncertainty [79] and that our model for the electrode neglects some known contributions to the capacitance due to the electronic structure of the surface, we may regard this model result for the equilibrium charge and potential as being in satisfactory agreement with what is now known experimentally.

These calculations of the equilibrium potential illustrate a generic challenge for modeling of electrochemical interactions. Note, for example, in Table III that the estimated equilibrium potential, of the order of a few tenths of volts relative to the point of zero charge, must be calculated as the difference between quantities as large as several tens of eV. Fortunately the ionization potentials are certainly known with the needed accuracy, but the work functions of electrodes are not. We discussed further evidence that inaccuracies in estimated work functions and points of zero charge are the most likely source of error in estimates of equilibrium potentials in Ref. 44. For the purposes of calculating activation barriers, the accuracy of the estimates of the equilibrium potential are adequate to assure for the cuprous–cupric problem that errors in the charge on the electrode at equilibrium have a smaller effect than the effects of other uncertainties in the calculation. On the other hand, to actually calculate voltammagrams, for example, from first principles would require another order of magnitude in accuracy in calculation of the energies. We do note, however, that on the other hand for the cuprous–cupric electron transfer and similar problems it is important to do at least as well as we have done in calculating the charge on the electrode at equilibrium. In particular it is not sufficient, as some authors have done, to simply calculate barriers at zero charge on the electrode, hoping that the effect of the electrode charge at equilibrium is insignificant. Furthermore, it is important to assure that simulation models being used to calculate charge transfer barriers do at least as well as Table II in reproducing experimental solvation energies, in order to assure adequate estimates of the equilibrium field.

We calculated activation free energies using the molecular dynamics model described above under the alternative assumptions that the reaction was taking place in the diabatic and in the adiabatic limit. For each calculation we required knowledge of the matrix element $\Lambda(z)$ describing mixing of the electronic states of the electrode with states that describe the electron on the Cu^{2+} acceptor in solution. $\Lambda(z)$ is the least well characterized aspect of our theoretical model. Larry Curtiss used the extended Hückel (EH) molecular orbital method [80] to investigate the magnitude of the electronic matrix element $\Lambda(z)$. For the matrix element calculation, the Cu(100) surface was modeled by a 25-atom copper atom cluster with 16 atoms in the first layer and 9 atoms in the second layer. Structural parameters were taken from bulk crystal data for face centered cubic copper. $\Lambda(z)$ for electron transfer from a hydrated cupric cation to the copper surface was investigated [81] by obtaining the activated state by adjustment of the ligand atoms around the reacting center until the interacting orbitals are equally distributed on the reacting centers and the electrode. The energy difference between the donor/acceptor orbitals then gives $\Lambda(z)$. Such a calculation of $\Lambda(z)$ is quite approximate. An "apex" approach of the hydrated ion to the copper surface was used. The water molecule on the surface was at a distance of 2.0 Å (O–surface distance) in a onefold site on a Cu(100) surface. The calculated $\Lambda(z)$ as a function of the distance of the hydrated ion from the surface is given in Ref. 69. $\Lambda(z)$ was found to be exponential in z, and it was described quite well with the equation $\Lambda(z) = \Lambda^0 e^{-z/a}$, where $\Lambda^0 = 4 \times 10^7 \, cm^{-1}, a = 0.54$, and z is the distance of the ion from the surface in Å. Other types of configurations could allow for closer approach of the ion to the surface and a larger $\Lambda(z)$ value. Possibilities include a configuration in which one of the waters of the hydration shell of the ion is replaced by a water of the monolayer of water on the metal surface or a configuration where a face (instead of an apex) of the octahedral solvation shell of the ion approaches the water monolayer on the surface. This latter configuration was found to give larger $\Lambda(z)$ values and a closer approach in the case of homogeneous ferric–ferrous electron transfer [81]. This calculation of the matrix element is the least satisfactory feature of our calculations of the electron transfer rate, because it was not possible to fully explore the effect of all the water or to make calculations for a comprehensively large number of atomic configurations. A more satisfactory approach would calculate the electron wave functions in the transition state continuously as the water molecules move in a molecular dynamics calculation. This has not proved possible so far, so the approach described here is a kind of compromise. We discuss this somewhat further below.

The diabatic limit for the electron transfer will obtain in the case that $\Lambda(z)$ is much less than $k_B T$ at separations of the copper acceptor or donor at which the maximum transfer rate occurs. If that diabatic condition holds, then the rate may be calculated as described in Ref. 46:

$$\text{rate} = \frac{2\pi}{\hbar} \langle |\Lambda(\{R\})|^2 \delta(V_i(\{R\}) - V_f(\{R\})) \rangle_i, \tag{10}$$

where we use the notation $\langle f \rangle_a$ to indicate an equilibrium average of the quantity f in a system with a potential energy function corresponding to state a. In Eq. (10), V_i and V_f (denoted V_i' and V_f' in Ref. 46) are total potential energies of the classical molecular dynamics system when the ion is in the initial and final charge state, respectively. They contain constants taking account of the ionization energy of the cuprous and cupric ions and of the work function of the copper surface. The coordinate z describing the distance of the ion from the electrode can be separated out of the last equation which then becomes

$$\text{rate} = \int dz k(z) C_i(z) \tag{11}$$

in which

$$\kappa(z) = \frac{2\pi}{\hbar} |\Lambda(z)|^2 \langle \delta(V_i(\{R\}) - V_f(\{R\})) \rangle_{i,z}, \tag{12}$$

where $\langle \ldots \rangle_{i,z}$ is an equilibrium average under the initial stage dynamics with the constraint that z is fixed. $C_i(z)dz$ is the probability of finding the initial-stage ion in the range of distances z to $z + dz$ from the electrode. Equation (11) will, of course, arise on general phenomenological grounds in both the adiabatic and diabatic limits.

In the canonical ensemble, $C_i(z)$ takes the form

$$C_i(z) = \frac{\int d\Omega e^{-\beta V_i}}{\int dR e^{-\beta V_i}}, \tag{13}$$

where the integral $\int d\Omega \ldots$ is over all coordinates $\{R\}$ except z. Noting that $C_i(z)$ will vanish as $1/L$, where L is the sample depth in the z-direction, we can produce a finite quantity in the limit $L \to \infty$ and make contact with the experimentally measured electrochemical rate constant κ_{ec} (which has units of velocity) by rewriting Eq. (11) as

$$\kappa_{ec} = \lim_{L \to \infty} \text{rate} = \int dz k(z) C_i'(z) \tag{14}$$

in which

$$C_i'(z) = \lim_{L \to \infty} LC(z)$$

can be interpreted as the ratio of the reactant ion density at z to the reactant ion density in the bulk of the electrolyte.

We write $C_i'(z)$ in terms of a free energy $F_z(z)$ a follows:

$$C_i'(z) = \frac{\int d\Omega' dz' e^{-\beta V_i(\{\Omega'\}, z')} \delta(z - z')}{\int d\Omega' dz' e^{-\beta V_i(\{\Omega'\}, z')} \delta(z_{bulk} - z')} \equiv e^{-\beta F_z(z)} \tag{15}$$

with z_{bulk} a position in the bulk of the sample (in simulation, we take z_{bulk} to be the center of the simulation cell). The derivative

$$\frac{\partial F_z}{\partial z} = -k_B T \frac{\partial}{\partial z} \ln \left[\frac{\int d\Omega' dz' e^{-\beta V_i(\{\Omega'\}, z')} \delta(z - z')}{\int d\Omega' dz' e^{-\beta V_i(\{\Omega'\}, z')} \delta(z_{bulk} - z')} \right]$$

$$= -k_B T \frac{\int d\Omega' dz' e^{-\beta V_i(\{\Omega'\}, z')} \frac{\partial}{\partial z} \delta(z - z')}{\int d\Omega' e^{-\beta V_i(\{\Omega'\}, z')}} \tag{16}$$

can be integrated by parts to yield

$$\frac{\partial F_z}{\partial z} = -k_B T \frac{\int d\Omega' dz' e^{-\beta V_i(\{\Omega'\}, z')} \delta(z - z') \left(-\beta \frac{\partial V_i}{\partial z'} \right)}{\int d\Omega' e^{-\beta V_i(\{\Omega'\}, z')}} = \left\langle \frac{\partial V_i}{\partial z} \right\rangle_{i,z}, \tag{17}$$

Thus the free energy associated with $C'(z)$ can be obtained from simulation by constructing the potential of mean force exerted on the ion in its initial state while fixed at various positions z.

We write $\kappa(z)$ in terms of a free energy at a fixed value ΔE of the difference $V_i - V_f$ by defining

$$F_{i,z}(\Delta E) = -k_B T \ln \left[\frac{k_B T \int d\Omega e^{-\beta V_i} \delta(\Delta E - V_i + V_f)}{\int d\Omega e^{-\beta V_i}} \right]. \tag{18}$$

An arbitrary factor of $k_B T$ is inserted inside the definition in order to make the units of $F_{i,z}(\Delta E)$ those of energy. (The result in the next equation is independent of this factor.) Then the rate constant can be written as

$$\kappa_{ec} = \frac{2\pi}{\hbar k_B T} \int dz |\Lambda(z)|^2 e^{-\beta F_{i,z}(\Delta E = 0)} C'(z) \tag{19}$$

or, incorporating the free energy associated with $C'(z)$,

$$\kappa_{ec} = \frac{2\pi}{\hbar k_B T} \int dz |\Lambda(z)|^2 e^{-\beta F_i(\Delta E = 0, z)} \tag{20}$$

with the total free energy

$$F_i(\Delta E, z) \equiv F_{i,z}(\Delta E) + F_z(z). \tag{21}$$

The last two equations show that the barrier height for the rate has two con-tributions: One, $F_{i,z}(\Delta E = 0)$, is due to solvent rearrangement and is essen-tially the contribution which is assumed to dominate in most formulations of Marcus theory. The second term, $F_z(z)$, is the contribution to the barrier aris-ing because the reactant ion must overcome a repulsion, partly arising from the charge on the electrode, in order to approach the electrode closely enough so that the coupling constant becomes significant. We sometimes refer to this as the approach free energy. In our model, the approach free energy dominates the barrier for the cuprous–cupric electron transfer. (We did not assume this. It emerges from the calculation.) The larger point is that, in all electron trans-fer problems at electrodes, there will be contributions from solvent rearrange-ment and from approach, and the question of which one will dominately determine the barrier is one of quantitative detail. This point has also been emphasized by Schmickler and coworkers [66].

For the cuprous–cupric interaction, calculations using the diabatic formal-ism (reported in more detail below) showed that the maximum transfer rate occurred at separations of the copper ion from the electrode at which the diabatic approximation was not justified (and gave too large an activation energy, relative to experimental values). Accordingly, we constructed an adiabatic free energy surface by use of the molecular dynamics simulations in which the forces on the atoms were determined from the potential energy function given by mixing the potential energy functions associated with the product and reactant states through the mixing matrix element $\Lambda(z)$—that is, by diagonalizing the matrix

$$V = \begin{pmatrix} V_i & \Lambda \\ \Lambda & V_f \end{pmatrix}. \tag{22}$$

Free energy surfaces were constructed analogously to the diabatic case at fixed z using

$$F_{-,z}(\Delta E) = -k_B T \ln k_B T \langle \delta(\Delta E - V_- + V_+) \rangle_{-,z} \tag{23}$$

with $V_\pm$ the upper and lower diagonalized energy surfaces. (For clarity of presentation and comparison with the diabatic case, in the plots we use the coordinate $\Delta E' = V_- - V_+ - 2\Lambda(z)$ which goes to zero at the minimum energy gap between the upper and lower energy surfaces, and we also choose the sign of $\Delta E'$ to match the sign of ΔE in the diabatic case.) The total free energy $F_-(\Delta E, z)$ is constructed as was done for the diabatic case using Eq. (21) and (23). The numerical computation of $F_i(\Delta E, z)$ is rendered numerically practical [46] by the introduction of a new potential energy function which mixes the potentials associated with the initial and final states using a scalar mixing parameter λ which takes values (chosen for computational convenience) between 0 and 1

$$V_\lambda \equiv (1 - \lambda)V_i + \lambda V_f \tag{24}$$

in the diabatic case or

$$V_\lambda \equiv (1 - \lambda)V_- + \lambda V_+ \tag{25}$$

in the adiabatic case in the expression for $F_i(\Delta E, z)$ as follows:

$$\frac{\int d\Omega e^{-\beta V_{i,-}} \delta(\Delta E - V_{i,-} + V_{f,+})}{\int d\Omega e^{-\beta V_{i,-}}}$$
$$= \frac{\int d\Omega e^{-\beta(V_\lambda + \lambda \Delta E)} \delta(\Delta E - V_{i,-} + V_{f,+})}{\int d\Omega e^{-\beta V_{i,-}}}$$
$$= e^{-\lambda \Delta E \beta} \frac{\int d\Omega e^{-\beta V_\lambda} \delta(\Delta E - V_{i,-} + V_{f,+})}{\int d\Omega e^{-\beta V_\lambda}} \frac{\int d\Omega e^{-\beta V_\lambda}}{\int d\Omega e^{-\beta V_{i,-}}}. \tag{26}$$

Thus

$$F_{i,-}(\Delta E, z) = -k_B T \ln k_B T \langle \delta(\Delta E - V_{i,-} + V_{f,+}) \rangle_\lambda - \lambda \Delta E + F_\lambda - F_{i,-}, \tag{27}$$

where $\langle \cdots \rangle_\lambda$ is an average in the ensemble defined by the potential function V_λ and the last two terms are

$$F_\lambda - F_{i,-} = -k_B T \ln \frac{\int d\Omega e^{-\beta V_\lambda}}{\int d\Omega e^{-\beta V_{i,-}}}. \tag{28}$$

These expressions are valid in principle for any λ. In practice, they are useful because λ can be varied so that, for a small range of values of ΔE, a specific λ

will result in an average value of $V_{i,-} - V_{f,+}$ that lies within this small range of ΔE. Then when a simulation is made to compute the first term in Eq. (27), many of the points on the trajectory will satisfy the delta function and sampling statistics are good. (This is a version of the well-known method of importance sampling.) The result of a series of such calculations of the factor $-k_B T \ln \langle \delta(\Delta E - V_{i,-} + V_{f,+}) \rangle_\lambda$ can in practice often be used to estimate the term $F_\lambda - F_{i,-}$ simply the choosing $F_\lambda - F_{i,-}$ so that $F_{i,-}(\Delta E, z)$ is independent of which λ value is used to evaluate it. We have taken several other approaches to the computation of $F_\lambda - F_{i,-}$. The most direct is simply to use Eq. (28) in the form

$$F_\lambda - F_{i,-} = \int_0^\lambda \langle (V_{f,+} - V_{i,-}) \rangle_\lambda d\lambda \qquad (29)$$

[with $\lambda = 1$ this is Eq. (9)]. Using the combination of parameters corresponding to the calculated equilibrium electrode charge of $\sigma_{eq} \equiv \sigma = 5\mu C/cm^2$ (see above), we evaluated the diabatic free energy surfaces $F_{i=Cu^{2+}}(\Delta E, z)$ and $F_{f=Cu^{1+}}(\Delta E, z)$ and evaluated the adiabatic surface $F_-(\Delta E, z)$ at room temperature with the matrix element corresponding to the apex configuration of the solvation shell with one intervening water molecule between the shell and the electrode with results shown in Figs. 22 and 23.

In Fig. 22, the lower of the each of the two diabatic surfaces in shown. The free energy as a function of ΔE at fixed z is close to parabolic for each surface (due to the approximately gaussian sampling of ΔE), and the curvature and

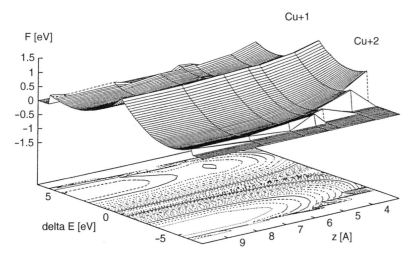

Figure 22. Free energy surfaces in the diabatic limit.

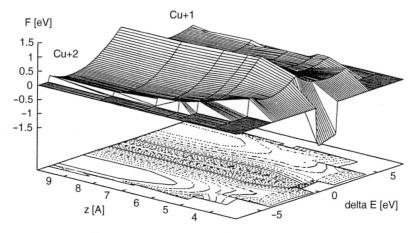

Figure 23. Free energy surfaces in the adiabatic case.

the position of the minimum with respect to ΔE depend weakly on z. The Cu^{2+} and Cu^{1+} states correspond to negative and positive values of ΔE, respectively. The height of the "solvation fluctuation" barrier at $\Delta E = 0$ is 1.1 eV near the center of the box. The surface rises as z decreases, reflecting a combination of effects due to both (a) the positive charge on the electrode repelling the positive ions and (b) a loss of solvation energy. The rise is greater on the Cu^{2+} side than for the Cu^{1+}. A strictly diabatic interpretation of this free energy surface (which is incorrect in the regions of large mixing matrix element, i.e., $z \gtrsim 5$ Å) yields the position z_{max} of maximum contribution to the rate when the integrand of Eq. (20) is maximized. This corresponds to

$$\frac{\partial}{\partial z}\left(\ln\frac{|\Lambda(z)|^2}{\Lambda_0^2} - \beta F_i(\Delta E = 0, z)\right) = 0. \tag{30}$$

Taking account of the numerical uncertainties, we estimate that the solution to Eq. (30) occurs near $z_{max} \sim 4$–5 Å. The approximate activation energy in this diabatic interpretation is

$$F_{act} \equiv -\frac{\partial}{\partial \beta} \ln \kappa(z_{max}) = F_i(\Delta E = 0, z_{max}) = 1.3 \pm 0.1 \, \text{eV}, \tag{31}$$

which is much higher than the experimentally reported activation energy $(0.33 \pm 0.5 \, \text{eV})$

TABLE IV
Results for the Kinetic Barrier for the Cuprous–Cupric Electron Transfer Reaction

Quantity	Values (eV $\pm$ 0.1 eV)	(kJ/mol $\pm$ 10 kJ/mol)
Diabatic barrier	1.30	125
Adiabatic barrier, first Λ	0.43	41
Adiabatic barrier, second Λ	0.17	16
Experiment	0.33 $\pm$ 0.05	31.7 $\pm$ 4.8

The lower adiabatic free energy surface $F_-(\Delta E, z)$ is shown in Fig. 23. (Zero values correspond to no data collected.) For $z \gtrsim 5\text{Å}$ this surface is the same as shown in the diabatic case. However, as the mixing becomes larger with decreasing z, qualitative differences appear. The solvation fluctuation barrier along the coordinate ΔE decreases, while the "approach free energy" F_z reaches a maximum and falls precipitously as the exponential matrix element grows. While the plunge of the free energy surface toward negative values as z becomes very small is not described correctly in our classical model (which neglects fine structure of the electrode surface and the complex dependence of the matrix element on $\{\Omega\}$ including electrode coordinates), the dynamics in this "black box" region at very small z are relatively unimportant for determining the activation energy if the saddle point in ΔE and z on the adiabatic surface at the onset of the plunge corresponds to the activated state of the Cu^{2+}/Cu^{1+} reaction. This occurs (for this choice of $\Lambda(z)$) at $F_-(\Delta E, z) = 0.4 \pm 0.1$ eV, which is in much closer agreement with experiment than the diabatic result. As we discuss below, taking account of the uncertainty in $\Lambda(z)$ increases the uncertainty in the barrier value, but not enough to change the conclusion that the adiabatic model gives better agreement with experiment than the diabatic one. We summarize results for the activation energies in Table IV.

Of the uncertainties in the calculated numbers, we believe the most important to arise from the matrix element $\Lambda(z)$. As discussed above, the matrix element was calculated in the "apex" configuration, which involves an intermediate water molecule between the wall and the solvation shell. Similar previous calculations for the case of iron indicated that the matrix element could be larger in other configurations. In particular, the matrix element for a similar geometry but without the extra water molecule had similar exponential behavior in z but with a larger prefactor corresponding to approximately 1 Å shift in the $\Lambda(z)$ curve. To explore the effects of such differences in the matrix element, we ran adiabatic simulations at the previously determined equilibrium charge $\sigma = 5 \, \mu C/cm^2$ with just this prefactor and found that the

TABLE V
Calculational Estimates of the Transfer Coefficient

σ (μC/cm^2)	F_{act} (eV)	δF_{act} (eV)	$\delta \Phi$ (eV)	β
0.0	0.37	0.06	0.25	0.24
5.0	0.43			
9.0	0.58	0.15	0.81	0.19

effect of the larger matrix element is a reduction in the approach barrier by 0.26 eV (also shown in Table IV).

To obtain the transfer coefficient from this simulation, we require a calculation at fields different from the equilibrium field. We imposed two ionic screening charges of 9μC/cm^2 and 0μC/cm^2 different from the estimated charge at the equilibrium potential and recalculated the barrier height. Results are summarized in Table V.

From Table V one sees that we are obtaining a transfer coefficient of about 0.2 ± 0.5, which is much lower than the experimental value of 0.6 as described above. The calculated transfer coefficient is probably low due to the fact that in this classical simulation, the vacuum region between the infinitely thin surface charge and the first water layer contains a significant fraction of the net potential drop across the cell. (See Fig. 12. The data in Fig. 12 are from Ref. 46, but similar results were obtained for the cuprous–cupric simulation.) In contrast, in results from the calculations described in the previous section and in Ref. 36 for the electrode electronic structure, the quantum mechanically treated electrons on the electrode have a finite spatial extent and lower the field at the interface. On the other hand, since the barrier in this simulation is not very sensitive to the net potential drop, the result for the barrier height is not as sensitive as the transfer coefficient to this problem.

Our calculations of the activation free energy barrier for the cuprous–cupric electron transfer were not precise enough to permit a very accurate estimate of the absolute value of the exchange current for comparison with experiment. In principle, a determination of the absolute rate from the activation energy requires a calculation of the relevant correlation function [82] when the ion is in the transition region within the molecular dynamics model. We have not carried out such a calculation, but can obtain some information about the amplitude by comparing experiments with the transition state theory expression [84]

$$k_{ec,TST} = \sqrt{k_B T / 2\pi m_r} \exp(-\beta F_{act}), \qquad (32)$$

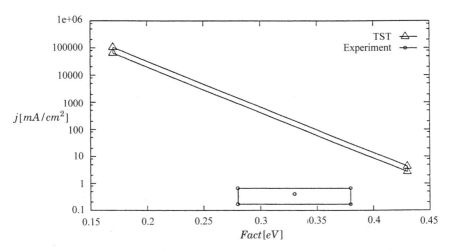

Figure 24. Transition state theory estimates of the exchange current. From Ref. 69, by permission.

where m_r is a mass associated with the dynamics of the reaction coordinate and F_{act} is the activation energy that we have calculated. This expression does not depend sensitively on m_r in the present case. We evaluated Eq. (32) for the two choices that m_r is the mass of a copper ion plus 6 water molecules and that m_r is the mass of just the copper ion. Figure 24 shows the exchange current density deduced from Eq. (32) (with a cupric ion bulk concentration of 0.1 mol/L) as a function of F_{act} at room temperature for these two choices of m_r. We also indicate the calculated values of F_{act} for the matrix element functions used as well as the range of experimental exchange currents and activation energies determined by the Argonne experiments [69]. From these results we concluded that the calculated activation energies are consistent with the experiment within the uncertainties but that transition state theory overestimates the reaction rate by about three orders of magnitude. Such deviations of observed rates from transition state theory are well known and are usually attributed to effects of damping on the motion of the reactant. A qualitative understanding can be achieved from Kramers' theory [83]. In the present case, the overdamped limit of Kramers theory is relevant. Following the discussion of Ref. 83 the reduction factor, relative to transition state theory, is expected to be of order λ/l_{TS}, where λ is the mean free path of the reactant and l_{TS} is the size of the transition region. We estimated from data in Fig. 23 that $l_{TS} \approx .5$ Å so that a reduction of three orders of magnitude relative to the transition state theory would require a mean free path for the reactant of order $\lambda \approx 0.5 \times 10^{-3}$ Å. This is of order of the motion during a few time steps of the simulation and is not unreasonable. A more detailed calculation of the relevant correlation function within the molecular dynamics model is possible.

The Argonne experimental results [69] were consistent with an Arrhenius temperature dependence of the cuprous–cupric rate, though the uncertainties were substantial regarding this conclusion. We only performed simulations at one temperature and so did not check whether the model predicts a strictly Arrhenius behavior. The most obvious case in which the model would lead to a substantial non-Arrhenius behavior would arise if a large part of the barrier free energy were entropic in nature. In previous work on the ferrous–ferric interaction [46], such entropic effects were found to be very small. In the cuprous-cupric case, however, both the charges on the ions and the charge on the electrode is smaller, leaving open the possibility of a larger entropic contribution.

We should emphasize that this study of the cuprous–cupric electron transfer reaction has led to some rather unconventional conclusions, particularly with regard to the nature of the kinetic barrier, which we believe is an approach barrier and not the solvent rearrangement barrier of Marcus theory. Though solvent rearrangement is present in the model it does not dominate the barrier. We are quite confident of this conclusion because we were forced to it by constraining our model quite strongly by linking it quantitatively to first principles potentials and confirming that they gave correct solvation behavior and then finding that such a constrained model could not account for the barrier observed in the carefully controlled experiments by our collaborator Zoltan Nagy unless the reaction was adiabatic and the barrier had an approach character. Thus theory, simulation, and experiment were closely linked in this study.

Several other groups have made similar calculations of electron transfer rates at electrode–electrolyte interfaces. These include Rose and Benjamin [84, 85] and Straus and Voth [87], who made calculations of the ferrous ferric rate. Rose and Benjamin took the effects of charge on the electrode into account, whereas Straus and Voth did not. Straus and Voth concluded that the ferrous–ferric transition was adiabatic, whereas in our study [46] of the ferrous-ferric rate we concluded that it was diabatic. Through our study did include fields, the calculation of the ferrous–ferric rate reported reported in Ref. 46 was inadequate in two respects: The force fields used to describe the iron–water interactions did not reproduce the solvation energies at the same level of accuracy which we achieved in the cuprous–cupric case (Table II and Ref. 69). As emphasized earlier, an error in the solvation energy propagates to an error in the estimated charge on the electrode at equilibrium. Because the solvation energies are so large, a small percentage error in them leads to big errors in the charge and possibly to large, even qualitative errors in the calculation of the kinetic parameters. Second, as mentioned earlier and elsewhere, there was a double counting error in the electrostatics arguments in Ref. 46. For these reasons we think that the question of the nature of the ferrous–ferric electron transfer at a metal electrode (specifically gold) is not settled. Despite the various caveats just cited, one can trace the different conclusions of our paper [46] and of Straus et al. [86] concerning the adiabaticity of the ferrous–ferric electron transfer to the different iron–electrode separations at which it was concluded that most of the electron transfer takes place at equilibrium (roughly 7 and 5 Å, respectively). The larger distance that we used reduces the coupling and leads to a diabatic reaction, whereas if we had obtained a distance comparable to the one in Ref. 86 we would also have concluded that the reaction was adiabatic. However, we emphasize that one cannot make a quantitative estimate of this distance, as opposed to a physical guess, without including effects of the fields and going through an argument like that which we have reviewed in the cuprous–cupric case [see particularly Eq. (30)]. We attempted to do that in Ref. [46] but, because of the problems just cited, cannot be sure of the conclusion.

Though our focus here is mainly on metal–electrolyte interfaces, we should mention the very extensive simulations that are being carried out by Barton Smith [87] on electron transfer at semiconductor–electrolyte interfaces. As mentioned earlier, many aspects of the physics of semiconductor–electrolyte interface are quite different because of the very long screening lengths in the electrode. Smith's simulations have mainly sought qualitative insight, particularly concerning how localization of the wave function occurs as the electron transfers from the semiconductor to the metal, though his models are quite realistic. He uses innovative wave packet propagation methods for these studies.

V. DISCUSSION AND CONCLUSIONS

Here we have sought to show how currently accessible simulation techniques can be brought to bear on qualitative and quantitative issues in the study of electrode–electrolyte interface. In our view, which we hope is exemplified by the cases described in detail, the firmest and most enlightening conclusions can be drawn when there are very tight links between theory, simulation, and experiment. One hears it said, for example, that there is no need for more theory in electrochemistry because "we have the theory." But in the examples cited, we have seen that studies in which careful attention is paid to making simulations quantitatively realistic, qualitative conclusions can emerge which are not part of the currently accepted theoretical picture. This occurred in our studies of the fields near electrodes and also in our discovery of the importance of approach free energy in the kinetic barrier for the cuprous–cupric electron transfer.

Another general theme that we intend to illustrate by these examples is the linking of length and time scales in theory and simulation. We believe that this theme has wide application. It is well known that continuum theories (hydrodynamics, Debye–Hückel / Gouy–Chapman, macroscopic dielectric continuum theory, elasticity, etc.) work down to very small length scales, often around a nanometer, but not to atomic scales. This means that it is not useful except in some exceptional cases, such as biological systems where there is no continuum theory, to attempt huge simulations with microscopic specificity, such as molecular dynamics. At best, such multimillion particle simulations reproduce known results of continuum theory at large expense. What microscopic simulation can contribute is boundary conditions which always must be postulated for continuum calculations. These can be determined on nanometer scales that molecular dynamics find cogenial. The calculation of the capacitance of the copper–water interface which we described here was intended in part to illustrate how such a calculation goes. Similar approaches should work for determining boundary conditions for other continuum theories.

The linking of electronic structure determination with atomic dynamics is the other scale linking problem that we have discussed. In electrochemistry, we find that the most straightforward approach, in which parameters for force fields are determined from first principles and passed to molecular dynamics simulations, has limited applicability (though we are still using it for kinetic problems). The new and much celebrated direct dynamics methods provide a way to go beyond parameter passing, and we have seen some successes in attempts by ourselves and others to implement them, but also some limitations. We are not yet able to carry out such simulations for a full description of transition metals or transition metal oxides, and our estimates [51] and attempts suggest that it may be some time before we can. In addition to the tight binding methods that we mentioned earlier as a possible way out of this impasse, ultrasoft potentials, used by some workers for large simulations of metal–oxide interfaces, may also prove useful.

Acknowledgments

Our first debt of gratitude and appreciation must be paid to Larry Curtiss and collaborators at Argonne, with whom we have worked for more than 15 years on these problems. Evidence of the Curtiss group's contributions to our studies is very clear throughout this chapter. We also are happy to acknowledge a collaboration with Zoltan Nagy's experimental electrochemistry group at Argonne which has lasted nearly as long. Others who have contributed substantially to the studies discussed in this review include J. Hautman, A. Rahman, Y.-J. Rhee, J. Rustad, P. Schelling, B. Smith, A. Mazzolo, N. Yu, Y. Zhou, and M. Zhuang. We have benefited from continuing interactions with the groups of A. Nozik (NREL) and A. Felmy (PNNL). This work was supported in part by the University of Minnesota Super Computer Institute, by the US Department of Energy grant DE-FG02-91-ER45455, and by the National Science Foundation grant DMR-9522286.

References

1. A. Volta, *Philos. Trans. Roy. Soc. London* **2**, 403–431 (1800).

2. M. Faraday, *Philos. Trans.*, 507–522, (1833); *Philos. Trans.*, 675–710, (1833).

3. H. L. F. von Helmholz, *Ann. Phys.* **89**, 211 (1853); *Ann. Phys.* **7**, 337 (1879).

4. K. Itaya, *Nanotechnology (UK)* **3**, 185 (1992); T. Hachiya, K. Itaya, *Ultramicroscopy (Netherlands)* 42A,445 (1992); C. C. Chen and A. A. Gewirth, *Phys. Rev. Lett.* **68**, 1571 (1992).

5. B. M. Ocko, J. Wang, A. Davenport, and H. Isaacs, *Phys. Rev. Lett.* **65**, 1466 (1990).

6. P. Fery, W. Moritz, and D. Wolf, *Phys. Rev. B* **38**, 7275 (1988).

7. D. Aberdam, Y. Gauthier, R. Durand, and R. Faure, *Surf. Sci.* **306**, 114 (1994).

8. V. L. Shannon, D. A. Koos, and G. L. Richmond, *J. Chem. Phys.* **87**, 1440 (1987); G. G. L. Richmond, J. M. Robinson, and V. L. Shannon, *Prog. Surf. Sci.* **2**, 1 (1988).

9. D. Grahame, *Chem. Rev.* **41**, 441 (1947).

10. G. Gouy, *J. Phys. Radium* **9**, 457 (1910); *Compt. Rend.* **149**, 654 (1910); D. L. Chapman, *Phil. Mag.* **25**, 475 (1913).

11. O. Stern, *Z. Elektrochem.* **30**, 508 (1924).

12. N. F. Mott, *Proc. Roy. Soc. (London)* **A171**, 27 (1939).

13. W. Schottky, *Z. Phys.* **113**, 367 (1939); **118**, 539 (1942).

14. For example, *Local Density Approximations in Quantum Chemistry and Solid State Physics*. Proceedings of a Symposium on Local Density Approximations in Quantum Chemistry and Solid State Theory, Copenhagen, Denmark, 10–12 June 1982, edited by J. P. Dahl, J. Avery, Plenum, New York, 1984.

15. M. M. Francl et al. *J. Chem. Phys.* **77**, 3654 (1982).

16. N. P. Allen and D. J. Tildesley, *Computer Simulation of Liquids*, Clarendon Press, Oxford, 1987.

17. R. Car and M. Parrinello, *Phys. Rev. Lett.* **55**, 2471 (1985).

18. M. T. M. Koper, A. P. J. Jansen, R. A. van Santen, J. J. Lukkien, and P. A. J. Hilbers, *J. Chem. Phys.* **109**, 6051 (1998); M. T. M. Loper, J. J. Lukkien, A. P. J. Jansen, and R. A. van Santen, *J. Phys. Chem. B* (1999) (in press).

19. D. L. Price, Ph.D. thesis, University of Minnesota (1984), unpublished.

20. L. I. Daikhin, A. A. Kornyshev, and M. Urbakh, *Phys. Rev.* **E53**, 6192 (1996); *J. Chem. Phys.* **108**, 1715 (1998); M. Urbakh and L. I. Daikhin, *Phys. Rev.* **B49**, 4866 (1994); L. I. Daikhin and M. Urbakh, *Langmuir* **12**, 6354 (1996); L. I. Daikhin and M. Urbakh, *Phys. Rev.* **E59**, 1821 (1999).

21. A. K. Theophilos and A. Modinos, *Phys. Rev.* **B6**, 801 (1972).

22. J. W. Halley and D. Price, *J. Electroanal. Chem.* **150**, 347 (1983).

23. OK Rice. *J. Phys. Chem.* **30**, 1926 (1926).

24. R. J. Watts-Tobin. *Philos. Mag.* **6**, 133 (1961).

25. W. R. Fawcett. *J. Phys. Chem.* **82**, 1385 (1978).

26. R. Guidelli, *J. Electroanal. Chem.* **123**, 59 (1981).

27. G. Valette, *J. Electroanal. Chem.* **122**, 285 (1981); **138**, 37 (1982).

28. A. Hamelin, L. Stoicoviciu, and Silva. *J. Electroanal. Chem.* **229**, 107 (1987).

29. J. W. Halley, B. Johnson, D. Price, and M. Schwalm. *Phys. Rev.* **B31**, 7695 (1985).

30. J. W. Halley and D. Price, *Phys. Rev.* **B35**, 9095 (1987).

31. D. L. Price and J. W. Halley, *Phys. Rev.* **B38**, 9357 (1988).

32. W. Schmickler and D. Henderson, *J. Chem. Phys.* **80**, 3381 (1984).

33. J. P. Badiali, M. L. Rosinberg, and J. Goodisman, *J. Electrochem. Chem.* **143**, 173 (1983); **150**, 25 (1983).

34. A. Kovalenko and F. Hirata, *J. Chem. Phys.* **110**, 10095 (1999).

35. D. L. Price and J. W. Halley, *J. Chem. Phys.* **102**, 6603 (1995).

36. S. Walbran, A. Mazzolo, and J. W. Halley, *J. Chem. Phys.* **109**, 8076 (1998).

37. J. B. Strauss and G. A. Voth, *J. Chem. Phys.* **97**, 7388 (1993).

38. S. Izvekov, M. R. Philpott, and R. I. Eglitis, unpublished data.

39. M. Yamamoto, M. Kinoshita, and T. Kakiuchi, unpublished data.

40. R. Car and M. Parrinello, in: *Simple Molecular Systems at Very High Density*, edited by A. Polian, P. Loubeyre, and N. Boccara, Plenum, New York; *NATO Ad. St. Inst. Ser. B* **186**, 455 (1988); D. M. Wood and A. Zunger, *J. Phys. A* **18**, 1343 (1985).

41. L. Verlet, *Phys. Rev.* **159**, 98 (1967).

42. P. Zapol, C. Naleway, P. Deutsch, and L. A. Curtiss, unpublished data.

43. K. Heinzinger and E. Spohr, *Electrochim. Acta* **34**, 1849 (1989); E. Spohr and K. Heinzinger, *Chem. Phys. Lett.* **123**, 218 (1986).

44. J. W. Halley, S. Walbran, and D. L. Price, In: *Interfacial Chemistry*, edited by A. Wieckowski, Dekker, New York, 1999, pp. 1–18.

45. D. Henderson and L. Blum, *J. Chem. Phys.* **69**, 5441 (1978).

46. B. Smith and J. W. Halley, *J. Chem. Phys.* **101**, 10915 (1994).

47. I. R. Zhang, H. S. White, and H. Ted Davis, *The Electrochemical Society Proceedings*, Vol. 93-5, *Microscopic Models of the Electrode–Electrolyte Interface*, edited by J. W. Halley and L. Blum, the Electrochemical Society, Pennington, NJ, 1993, pp. 106–111.

48. G. Rickayzen, *The Electrochemical Society Proceedings*, Vol. 93-5, *Microscopic Models of the Electrode–Electrolyte Interface*, edited by J. W. Halley and L. Blum, the Electro- chemical Society, Pennington, NJ, 1993, pp. 143–153.

49. M. S. Fleischman and E. Pons, *J. Electrochem.* **261**, 301 (1989).

50. Y. Zhou, A. Mazzolo, J. W. Halley, and D. L. Price, *Int. J. of Thermophys.* **19**, 663 (1998).

51. S. Walbran, Ph.D. thesis, University of Minnesota, 1998, unpublished data.

52. D. L. Price, *J. Chem. Phys.*, **112**, 2973 (2000).

53. N. Troullier and J. L. Martins, *Phys. Rev.* **B43**, 1993 (1991).

54. L. Kleinman and D. M. Bylander, *Phys. Rev. Lett.* **48**, 1425 (1982).

55. S. G. Louie, S. Froyen, and M. L. Cohen, *Phys. Rev.* **B26**, 1738 (1982).

56. R. Naneva, V. Bostanov, A. Popov, and T. Vitanov, *J. Elec. Interfac. El. Chem.* **274**, 179 (1989).

57. P. K. Schelling, J. W. Halley, and N. Yu, *Phys. Rev.* **B58**, 1279 (1998).

58. N. Yu and J. W. Halley, *Phys. Rev.* **B51**, 4768 (1995).

59. M. Elstner, D. Porezag, G. Jungnickel, J. Elsner, M. Haugk, Th. Frauenheim, S. Suhai, and G. Seifert, *Phys. Rev.* **B58**, 7260 (1998).

60. W. J. Shack, T. F. Kassner, P. S. Maiya, J. Y. Park, and W. E. Ruther, *Nucl. Eng. Des.* **108**, 199 (1988).

61. F. Williams and A. Nozik, *Nature* **311**, 21 (1984).

62. *Proceedings, Symposium on Batteries and Fuel Cells for Stationary and Electric Vehicle Applications*, edited by A. R. Landgrebe and Z. Takehara, Electrochemical Society, Pennington, NJ, 1993.

63. R. Marcus, *J. Chem. Phys.* **24**, 966, 979 (1956); R. Marcus, *J. Chem. Phys.* **26**, 867, 872 (1957); R. Marcus, *Can J. Chem* **37**, 155 (1959); R. Marcus, *Trans. Symp. Electrode Processes, Philadelphia* **1959**, 239 (1957); R. Marcus, *Trans. N. Y. Acad. Sci.* **19**, 423 (1957); R. A. Marcus, *Dis. Faraday Soc.* **29**, 21 (1960); R. Marcus, *J. Phys. Chem.* **67**, 853, 2889 (1963); R. Marcus, *J. Chem. Phys*, **43**, 679 (1965).

64. V. G. Levich and R. R. Dogonadze, *Coll. Czech. Chem. Comm.* **26**, 193 (1961).

65. N. S. Hush and J. Ulstrup, in *Proceedings of the Conference Electron and Ion Transfer in Condensed Media, Theoretical Physics for Reaction Kinetics*, edited by A. A. Kornyshev, M. Tosi, and J. Ulstrup, World Scientific, Singapore, 1997, pp. 1–24.

66. M. T. M. Koper and W. Schmickler, *Chem. Phys.* **211**, 123 (1996).

67. M. T. M. Koper, J.-H. Mohr, and W. Schmickler, *Chem. Phys.* **220**, 95 (1997).

68. A. M. Kuznetsov, *Charge Transfer in Physics, Chemistry and Biology*, Gordon and Breach, Reading, PA, 1995.

69. J. W. Halley, B. B. Smith, S. Walbran, L. A. Curtiss, R. O. Rigney, A. Sujianto, N. C. Hung, R. M. Yonco, and Z. Nagy, *J. Chem. Phys.* **110**, 6538 (1999).

70. K. Toukan and A. Rahman, *Phys. Rev.* **B31**, 2643 (1985).

71. J. Hautman, J. W. Halley, and Y.-J. Rhee, *J. Chem. Phys.* **91**, 467 (1989).

72. L. A. Curtiss, J. W. Halley, J. Hautman, and A. Rahman, *J. Chem. Phys.* **86**, 2319 (1987).

73. L. A. Curtiss and R. Jurgens, *J. Phys. Chem.* **94**, 5510 (1990).

74. Y. Marcus, *Ion Solvation*, John Wiley & Sons, Chichester, 1985.

75. D. R. Lide (ed.), *Handbook of Chemistry and Physics*, 78th ed., Chemical Rubber, Cleveland, 1996, pp. 12–122.

76. J. Lipkowski and P. N. Ross, *Structure of Electrified Interfaces*, VCH Publishers, New York, 1993, p. 213.

77. J. Lecoeur and J. P. Bellier, *Electrochim. Acta* **30**, 1027 (1985).

78. D. R. Lide, ed. *Handbook of Chemistry and Physics*, 77th ed. Chemical Rubber, Cleveland, 1997, pp. 10–214.

79. S. Hartinger and K. Doblhofer, *J. Electrochem.* **380**, 185 (1995).

80. R. Hoffmann, *J. Chem. Phys.* **39**, 1397 (1963).

81. L. A. Curtiss, J. W. Halley, J. Hautman, Z. Nagy, Y.-J. Rhee, and R. M. Yonco, *J. Electrochem. Soc.* **138**, 2032 (1991).

82. D. Chandler, *J. Chem. Phys.* **68**, 2959 (1978).

83. H. Frauenfelder and P. G. Wolynes, *Science* **229**, 337 (1985).

84. D. A. Rose and I. Benjamin, *Chem. Phys. Lett.* **234**, 209 (1995).

85. D. A. Rose and I. Benjamin, *J. Chem. Phys.* **100**, 3545 (1994).

86. J. B. Straus, A. Calhoun, G. A. Voth, *J. Chem. Phys.* **102**, 529 (1995).

87. B. B. Smith and A. J. Nozik, *J. Phys. Chem.* **B101**, 2459 (1997).

AUTHOR INDEX

Numbers in parentheses are reference numbers and indicate that the author's work is referred to although his name is not mentioned in the text. Numbers in *italic* show the pages on which the complete references are listed.

SUBJECT INDEX